ADVANCES IN THERMAL MODELING OF ELECTRONIC COMPONENTS AND SYSTEMS

VOLUME 4

ASME PRESS SERIES ON ADVANCE IN THERMAL MODELING OF ELECTRONIC COMPONENTS AND SYSTEMS

Series Editors: Avram Bar-Cohen and Allan D. Kraus

Volume 1 Hemisphere Publishing Corp., New York, 1988
Volume 2 ASME Press, New York, 1990
Volume 3 ASME Press/IEEE Press. New York, 1993
Volume 4 ASME Press, New York, 1998

ASME PRESS SERIES

ADVANCES IN THERMAL MODELING OF ELECTRONIC COMPONENTS AND SYSTEMS

VOLUME 4

Editors

Avram Bar Cohen
Department of Mechanical Engineering
University of Minnesota
Minneapolis, Minnesota

Allan D. Kraus
Allan D. Kraus Associates
Aurora, Ohio

ASME PRESS
The American Society of
Mechanical Engineers
New York

ISBN 0-7918-0065-2
ISSN 0897-7453

CONTENTS

PREFACE

In the late 1990's, thermal management is again on the critical path in the velopment and commercialization of electronic products. Power dissipation is rging and available volume and allowable weight continues to shrink. And, spite Moore's law and improvements in performance and functionality, the price ints–for most electronic products, as well as the requisite thermal packaging rdware–remain fixed. Thus, air cooling will continue to play a central role the thermal packaging of electronic systems, but significant improvements in chniques and design methodology will be needed if the range of air cooling chnology is to extend beyond its traditional bounds.

Volume Four of the ASME Press Series "Advances in Thermal Modeling of ectronic Components and Systems," is focused on air cooling technology. In e first chapter of this volume, "Thermal Analysis of Natural Convection Elec-onic Systems: Status and Challenges," Clemens Lasance (Philips-Eindhoven, etherlands) and Yogendra Joshi (University of Maryland) assess the state-of-e-art and future promise of predictive modeling techniques which incorporate merical solutions of the governing momentum and energy equations.

Chapter Two, "Thermal Modeling of Air Coooled Components Mounted on inted Circuit Boards," by Sorin Witzman [prepared while he was associated th Nothern Telecom Ltd (Nortel)] describes an experimentally validated thermal sign methodology which relies on the superposition of the locally-determined mponent temperature rise on the globally-induced adiabatic temperature.

In "Governing Relations and Performance Limits in Air-Cooled Heat Sinks," llan D. Kraus (Allan D. Kraus Associates) and Avram Bar-Cohen (University of innesota) explore the design and optimization of multiple fins clustered and ar-yed in various configurations. In Chapter Four, Robert E. Simons (consultant) oses this volume with an annotated "Bibliography of Heat Transfer in Elec-onic Equipment: 1990-1994," in which he covers 400 selected papers, articles d published patents.

The editors are very grateful to the authors for providing ASME with this op-rtunity to meet the education and needs of engineers, researchers and managers the electronic industry. Volume Three of "Advances in Thermal Modeling of ectronic Components and Systems" found widespread acceptance in the elec-onic packaging community and a similar reception for Volume Four is antici-ted. It is the editors' expectation that this volume will be of particular interest to thermal specialists and software development engineers, as well as product d velopment teams throughout the electronic industry. Moreover, it is expect that technology developers and researchers struggling to extend the limits of a cooling technology will find this volume an invaluable source of data, referenc and thermal analysis/design techniques.

AVRAM BAR-COHEN

ALLAN D. KRAUS

CHAPTER 1

THERMAL ANALYIS OF NATURAL CONVECTION ELECTRONIC SYSTEMS: STATUS AND CHALLENGES

Clemens J. M. Lasance
Philips Research Laboratories
Eindhoven, the Netherlands
and
Yogendra Joshi
CALCE ELECTRONIC PACKAGING
RESEARCH CENTER
University of Maryland
College Park, Maryland

1.1 INTRODUCTION

The tremendous increase in computer power, combined with the increasing "user-friendliness" of Computational Fluid Dynamics (CFD) codes has finally brought numerical analysis, at the system level, to the desk of the application engineer. The last five years reveal an increasing number of papers devoted to the numerical study of real electronic systems, including transient effects and the use of relatively simple turbulence models. Due to heavy time pressure on electronic system designers, there is currently a tendency to make a full 3-D model of the system under development. The model is then "validated" by some "correction" factors, usually based on expert knowledge, by data from simple flow visualization experiments and/or temperature measurements.

Suppose that a system level packaging engineer, who is responsible for the thermal design of a new product, is asked by the product manager to estimate the critical temperatures, such as the die temperature of an active component, the coil temperature of a transformer or the peak temperature inside an electrolytic capacitor. If the objectives are to approximate the worst case situation or to support a feasibility study, or to derive some Figure of Merit to compare different designs, the current state-of-the-art provides sufficient information to perform a highly useful parametric study. However, if the objective is to give an estimate of the expected reliability of the complete system, the packaging engineer is in trouble. It is now well known that reliability of individual components shows a dependence not only on the temperature itself but, often of more importance, on its spatial and time gradients. Indeed, the level of accuracy needed in thermal simulations is highly dependent on the application.

The state-of-the-art in 1996 with respect to the thermal characterization of complex electronic systems indicates that enormous progress has been made over the last five years in grasping the essential features of all heat transfer mechanisms acting concurrently in real systems. These tools can advantageously be used in the early phases of the Product Creation Process and, as a result, the number of prototypes needed to arrive at a reliable design can be reduced considerably and First-Time-Right designs become feasible.

Still, the era of experimental work is not yet over; only the emphasis has been shifted to different aspects. The focus is now on concurrent use of experimental and numerical analysis techniques in which the experimental effort is required to support the numerical analysis, in other words, in the validation of numerical "experiments." If the objective of the analysis is to study those phenomena that are related to the mean characteristics of only the flow, the current state-of-the-art is well able to provide the right answers and, in theory, no substantial differences should be found between numerical and experimental results. However, if the temperatures of solid objects are also required, the problem of validation increases to such an extent that an unrealistic amount of effort has to be put into the design of the test facility to guarantee the required accuracy. Because the boundary layers and the conjugate effects must also be captured in sufficient detail, many designers believe that increasing the accuracy is merely a function of the number of grid cells, or, in other words, that buying a bigger computer solves the problem. The reason for the origin of this belief might well be the unprecedented increase in computer power.

When CFD was initially launched as a user-friendly tool for designers of electronic systems, the emphasis was on gaining physical insight into flow patterns, local air temperatures and average velocities as well as on performing many parametric studies in a very short time. While it is currently possible to faithfully simulate the thermofluid behavior of a realistic, though simple, configuration, in general, it is not possible to capture all of the fine details of the fluid flow in a complex realistic system. The only way to tackle this problem is to use "external" data, such as imposed heat transfer coefficients and thermal resistances, to solve for temperatures within solid regions and to get some idea of junction temperatures. Realizing this, the CFD world could easily be divided into two parts:

- **Real-physics simulation:** 2-D, geometrically simple, laminar or sophisticated turbulence models, steady/transient and simple boundary conditions.

- **Real-life simulation:** 3-D, geometrically complex, laminar or simple turbulence models, steady state and complex boundary conditions.

And never the twain shall meet............Unfortunately (in a sense), it is currently possible to work with sufficiently fine meshes to model some realistic systems. As a result the two territories which, formerly, were nicely separated, are beginning to overlap. Azar and Russell (1991) were able to visualize the overwhelming complexity resulting from non-uniformly positioned and dimensioned components on a circuit board with water as a cooling agent. On the basis of this and more recent evidence, one of the authors (Azar) appears to advocate the idea that the fluid flow phenomena that are observed in experiments are so complex that no existing CFD code is able to predict correctly even the direction of all local velocities. With increasing capabilities of CFD codes, such challenges may, in fact, be met by code vendors in the future.

The reader will notice after some reading that here, very few correlations and tables are presented. In contrast to the characterization of simple geometries, the authors consider the search for generalization in complex systems as counterproductive. Hundreds of runs are usually needed to come up with some correlation. Due to the multitude of dimensions and properties, every change in one of the parameters for which account was not taken renders the correlation useless. This should not be considered as a serious drawback because with modern analysis tools and appropriate experimental backup, every "new" system resulting from a major change should be treated as an entity in a catalogue of electronic systems. Instead of correlations, the physical, experimental and numerical insight gained from previous experience should be fostered and documented because such is of the utmost value in shortening the Time-to-Useful-Results. However, when the designer is fully aware of these limitations, worthwhile information might be extracted from some selected correlations.

In an interesting article, Oktay et al (1986) compared the heat fluxes in electronic devices with those in a number of natural and man-made events. The overall conclusion was that, while the heat fluxes at the chip level were approaching highly energetic processes such as the heat flux at the surface of the sun, the requirement to keep operating temperatures typically within 85° to 125°C made electronics cooling a particularly challenging problem. In response to this identified need, considerable attention and research continues to be devoted to high heat flux processes such as liquid immersion cooling and various thermal enhancements of forced convection cooling.

While forced convection cooling in a vented system is ultimately bounded by external factors, natural convection cooling of (vented) enclosures is essentially limited by the area available. As a consequence of safety reasons, the touch temperature of most enclosures is limited to about 60°C. With an ambient temperature of 30°C and a total heat transfer coefficient of $10\,W/m^2$-K, the heat flux at the boundary of a completely sealed enclosure can never be higher than $300\,W/m^2$. Heat fluxes and volumetric generation rates for several electronic systems commonly in use are compared in Table 1.1.

Table 1.1. Typical power/area and power/volume data for some sealed enclosures.

	Dissipation W	**Outer Surface Area** m^2	**Heat Flux** W/m^2
Telecom	2000	6	333
TV Set	200	0.5	400
Audioset	100	0.3	333
Notebook (open)	10	0.05	200
Cellular phone	2	0.01	200
Lamp	3	0.01	300

	Volume $/m^3 \times 10^3$	**Volumetric Power Generation** $W/m^3 \times 10^3$
Telecom	1000	2
TV Set	80	2.5
Audioset	3	30
Notebook (open)	1	10
Cellular phone	0.05	40
Lamp	0.03	100

Some interesting conclusions can be drawn from these data. For example, some systems are quite near the practical limit of 300 W/m^2. However, the electronic lamp has a much higher surface temperature, as becomes painfully evident when one tries to replace a failed lamp immediately after a burnout. It is also evident that the telecom equipment and the TV in the example cannot function without a vent. Vented systems allow a dissipation of as much as 2 W/cm^2 maximum, which is about two orders of magnitude higher than the dissipation in a closed area. The reason that the audioset can dissipate such a high power is due to the external heat sink. In contrast to telecom equipment, for which the power dissipation roughly scales with the volume, or rather the number of boards, such a metric cannot be found for other equipment. Finally, it becomes

Table 1.2: Value of Products for Basic Markets - U.S. 1994 (Source: The National Technology Roadmap for Electronic Interconnections).

Market Segment	**Percentage**	**1994 U.S. Value in billions of dollars**
Automotive	4	10.99
Business/Retail	5	13.72
Computer	30	82.44
Consumer	3.8	10.44
Industrial	9.3	25.56
Instrumentation	8.7	23.91
Military/Aerospace	16.6	45.62
Telecommunications	22.6	62.13

clear that notebook computers and cellular phones are operating near the limits of natural convection cooling.

A significant shift in the overall electronic product market has begun to occur since the inception of the 1990's. Table 1.2 shows the 1994 market share for various electronic products in the United States as compiled by the Institute for Interconnecting and Packaging Electronic Circuits.

While computing products still account for slightly less than a third of the total, electronic products are making rapid inroads in a number of other areas. Even within computing products, an increasing share of the market is being taken up by compact, portable equipment. This trend toward smaller sizes, portability and increased functionality applies equally to other product categories, including telecommunications, consumer and business/retail applications. Automotive electronics is another rapidly growing application area as cost-effective and reliable electronic packaging schemes are developed to handle the harsher environments encountered by such products. The military/aerospace market is relying more and more on the use of commercial parts to reduce costs without compromising on product reliability. Concepts such as "ruggedization" and "commercial off-the-shelf component" describe these trends.

The foregoing changes have brought about a resurgence of interest in low cost, passive thermal management schemes such as natural convection. Currently, the cooling fluid is almost exclusively air, but dielectric-liquid natural-convection cooling is not to be excluded as a future alternative. These schemes are a must for portable equipment because they do not compete for battery power. They are also of interest in many other cost-sensitive and high-reliability applications. The availability of many advanced materials such as composites, employing a variety of high-thermal conductivity additives, has made the availability of high-thermal conductivity substrates and heat sink materials possible. The use of these promises the extension of natural convection cooling to many other applications.

Another significant development is the current interest in high temperature electronic devices and systems. Efforts are currently underway to identify packaging materials and device technologies which will function reliably above the current conventional maximum junction temperature limits. Once these become commonly used, they will extend natural-convection cooling to many new applications.

Increasing global competition is strongly impacting the design of modern electronic products. While the product development time for a mainframe computer of the 1980's involved several years, portable computing and consumer products have a time-to-market of several months. Further reductions in cycle time are continually being sought by companies striving for a greater share of the market. Rapid times-to-market do not leave room for expensive trial and error approaches to product design. Experimentally validated computational modeling has become the method of choice for rapidly carrying out numerous "what if?" parametric studies during design. This increasing reliance on the predictive capabilities of system level computational characterization poses a considerable challenge to the thermal analyst.

The main topic of this chapter is the assessment of the current state-of-the-art in the thermal characterization of natural convection cooled electronic systems and identification of research needs in this area. Studies dealing with air and liquid cooling are described. Because the emphasis is on electronic systems within the overall packaging hierarchy, characterization at the component and board levels are addressed in as much as they relate to the system. By characterization, both experimental and computational approaches that are used to determine the thermal fields within an electronic system are implied.

The choice of material has been dictated by the targeted application of electronics cooling. As such, a number of interesting and practically relevant natural convection topics have been left out. These include transport at Rayleigh numbers above 10^{10}, effects of non-atmospheric pressures and cryogenic temperatures. The primary focus of the chapter is on the predictive modeling of steady state natural convection transport in geometrically complex enclosures.

The definition of "complex" is somewhat subjective. It has been assumed to include protruding components and circuit card assemblies within fully or partially enclosed air or liquid filled spaces. Because most electronic equipment boxes are rectangular in shape, any discussion of natural convection in other geometries, including internal flows within cylinders, between annuli of concentric cylinders and external flows around cylinders, spheres and tube bundles has been excluded. The detailed discussion of transport features such as flow and temperature patterns in the fluid is not included as the primary emphasis is on the prediction of chip and package temperatures in the conjugate heat transfer analyses usually required in such models. In the past, natural convection cooled electronic enclosures have often been analysed using a thermal resistance network type approach. In this chapter, only the methodologies incorporating numerical solutions of the entire set of differential conservation equations are considered.

This chapter begins with a survey of the existing literature on natural convection air and liquid cooled electronic systems. The challenges in predicting the thermal characteristics of such systems are illustrated by reviewing, in Appendix A, the research on geometrically simpler enclosures that have been investigated in considerable detail and, in Appendix B, the intriguing complexity of the heat transfer phenomena above heated horizontal plates.

Three examples of natural convection cooled systems are analyzed next to illustrate the combined experimental/computational approach espoused in this chapter. The first addresses natural convection within a small cubic enclosure 36 mm on a side containing a single component and a printed circuit board (PCB). Challenges posed by the small size, multiple heat transfer modes and orientation effects are clearly brought out even for this simple geometry. The second example considers a more realistic application of natural convection air cooling of a television set. The complexity of this example is handled by using a computational fluid dynamics approach in conjunction with experimental specification of surface heat transfer coefficients. Such an approach, with appropriate enhancements in global/local coupling are likely to form the basis of analysis for a wide array of current and future systems. The final example is a study of natural convection cooling of a three-by-three array of protrusions in a dielectric liquid. The significantly larger density and specific heat of liquids compared to air result in superior thermal performance in such systems which is comparable to forced convection air cooling.

The description of the state-of-the-art is followed by an identification of the challenges faced by the thermal analyst involved in the characterization of natural convection cooled systems. Significant impediments to achieving a predictive capability in accurate modeling of such systems are discussed. Recommendations for future work in this area form the conclusions to this chapter.

1.2 LITERATURE SURVEY

1.2.1 AIR COOLED SYSTEMS

To present an up-to-date literature survey on the subject of natural convection cooled systems is not a trivial task. As indicated in the general introduction, only a subset of all possibly relevant topics related to natural convection in closed or vented enclosures is treated in this chapter. For example, the natural convection behavior of geometrically simple enclosures, either with uniform boundary conditions or with discrete heat sources, is not discussed but with a few exceptions. However, to decide to which subset a paper belongs is a matter of personal opinion, and an apology is due those authors whose papers are not covered. Furthermore, papers that are not older than ten years have been chosen for review. Older work can easily be traced by consulting the references in the general reviews cited.

The papers surveyed are grouped into a number of categories: General Reviews, 2-D and 3-D Numerical Analysis of Idealized Systems, Experimental Analysis, and Analysis of Real Systems.

1.2.1.1 GENERAL REVIEWS

Early references are covered by Ostrach (1972) and Catton (1978). Their focus is heavily on geometrically simple enclosures. Bejan (1984) devotes Chapter Five in his book *Convection Heat Transfer* to enclosures.

Yang and Lloyd (1985) discussed the turbulent flow in vented simple and complex geometries. Although the Rayleigh numbers beyond which real turbulence occurs are not likely to be encountered in electronic equipment, with the possible exception of large telecommunication systems, the review presents useful information about the problems associated with non-laminar flow.

Yang (1987) presented a general overview on single-phase natural convection phenomena in enclosures. Amongst other things, laminar, turbulent, transient and unsteady effects are covered, as well as the flow in rectangular enclosures, horizontal annuli and three-dimensional bodies in enclosures. However, no reference is made to more complex geometries.

Gebhart et al (1988) devote Chapter Fourteen in their book to buoyancy-induced transport in enclosures and partial enclosures. Incropera (1988) formulated a comprehensive review of convection cooling options. In the natural convection section, parallel plates, both vertical and inclined, fin arrays, discrete heat sources and enclosures are covered. He stressed the need to clarify the significance of 3-D effects of the limitations associated with 2-D results. However, no complex geometries were considered.

Peterson and Ortega (1990) presented a table with studies of buoyancy-induced flow and heat transfer in vertical open channels in their chapter on *Thermal Control of Electronic Equipment.* Part of the chapter deals with vertical arrays, and discrete thermal sources on vertical plates. A few pages are devoted to discrete sources in enclosures.

Aihara (1991) reviewed natural convection around finite plates, non-circular cylinders, bodies of complex shape, non-smooth surfaces and the interaction of multiple heat sources. Although the flow in

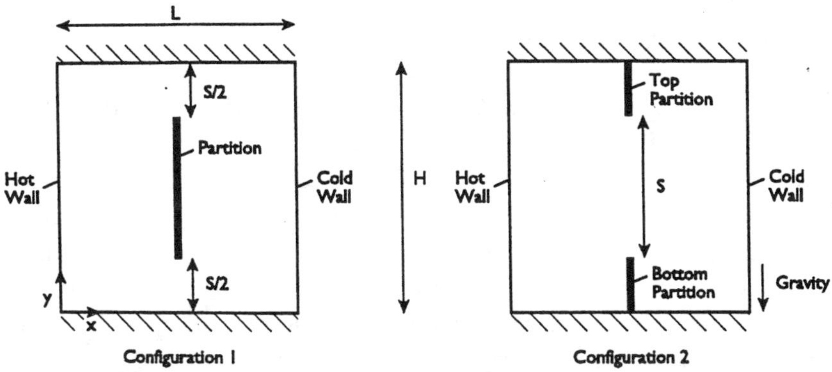

Figure 1.1: Two different partition arrangements [Kelkar and Patankar (1990)].

enclosures was given only three pages, the emphasis is on complex geometries and complicated thermal-boundary conditions.

1.2.1.2 SELECTED 2-D AND 3-D NUMERICAL ANALYSES OF IDEALIZED SYSTEMS

In addition to discrete heating, electronic systems possess geometrical complexity due to the presence of vents, partitions and heat sinks. Quite a number of papers are devoted to enclosures with horizontal or vertical partitions. Shaw et al (1988) described the influence of the position of a vertical adiabatic plate with discrete heat sources in a square enclosure. The effect of the distance between plate and top wall was also investigated. Kelkar and Patankar (1990) analyzed square enclosures with partitions in the laminar regime. Figure 1.1 shows the two different vertical partition arrangements covered.

It was observed that the overall heat transfer rate and through-flow are fairly insensitive to the decreasing gap width until half of the total height is blocked. Kangni et al (1991, 1995) calculated the influence of multiple vertical conducting partitions while Hawkins et al (1992) modeled two chambers separated by a horizontal partition with an opening. Beyond a certain aspect ratio, they found negligible heat exchange between the two chambers. Xia and Zhou (1992) studied a partially open cavity with a vertical divider. Results show that the presence of the divider significantly influences the flow and heat transfer in the cavity. The influence of both the divider height and the opening size becomes stronger at larger Rayleigh Numbers. Ten typical flow patterns have been found. Ntiburufata et al (1993) treated the effect of partitions in an enclosure with localized heating from below. Higher Rayleigh numbers (10^{10}-10^{12}) were covered in a recent study by Hanjalic et al (1996). These values are just sufficient to promote transition to turbulence along the walls causing a substantial increase in heat transfer coefficient. The phenomenon of transition is essentially unsteady and not reproducible, causing the experimentally obtained heat transfer correlations to differ considerably. Of particular relevance could be the observed jet-like horizontal flow at the partition tip across the partially divided enclosure. Usually, however, these high Rayleigh numbers are not encountered in electronic systems.

A few recent publications deal with protrusions and fins. Du and Bilgen (1990) performed a 2-D numerical study of three protruding elements on a PCB in a sealed enclosure, simulating an idealized case of an Electro-Magnetic-Compatibility-tight box in a larger system. Heat transfer by fins in slender cavities was studied by Frederick (1992). Facas (1993) presented results for an enclosure with fins attached to both vertical walls. It was observed that beyond a certain fin length, the fluid flow breaks down into secondary air circulations, thereby enhancing the heat transfer. Amin (1993) treated a rectangular enclosure with adiabatic sidewall and top wall, and one isothermal sidewall, fitted with a periodic array of protruding elements at the bottom, for Rayleigh Numbers between 10^2 and 10^5. An increase in heat transfer was observed when the phase shift of the protrusion (with respect to the side walls with non-symmetric boundary conditions) was equal to half its period, and more significant with increasing height of the protrusion. Up to 78 percent improvement has been found. Desai et al (1995) examined a fairly shallow enclosure provided with five protrusions, with an isothermal top wall and the other walls insulated. For Ra $< 10^7$, the flow is stable, while for Ra $> 3 \times 10^8$, a periodic flow was observed.

The influence of heat sources, vents or objects on the thermal behavior of enclosures has been the subject of various studies. Torok

(1984) discussed the application of tilted heat dissipating planes in an enclosure with and without vent openings. The addition of a vent only had a significant influence above a certain Rayleigh number. Yu and Joshi (1995) treated laminar convection in vented enclosures, in the Rayleigh Number range from 10^4 to 10^6. Four different vent locations were investigated and the vertical wall with conduction was provided with a flush-type heat source. It was concluded that the vent locations significantly affected the velocity and temperature fields. The combination of vents on the top and right walls provided the largest cooling effect. Ozoe et al (1988) calculated the temperature and flow fields in a rectangular enclosure with multiple point heat sources in the air for top-plane and side-wall cooling. House et al (1989) studied a vertical square enclosure containing a centered square obstruction. As was expected from the studies on differentially heated enclosures, the effect of the obstruction is not very significant as long as it is away from the boundary layers, in the relatively stagnant core. A centered object in a 2-D rectangular enclosure with conducting walls was also studied by Huang and Aggarwal (1995).

A cubic enclosure provided with an internal isolated heated vertical plate was the subject of an experimental and numerical study by Yang and Tao (1995). They concluded that in the low Rayleigh Number regime the plate average Nusselt number is quite close to that of an isolated vertical plate. Differences between the confined and the infinite space gradually become larger with increasing Rayleigh Number. This effect is attributed to the restriction of the plume originating at the trailing edge. Lacroix and Joyeux (1995) studied convection around horizontal vertically separated heated cylinders inside a cavity having finite wall conductance. Amongst other conclusions, it was found that heat transfer through the wall is significant even when its outer surface is insulated. Ahmed et al (1994) described the 2-D transient heat transfer behavior of flat plate heat sinks as frequently employed in TV sets. The basic features of a component attached to the heat sink were also taken into account.

Heat transfer in a 2-D shallow enclosure, representing an idealized notebook computer, with one to three heat sources at the bottom wall, was considered by Papanicolaou and Gopalakrishna (1995). One conclusion is that the heat transfer from a centrally located source is greatly affected by the presence of its neighbors, as compared to a single source. The results could help the designer in determining whether heat transfer will be dominated by conduction or convection. Adams et al (1995) and Vikram and Joshi (1996) calculated the conjugate heat transfer of a 3×3 array of electronic packages mounted on a horizontal board in a narrow-aspect ratio enclosure. Adams et al (1995) simulated a notebook computer, in the Rayleigh Number range from 10^4 to 10^7 corresponding to device power levels of 0.5 to 500 mW. For the lower power level, the heat transfer was conduction-dominated. For the highest power, conduction-only analysis underpredicted the heat transfer up to a factor of 4.4.

Despite the emphasis on complex geometries, it is wise to include some recent work on 2-D and 3-D numerical simulation of enclosures with discrete flush-mounted heat sources, without going into detail, and without striving for completeness. The following papers might be consulted: Keyhani et al (1988), Kim et al (1989), Shen et al (1989), Lee and Yovanovich (1989), Refai Ahmed and Yovanovich (1992), Hasnoui et al (1992), Hadim and Ramot (1993), Chen and Li (1993), Perna et al (1994) and Soylemez and Unsal (1994).

1.2.1.3 COMBINED CONDUCTION, NATURAL CONVECTION AND RADIATION

While many investigations have been carried out with regard to natural convection in closed enclosures, only a handful of studies exist on transport in vented enclosures including combined modes. Moutsoglou et al (1988, 1992) investigated the convection-radiation interaction in buoyancy-induced channel flows with and without venting. In their computations, two-dimensional vertical laminar flow was considered. An asymmetric, non-uniform, heat flux was prescribed on the inside of each of the channel walls. It was found that radiation tended to bring the temperatures of the two channel walls closer together. The radiation enhanced the heat transfer compared to a purely natural convection process. Beckermann et al (1994) reported a two-dimensional thermal analysis of a vertically oriented multi-board electronic module. Radiation was found to enhance heat transfer over natural convection and three-dimensional effects were significant.

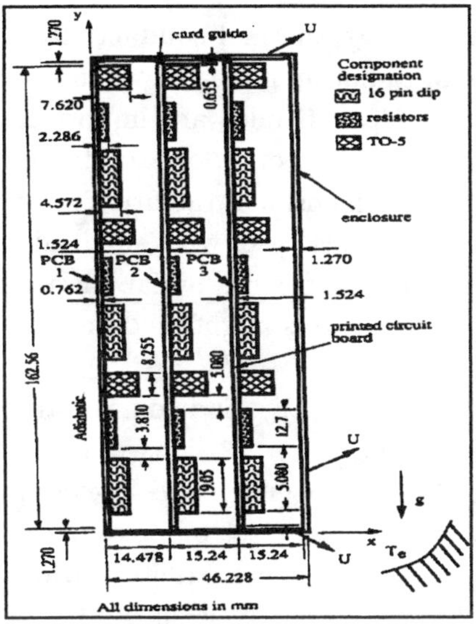

Figure 1.2: Enclosure and printed circuit board layout [Smith et al (1991)].

Balaji and Venkateshan (1993, 1994a, 1994b) conducted a series of studies on combined natural convection and radiation in closed and vented square enclosures. The results for unvented enclosures showed that radiation decreased the convection contribution to the local Nusselt number, but it either increased or decreased the overall heat transfer across the enclosure. For the vented enclosure, radiation from the left wall caused a temperature increase on the right wall resulting in a more vigorous air flow along this wall. Kassemi (1993) performed an analysis of radiation-natural convection interaction in a differentially heated square enclosure using the discrete exchange factor method. The computational results were obtained for ground-based and space-based environments. Heat transfer was found to be enhanced by radiation over natural convection.

Smith et al (1991) investigated combined conduction, natural convection and radiation heat transfer in an electronic chassis. As shown in Fig 1.2, a subsystem consisting of three circuit boards was considered. Both 1-D and 2-D radiation models were investigated and it was found that when boards and the top/bottom walls had large temperature differences, the 2-D model was necessary. One interesting result was that natural convection was not pronounced for a circuit board of high thermal conductivity. It was also argued that the influence of radiation increases when the PCB's cannot lose heat via the card guides. Guides are important, but do not alleviate the detrimental effects of a low conductivity board. It was also shown that a one-dimensional radiation model does not provide sufficient accuracy. For this case, with small distances between the boards, conduction heat transfer is by far the most important mode of heat transfer. However, this might be different when the enclosure is vented.

Lage et al (1992) numerically studied natural convection and radiation in a 2-D vented enclosure. They used a simplified three-surface radiation model. Wu and Cengel (1994) examined radiation heat transfer between rectangular electronic components on a PCB and the walls of the enclosure using a Monte Carlo technique, both for diffuse and specular surfaces. It was shown that radiation can serve as an effective mechanism to cool components in a sealed enclosure cooled externally. Dehghan (1994) conducted a numerical and experimental investigation of natural convection in discretely-heated open-top two-dimensional cavities. The heat source was mounted vertically and conduction in the substrate was considered. Surface radiation was also included for some cases. Dehghan and Behnia (1996) found that the radiation heat exchange had a significant influence on the thermal and flow fields. However, radiation did not enhance the total heat transfer coefficient because of simultaneous weakening of the convective contribution.

Yu and Joshi (1996) presented a three-dimensional numerical model to investigate combined heat transfer in a side-vented enclosure. A flush discrete heat source was mounted on a vertical substrate. Initial computations revealed that the approximate boundary conditions used on the opening in many two-dimensional studies were inappropriate for three-dimensional simulations. The calculations were performed on an enlarged domain to handle the difficulties in defining convection and radiation boundary conditions on the exterior enclosure walls and the opening. Radiation was incorporated on the basis of the radiosity/irradiation formulation. The solid surfaces were assumed to be opaque, diffuse and gray. The opening was assumed to be black at the ambient temperature. The enclosure had six internal surfaces. These were divided into smaller regions called effective surfaces, each assumed to be isothermal. The interaction between radiation, natural convection and conduction was examined for different Rayleigh numbers (input power levels) and surface emissivities of the substrate. The relative contributions of radiation and convection to the overall heat transfer were presented in terms of Nusselt numbers and heat transfer coefficients. The effect of radiation on venting was also studied. Experimental data were included to show the validity of the numerical results. Inclusion of surface radiation in the computations was found to reduce the component temperature by as much as 35 percent. Radiative transfer within the enclosure resulted in the warming of the enclosure walls and an overall reduction in the convective effects. The inclusion of a vent increased the radiative loss from the enclosure more significantly than the enhancement in convective transport.

The foregoing brief review indicates the need for investigations on combined heat transfer process in three-dimensional vented enclosures because these are often found in natural convection cooled applications.

1.2.1.4 EXPERIMENTAL ANALYSIS

Some work of the 1970's and early 1980's was reviewed in a paper by Buller and Duclos (1982). Kennedy and Kanehi (1983) investigated the effect of the angle of inclination on the thermal performance of packages with and without ventilation. Results indicate that enhancements are negligible for sealed enclosures, but improvements of up to 25 percent were observed by tilting with ventilated ends. Guglielmini et al (1985) published an experimental analysis of the cooling of multicomponent circuit boards with localized heat sources in partially confined enclosures with appropriate vents with the aim of optimizing the heat transfer performance. In another related paper by Guglielmini et al (1990), card spacing and the insertion of metallic plates between the cards were varied. Ortega and Moffat (1985) measured the heat transfer rates of a regular array of protrusions mounted on an insulating plane wall, both with and without an opposing insulating wall. It was found that local array heat transfer coefficient data correlate well with the induced Reynolds number, following about the same correlation as for pure forced convection.

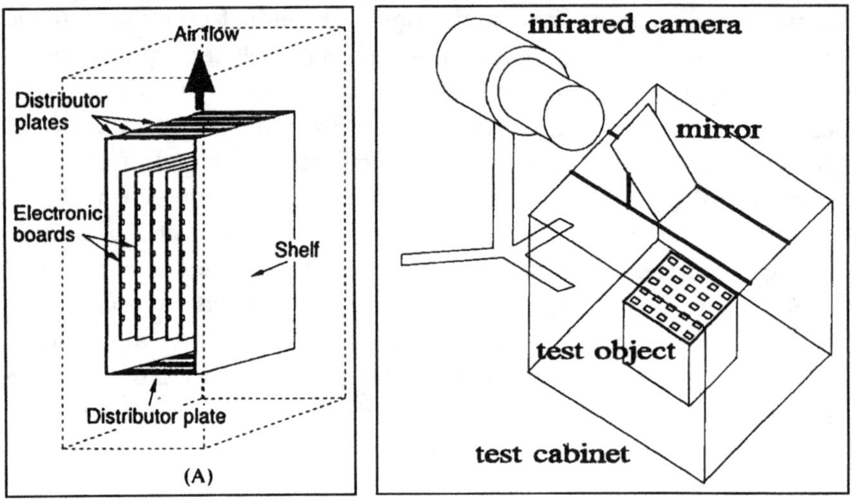

Figure 1.3: (a) An arrangement of electronic boards in an experimental apparatus [Kato et al (1991)] and (b) Schematic of the temperature measurement technique showing the test object with flush-mounted sources [Drabbels (1995)].

Kang and Jaluria (1990) treated the natural convection heat transfer characterization of a protruding 2-D source, either located on the bottom or on the side. Krane and Philips (1991) studied a square horizontal board in a low aspect ratio enclosure. The effect on the maximum temperature was measured by varying the enclosure geometry and the power settings. Chao et al (1989) presented the results of a numerical and experimental study of the effect of partial baffles as a function of the inclination. The maximum Rayleigh number in the numerical study was 6000 and in the experimental study 2×10^4. The Nusselt number was reduced by 20 percent as compared to the no-baffle situation for all angles. It was shown that the conductivity of the baffle played an important role in the resulting flow pattern and heat transfer rate.

Kato et al (1991) paid attention to the heat transfer between vertical boards populated with many small resistors as in Fig 1.3a. The open fraction of the vents at the top and bottom of the channel was varied between 0.08 and 1, and a significant influence was observed. For their case, it was concluded that the usual correlations for flat plate channel flow with constant heat flux provided a good approximation. A model of a digital transmission equipment containing vertically oriented boards with a number of resistors was studied by Ramos and Milanez (1996). The influence of board and shelf spacing was studied, as well as several kinds of baffles. Some design rules were presented.

Witzman et al (1992) performed an experimental study of the temperature fields for populated PCB's inside of open and hermetic enclosures. They concluded that the device temperature rise can be expressed as a summation of two terms: the temperature rise inside the thermal boundary layer and the temperature rise of the device surface. For boards with fairly good conductivity, it was argued that increasing the air velocity, that is, by changing to forced convection, did not provide the impact that people tend to expect. An interesting idea was presented concerning the use of non-contact heat sinks inside the device thermal plume to enhance the adiabatic heat transfer coefficient. Though not as effective as contact heat sinks, such heat sinks could be used in cases where no mechanical stress on the device is allowed.

Ortega et al (1993) performed experiments to characterize the combined conduction, convection and radiation heat transfer from a board in a shallow horizontal enclosure. Both simulated and actual enclosures were studied. Excellent agreement was found between a one-dimensional model description and the measured 3D results. The sensitivity of the maximum temperature to the accuracy of the heat transfer coefficient was found to be of minor importance. In a related paper concerning the calculation of the thermal "footprint" associated with a component on a board, Lall et al (1994) derived thermal design relationships for the determination of the peak temperatures of sources on a conducting board in the same type of enclosures.

Mayinger and Wang (1993) conducted a study of a vertical PCB array in a closed casing. It was found that the unstable thermal boundary layer on the top wall caused unsteady air circulation. The temperature fields were measured using holographic interferometry [Fig 1.4]. It was also found that the average heat transfer coefficient on the top wall was 3 to 5 times higher than on the bottom wall and that the heat transfer in the array was greatly influenced by the distance between the PCB's. In a second paper, Wang and Mayinger

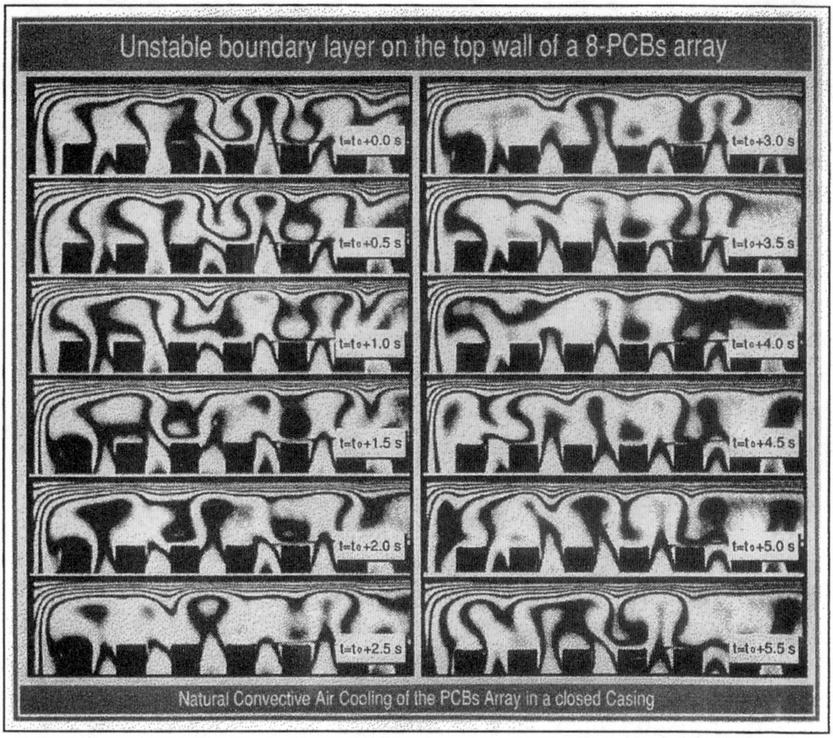

Figure 1.4: Time-sequence using holographic interferometry showing unstable boundary layers on the top wall of an array with vertical PCB's [Wang (1997)].

(1994) added inclination as a parameter. O'Meara and Poulikakos (1994) tested an array of vertical plates with uniform dissipation by varying the spacing and the distance between the plates and the top and bottom walls. The effect of a second row of plates, either in-line or staggered, was also studied. It was concluded that the results are very useful for providing engineering estimates for an optimal layout. In particular, increasing the bottom wall spacing beyond a certain value drastically reduced the negative effect of the presence of this wall.

Drabbels (1995) studied the behavior of a horizontal plate provided with a 5 × 5 array of flush-mounted heat sources (Fig 1.3b) both experimentally and numerically. One of the conclusions was that experimental verification of the numerical results was seriously hampered by the lack of sufficiently accurate values for the thermal

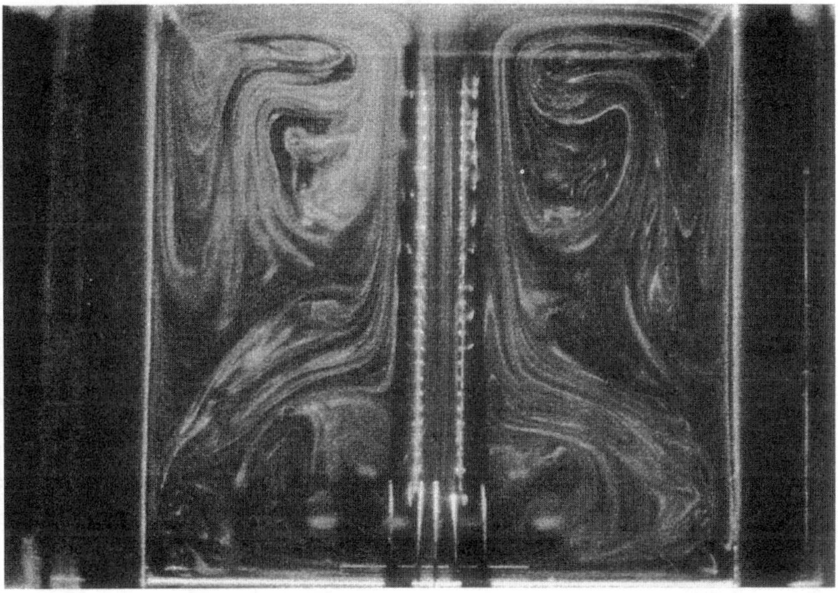

Figure 1.5: Flow visualization of two vertical boards in a large enclosure [Mergui et al (1990)].

conductivity and emissivity of the plate material. Another problem appeared to be the vizualisation of relatively low velocity air flows. Ortega and Lall (1996) did experiments to measure the heat transfer coefficients on the surface of a square flush heat source mounted at the center of a horizontal FR4 plate in a small enclosure. A rake of thermocouples, of which the position could be accurately controlled, was used to measure the temperature profile near the surface. It was concluded that the data were in significant disagreement with commonly available correlations for horizontal plates in infinite media and for enclosures with differentially heated walls.

1.2.1.5 ANALYSIS OF REAL SYSTEMS

Pioneering work focussing on the use of user-friendly CFD codes for the analysis of real systems was started in the early 1980's by a French consortium. This work resulted in several papers describing the temperature and velocity fields in telecommunication equipment [Le Jannou et al (1986), Cadre et al (1986, 1989a, 1989b, 1990), Mergui et al (1990)]. A typical result of their flow vizualization studies is reproduced in Fig 1.5.

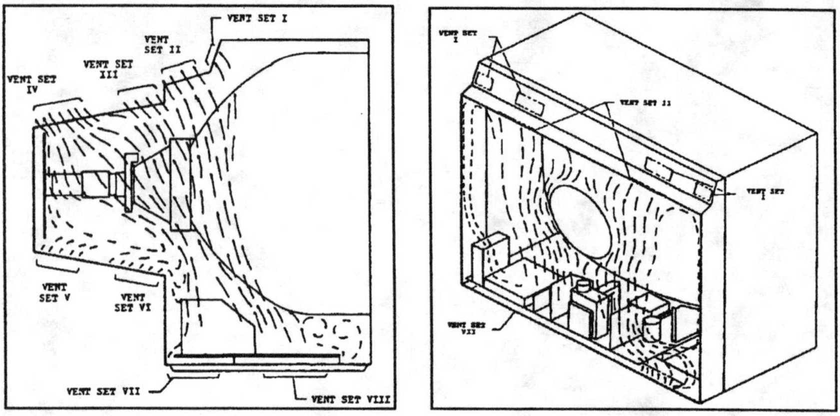

Figure 1.6: Typical flow patterns in a TV set with all vents open in two different views [Yuki et al (1990)].

During the early and mid 1990's, many papers were presented discussing the application of commercially available CFD codes for the analysis of natural convection cooled, complex electronic systems. Important design guidelines can be drawn from these publications. However, because the scope varies so widely, the interested reader is asked to consult the individual articles: Fuchs-telecom (1989), Nelemans et al-telecom (1989), Prakash-general (1990), Yuki et al-TV set (1990) [Fig 1.6], Krane and Phillips-TV set (1991), Lasance-TV set (1991), Drake et al-general (1994), Parry and Tatchell-general (1994), Lee et al-portables (1994), Garcia et al-telecom (1995), and Free et al-general (1995).

Associated with the study of vented systems are the papers that discuss the acquisition of the air resistance coefficients. Amongst others, these include the papers by Ishizuka et al (1987), Guglielmini et al (1985, 1987), and Misale (1993). In another paper, Misale (1994) compared four practical relationships currently used by designers.

Very recently, Azar wrote a chapter on "Limits of Air Cooling-A Methodical Approach" in a book edited by Kim and Lee (1996). He discusses some first-order approaches to calculate heat transfer from circuit boards and components in order to assess the potential limits of natural (and forced) convection cooling with air early in the design process.

To conclude this section, attention should be drawn to the valuable information to be found in the proceedings of the User Conferences organized by various vendors of CFD software. In these, the emphasis usually lies on very pragmatic approaches to tackle the challenging problems posed by the CFD analysis of complex systems. For example, typical telecom equipment contains 6 racks, each rack with 100 PCB's and each PCB with 300 components. However, only very few of these are critical. Typical examples of cases studied are shown in Fig 1.7

1.2.2 NATURAL CONVECTION LIQUID IMMERSION COOLING

Thermal control of electronic components using dielectric liquids has received considerable attention due to the inherently high heat removal capabilities of liquids compared with air. Candidate liquids for immersion cooling must have high dielectric strength, be chemically inert, environmentally safe and non-toxic. A family of such coolants is commercially available as Fluorinerts (1985) with a wide range of boiling points. Their availability has prompted a number of investigations of their heat transfer characteristics both for single phase and phase change schemes. Direct liquid immersion cooling has already been employed for thermal management of supercomputers. Single phase and phase change schemes for liquid immersion cooling have been reviewed by Bar-Cohen (1993) and Bergles and Bar-Cohen (1990).

The fluid circulation in immersion cooling is either forced or buoyancy induced. Natural convection cooling presents an attractive option for enclosed systems such as power supplies and certain electronic modules due to its design simplicity and inherent high reliability. Both experimental and computational investigations have been undertaken to characterize natural convection in configurations of interest in electronic packaging. The transport in such geometrically complex systems is often three-dimensional and includes discrete heating conditions and multiple heat transfer paths. Typically, the electronic components in the available studies have been simulated as discrete heat sources flush mounted on, or protruding from a substrate.

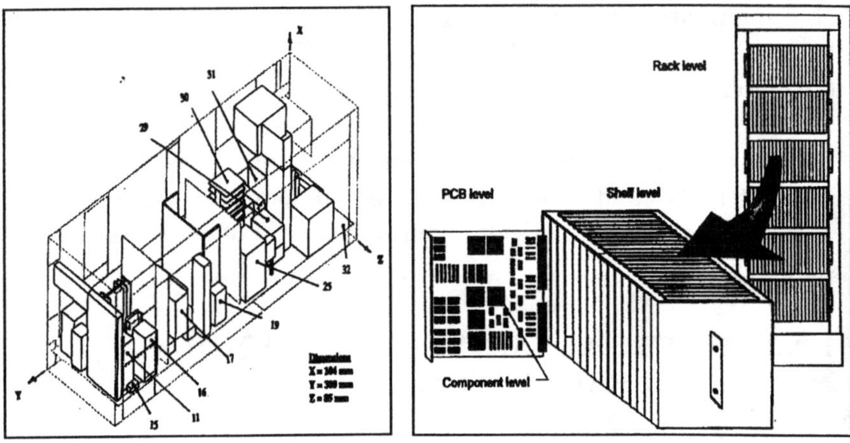

Figure 1.7: (a) Low voltage power supply [Parales, Flotherm user meeting (1993)] and (b) Typical telecom equipment with only very few thermally interesting components [Kordoban, Flotherm user meeting (1994)].

Experimental studies of natural convection liquid cooling have investigated arrays of simulated electronic components arranged

- on vertical surfaces [Baker (1972, 1973), Park and Bergles (1987), Joshi et al (1989a)],
- in vertical channels [Bar-Cohen and Schweitzer (1985), Joshi et al (1989b)] and
- in rectangular enclosures [Kelleher and Knock (1987), Keyhani et al (1988, 1991)), Joshi et al (1990), Heindel et al (1995a, 1995b)].

  Because the focus of the present chapter is on system level studies, only the last category is considered further.

Kelleher and Knock (1987) carried out an experimental study of natural convection from a single heated protrusion in a water-filled rectangular enclosure. Two-dimensional computations of transport in the same geometry were presented by Lee et al (1987). Keyhani et al (1988, 1991) and Chen et al (1988) experimentally investigated the two-dimensional transport resulting from a number of discrete protruding and flush heated elements placed within a rectangular enclosure filled with water and ethylene glycol.

Joshi et al (1990) examined natural convection from a 3×3 array of heated protrusions in a rectangular enclosure filled with Fluorinert FC 75. The heated elements geometrically simulated 20 pin Dual-In-Line packages (DIP's). They were mounted on one vertical wall of a rectangular enclosure with their largest dimension oriented horizontally. The top and bottom enclosure surfaces were maintained at uniform temperatures and the rest of the boundaries were insulated. Flow visualizations and heat transfer measurements were presented for a range of input power levels through each protruding element for a fixed enclosure width. The flows were seen to be strongly three-dimensional and often unsteady, requiring a time-dependent numerical simulation.

The enclosure boundary conditions, the liquid Prandtl number and the orientation of the heated components effect the heat transfer characteristics. Joshi et al (1994) examined these influences. Of specific interest were the effects of enclosure boundary conditions and width, component orientation, fluid thermophysical properties and partial powering of the array on natural convection heat transfer. Three combinations of top and bottom boundary temperatures were examined. The enclosure boundary condition effects were pronounced only at very low power dissipations. Measurements were obtained for two chamber widths and for three dielectric liquids to cover a Prandtl number range from about 30 to 1400. The data for various fluids were well correlated in the form of Nusselt number versus Rayleigh number. The component orientation effects were examined by comparing the data with the earlier study by Joshi et al (1990).

A three-dimensional experimental study of a single phase thermosyphon was presented by Polentini et al (1993). A 3×3 array of flush heaters was mounted on one wall of an enclosure, with the opposite wall cooled isothermally.

Numerical computations have been reported for some simple liquid-filled enclosures containing heat sources. Two-dimensional computations of transport for a protruding heat source in an enclosure were carried out by Lee et al (1987) and for three flush heat sources in a liquid filled enclosure by Prasad et al (1990). In both of these studies, the heat sources were simulated as uniform flux regions and the substrate was considered adiabatic. The fluorocarbon coolants used in liquid immersion cooling typically possess thermal conductivity values almost an order of magnitude smaller than conventional coolants such as water. This, in combination with the use of moderate thermal conductivity substrate materials such as ceramics, makes the conductive spreading within the substrate an important and sometimes dominant mode of heat removal from the electronic component. These effects were examined for a substrate-mounted protruding heat source in the two-dimensional computations by Sathe and Joshi (1991).

Three-dimensional computations of transport from a 3×3 array of rectangular protrusions in a liquid-filled enclosure were performed by Liu et al (1987a, 1987b). The liquid-exposed protrusion surfaces were considered to dissipate uniform heat fluxes, while the substrate was adiabatic. They found that there was little interference among chips and that the heat transfer was affected more by the stratified temperature field outside the boundary layer region than by the relative position within the boundary layer.

Long-time solutions were characterized by temperature oscillations on the faces. Substrate conduction effects were included by Wroblewski and Joshi (1993) in their study of natural convection from a substrate-mounted heat-generating protrusion (chip or electronic package) in a dielectric liquid filled cubical enclosure. The substrate forms one vertical wall of the enclosure. The opposite vertical wall was cooled and maintained at a constant temperature, while all other surfaces were insulated. A uniform volumetric heat generation was specified for the protrusion, and the problem of the conjugate conduction within the heat source and the substrate was solved in addition to the natural convection flow within the enclosure. Comparisons of the computations with the experimental results of Joshi and Paje (1991) for the same configuration were also provided. The effects of conduction were quite pronounced for the ceramic substrate since the ratio of substrate-to-fluid thermal ocnductivity was 575. The measured temperatures directly below the component were lower than the computed ones due to the assumption of a zero-thickness adiabatic boundary in contact with the back of the substrate. In the experiments, the back of the substrate was in concact with a lexan wall. This points to the importance of choosing realistic boundary conditions on surfaces in direct communication with the heat sources.

Computations of liquid immersion natural convection cooling were presented by Mukutmoni et al (1995) for a 3×3 array of heat generating, substrate-mounted, rectangular protrusions in an enclosure filled with dielectric liquid FC-75. The code was first validated with an experimental study of the same configuration [Joshi et al (1994)]. The effects of changing the substrate thermal conductivity were then investigated. The substrate was found to play a significant role in the heat removal from the protrusions. The flow was found to be unsteady at realistic power dissipations due to the large Rayleigh numbers. The coefficient of volumetric expansion for fluorocarbon liquids is almost three times as large as for water, resulting in these large Rayleigh numbers. The unsteadiness in the transport in the fluid was damped out by the thermal mass of the protrusions and measured temperatures on these surfaces were found to be steady.

Heindel et al (1995a, 1995b) presented computations, (1995a), and experiments, (1995b), for natural convection liquid immersion cooling of a 3×3 array of heaters flush-mounted to one vertical wall of a rectangular enclosure. For heater aspect ratios below 3, average Nusselt numbers increased with decreasing heater aspect ratio. For aspect ratios above 3, the two and three-dimensional predictions were within 5 percent of each other and results were approximately independent of the heater aspect ratio. Experimental measurements were in agreement with the three-dimensional predictions. Row averaged Nusselt numbers were well correlated with a Rayleigh number exponent of 0.25 based on component height, for Rayleigh Numbers below 1.2×10^8.

1.2.3 CLOSURE

In the open literature, there is an abundance of papers that treat relatively simple geometries and uniform boundary conditions, sometimes in astonishing detail. Unfortunately, the resulting correlations frequently cannot be used in electronic system design. Another observation is that the number of papers dealing with the analysis of complete systems is rapidly growing but again, results from these individual studies cannot be easily translated to other systems.

Many of the foregoing studies on complex systems attempted to characterize the flow and heat transfer behavior over a range of operating conditions for the selected configuration. However, due to the multiplicity of length scales and the large variability in the thermophysical properties of the various electronic packaging materials, the reported correlations are only applicable for limited conditions. Perhaps, their greatest utility is in providing a database against which numerical schemes can be validated. Indeed, a combined computational and experimental approach appears to be best suited for the typically complex electronic cooling applications. The computational technique employed must first be validated over the range of flow parameters investigated in the supporting experiments. This process provides important input on the appropriate grid sizes as well as on the time step necessary to resolve the observed transport behavior. The latter is particularly important in vigorous natural convection flows, typical in many Fluorinert liquids which have significantly higher volumetric expansion coefficients than water. Following a detailed validation, the code may be used as a convenient tool by the thermal designer to study the effects of varying thermophysical properties and geometric parameters, and to parametrically explore the configurations of interest.

1.3 SIMULATION OF NATURAL CONVECTION HEAT TRANSFER INSIDE ENCLOSURES: THREE EXAMPLES

Three examples of natural-convection driven electronic systems are discussed in some detail in this section. It is hoped that the reader will notice some common aspects in the analysis techniques that are recommended in investigating complex electronic systems. Because many parameters are usually unknown at the time the analysis is required, the designer benefits most from a judicious combination of experimental and numerical techniques.

The first example discusses such an approach for a simple, small enclosure with only one PCB and one component subject to essentially laminar flow. Despite its "simplicity", it will become clear that accurate temperature prediction is still hampered by many unknowns. The second example treats a much larger system, a TV set. Considering the current state-of-the-art, it is absolutely impossible to analyze all of the fine details of the temperature and velocity fields which, nevertheless, have an impact on the predicted temperature of critical electronic parts. A clever choice of some simple but highly efficient measurements can increase the predicted accuracy tremendously. On the other hand, especially in early design phases, absolute accuracy is not required and the main objective of the investigation is to perform parametric studies for which relative accuracies are sufficient. The final example considers the liquid cooling behavior of an idealized enclosure consisting of a 3×3 array of protruding elements. Liquid cooling using single-phase natural convection offers considerable enhancement potential compared to air cooling. The complexity in the modeling of heat transfer and fluid flow is illustrated for the selected model system. The large differences in the thermophysical properties of dielectric liquid coolants compared to air bring considerable challenges to the modeling.

Because the word "validation" is used frequently in different contexts, it is advantageous to make a distinction between *verification, validation* and *calibration.* The definitions proposed by Fisher and Rhodes (1994) will be followed.

- **Verification:** The procedure to ensure that the program solves the conservation laws correctly including the issue of grid independence.

- **Validation:** The procedure to validate the ability of models to predict complex flows, that is, the testing of predictive capabilities of physical models against detailed test data (surface pressure, velocity field and turbulence data).

- **Calibration:** The procedure to calibrate a code with a particular model in its ability to predict global quantities for realistic geometries of design interest.

Of these activities, verification is seen as the responsibility of the code developer and calibration is specific to a particular design application. The code is "tuned," as it were, to the specific application, preferably by using statistical techniques. It is generally rather dangerous to use the prediction capability outside the range of tested applications. The key activity, which is also the most demanding and expensive, is validation.

1.3.1 A SMALL ENCLOSURE WITH ONE COMPONENT AND ONE PCB

In a number of industrial cases, relatively small volumes (of order $50 \times 50 \times 50$ mm or less) are packed with electronics. The packaging engineer is confronted with a number of problems relating to the choice of the materials such as the optimum placement of the electronic parts and their position with respect to gravity. It is evident that for increasingly smaller systems, accurate experiments become difficult. At the same time, full numerical analysis becomes more feasible for two reasons. First, the number of nodes required for a given accuracy decreases with the third power of the dimension for given internal geometrical complexity. Second, the falling Rayleigh number indicates that laminar flow is a fact. In general, the physics of such a flow can be captured by solving the coupled mass, momentum and energy equations to any degree of accuracy provided that the input parameters are also known with the same required accuracy.

This subsection describes the heat transfer phenomena in an enclosure with dimensions $36 \times 36 \times 36$ mm. To gain insight into the basic physics, it was decided to provide the enclosure with only one component and one PCB. One objective of the research was to see if a commercially available user-oriented CFD code could be used to predict the essential features of the heat transfer inside the enclosure including conjugate effects and radiation, with a maximum inaccuracy of 5 percent in the absolute temperature. By ensuring that the surface of the component has a low emissivity, account of the radiation losses can be easily taken. Even if the emissivity or the view factors are not exactly known, their contribution to the overall heat transfer budget is minimized. Conduction losses via the wires can only be limited by reducing the dimensions. Obviously, this is a far greater challenge for the experiment than for the numerical analysis.

Because it is believed that the basic features underlying the treatment of this "simple" enclosure are representative of more complex

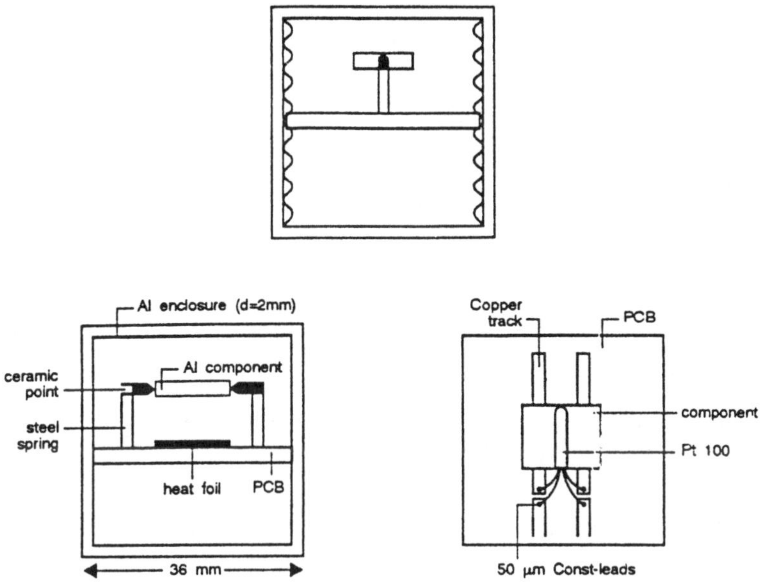

Figure 1.8: Sketch of the enclosure with one component and one PCB.

systems, the experimental and numerical analyses are discussed in some detail.

1.3.1.1 THE EXPERIMENTAL SETUP

The heart of the experiment consists of a component made of polished Aluminium, provided with a very small ceramic Pt100 element with a diameter of 0.6 mm which acts both as a heat dissipator and a temperature sensor. For an accurate determination of the heat transfer coefficient, say to within 5 percent, the heat transferred by convection must be accurately known, and the heat losses via the leads and by radiation must be minimized. It is not sufficient to know the heat losses via the leads, by whatever means, because the heat that flows via the leads into the PCB causes local heating of the air, thereby affecting the local heating by the component itself. Another problem is the provision of support for the component. Because the required tiny leads are not sufficient to support the component in all positions with respect to gravity, extra measures had to be taken to ensure stability. To this end, the component was held in place by two ceramic points glued to thin steel springs. These, in turn, were locked with Loctite in small holes in the PCB. Worst case estimation showed that the heat losses via the suspension would be less than a few percent. Two different sizes were manufactured: a rectangle measuring $10 \times 5 \times 2$ mm and a cylinder with a diameter of 15 mm and height 12 mm. Figure 1.8 shows a sketch of three views of the enclosure and the parts considered.

Minimization of Heat Losses Via the Leads.

One of the major problems to be addressed was the choice of the material and the dimensions of the leads in order to satisfy the requirement of the 5 percent heat loss limitation. It may be recalled that smaller components in a natural convection environment, depending on the thermal conductivity, lose as much as 50 to 80 percent of the dissipated heat via their 0.6 mm diameter leads to the PCB. The heat transfer path can be split into two parts in series: the path via the leads to the PCB, and the path via the PCB to the ambient air. To solve this problem, it is not necessary to use a full CFD or FEM analysis because analytical relationships can be derived, which have the advantage of enabling valuable parametric studies.

Thermal Resistance of the PCB.

Consider the heat transfer from the lead in the PCB to be similar to that in a circular fin. Then, the temperature rise [Kern and Kraus (1972)] is given by

$$\Delta T = \frac{q}{2\pi k d m r_1}\left[\frac{I_0(mr1)K_1(mr2) - I_1(mr2)K_0(mr1)}{I_1(mr1)K_1(mr2) - I_1(mr2)K_1(mr1)}\right] \tag{1.1}$$

where

$$m = \left(\frac{2h}{kd}\right)^{1/2}$$

Lasance (1986) shows that when q is divided by ΔT, the thermal resistance R_{pcb} can be reduced to

$$R_{\mathrm{pcb}} = \frac{1}{2hA_{pcb}} + \frac{\ln(A_{\mathrm{pcb}}/A_{\mathrm{lead}})}{4k\pi d} - \frac{\gamma}{2k\pi d} \tag{1.2}$$

Table 1.3: Percentage of heat losses for a lead length of 1 cm.

	Copper		**Titanium**		**P-Bronze**	
	h_{lead} (W/m^2-K)					
Diameter, μm	**5**	**50**	**5**	**50**	**5**	**50**
20	3	6	−11	−3	−2	1
50	17	22	−2	3	4	8
100	39	44	6	12	13	19

	Constantan		**Nichrome**	
	h_{lead} (W/m^2-K)			
Diameter, μm	**5**	**50**	**5**	**50**
20	−11	−1	−20	−5
50	0	4	−2	3
100	7	13	5	10

provided the inequalities

$$m\frac{A_{\text{lead}}}{\pi} \leq 0.2, \quad \frac{A_{\text{pcb}}}{A_{\text{lead}}} \geq 4 \quad \text{and} \quad m\frac{A_{\text{pcb}}}{\pi} \leq 3 \tag{1.3}$$

are also satisfied. Here γ is Euler's constant (≈ 0.57).

Thermal Resistance of the Leads.

The ideal lead metal should have a minimum thermal conductivity and a maximum electrical conductivity. Unfortunately, the design window is fairly small due to physical principles formulated by Wiedemann and Franz over a century ago. Because the relatively long and thin leads in this case also lose heat via convection and radiation, fin theory should be used. However, a simple fin relationhip is not appropriate because the "fin" loses heat from its tip to the PCB. Consider a spinal fin with a non-insulated tip. Its thermal resistance is given by Kraus (1982)

$$R = \frac{1 + k \tanh ml}{\frac{\pi}{4} k d^2 m \left(\frac{h_{\text{tip}}}{mk} + \tanh ml \right)} \tag{1.4}$$

If the heat transfer coefficient at the tip is replaced by the thermal resistance given by eq (1.2), the result, after some manipulation is

$$R_{\text{loss}} = \frac{R_{\text{pcb}} + \dfrac{\tanh ml}{kAm}}{1 + kAmR_{\text{pcb}} \tanh ml} \tag{1.5}$$

In this way, the entire system can be described by a parallel arrangement of three resistances. A simple program may be used to calculate the relative contributions of these resistances as a function of all relevant parameters. Some results for a specific situation are presented in Table 1.3 which shows the ratio of the heat losses to the heat convected in percentages. The estimated heat transfer coefficient for a bare wire is $h_{\text{lead}} = 50\,\text{W/m}^2\text{-K}$ and for a wire insulated with a loose sleeve, $h_{\text{lead}} = 5\,\text{W/m}^2\text{-K}$. Furthermore, the contribution of the Joule heating in and the voltage drop across the leads was calculated using the same code. The negative numbers in Table 1.3 are caused by excessive Joule heating. Theoretically it is possible, by using the right material, to let the self-heating of the leads just cancel the heat losses.

The few materials that seem to meet the specified requirements must also be solderable or weldable to the Platinum of the Pt100 and the Copper of the tracks on the PCB. After taking everything into account, a Constantan wire with a 50 μm thickness and provided with an insulating sleeve was finally chosen. Two PCB materials were chosen from the extreme ends of the practically useful thermal conductivity spectrum: hard paper and aluminium. The housing was also made of aluminium, having the advantage of creating virtually no temperature gradient, allowing the use of a single thermocouple. Its reading provided a direct estimation of the average heat transfer coefficient to the outside environment.

The last problem that remained to be solved was caused by the large discrepancy between the spatial dissipation in an actual system and the measurement setup. Two reasons emerge. The first concerns the fact that a real component loses a great deal of heat to the PCB via its leads and this path is no longer available in this measurment situation. In order to measure the same flow field caused by a "normal" dissipation, an extra power source simulating the heat losses due to the leads had to be provided. The second reason is that the mutual influence of the components on each other, either via the

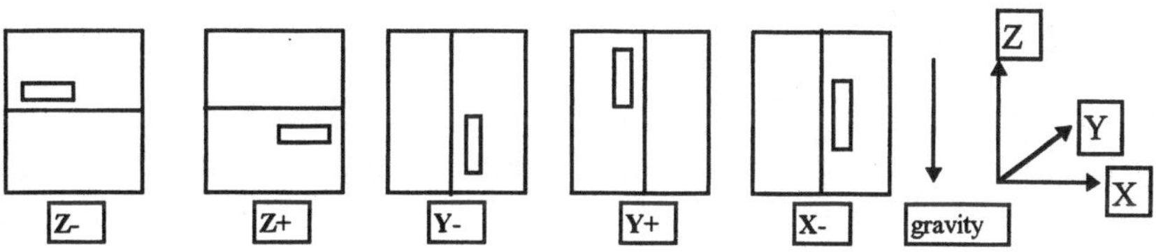

Figure 1.9: Parameters varied in the measurements.

PCB or via the air, is a very important phenomenon and must be taken into account. Both effects, to a first order approximation, were addressed by fixing a flexible heater foil to the PCB as shown in Fig 1.8

1.3.1.2 THE NUMERICAL ANALYSIS

The commercially available code, FLOTHERM, was used for the numerical analysis. In order to validate the numerical analysis, all relevent physical properties should be known with sufficient accuracy. For the present case, data was available concerning the following parameters:

- dimensions,
- emissivities,
- thermal conductivities,
- dissipations,
- the heat transfer coefficient at the outside of the enclosure and
- the influence of the numerical grid size.

These items will now be discussed in detail.

Dimensions: Because the CFD code used only allows for rectangular coordinates, almost all of the geometrical information from the experiment, except for the cylinder, can be readily used. To overcome this problem, the height of the polygon was adapted to ensure that the areas were, at least, identical. For steady state problems, this seems to be an appropriate method.

Emissivities: The most important emissivity is that of the component itself. This parameter has been measured using the method described by Lasance (1994b). The exact emissivities of the housing and the PCB are not very important because of the small ratio between the component and the other areas. The code calculates radiation between surfaces without taking account of reflections. A comparison of this method with a much more sophisticated method available in ANSYS shows that the simple approach introduced errors of less than 5 percent.

Conductivities: The effective conductivity of the hard paper presented the only problem. This value was retrieved by using the measurement discussed by Vanlaer and Lasance (1994).

Dissipations: The dissipations themselves do not present a problem because they can be measured with sufficient accuracy by means of the four wire technique.

External heat transfer: Because the internal and external flows are separated, a simple measurement of the temperature of the aluminium housing is sufficient to calculate an average heat transfer coefficient. This value is then used as a boundary condition in the numerical analysis.

Numerical grid: As far as the simulation of the boundary layers is concerned, the mesh was chosen is such a manner that further refinement showed no significant improvement. The number of mesh lines used was approximately $30 \times 30 \times 30$. In order to provide accurate simulation of the boundary layers, a non-uniform grid near every solid object was employed.

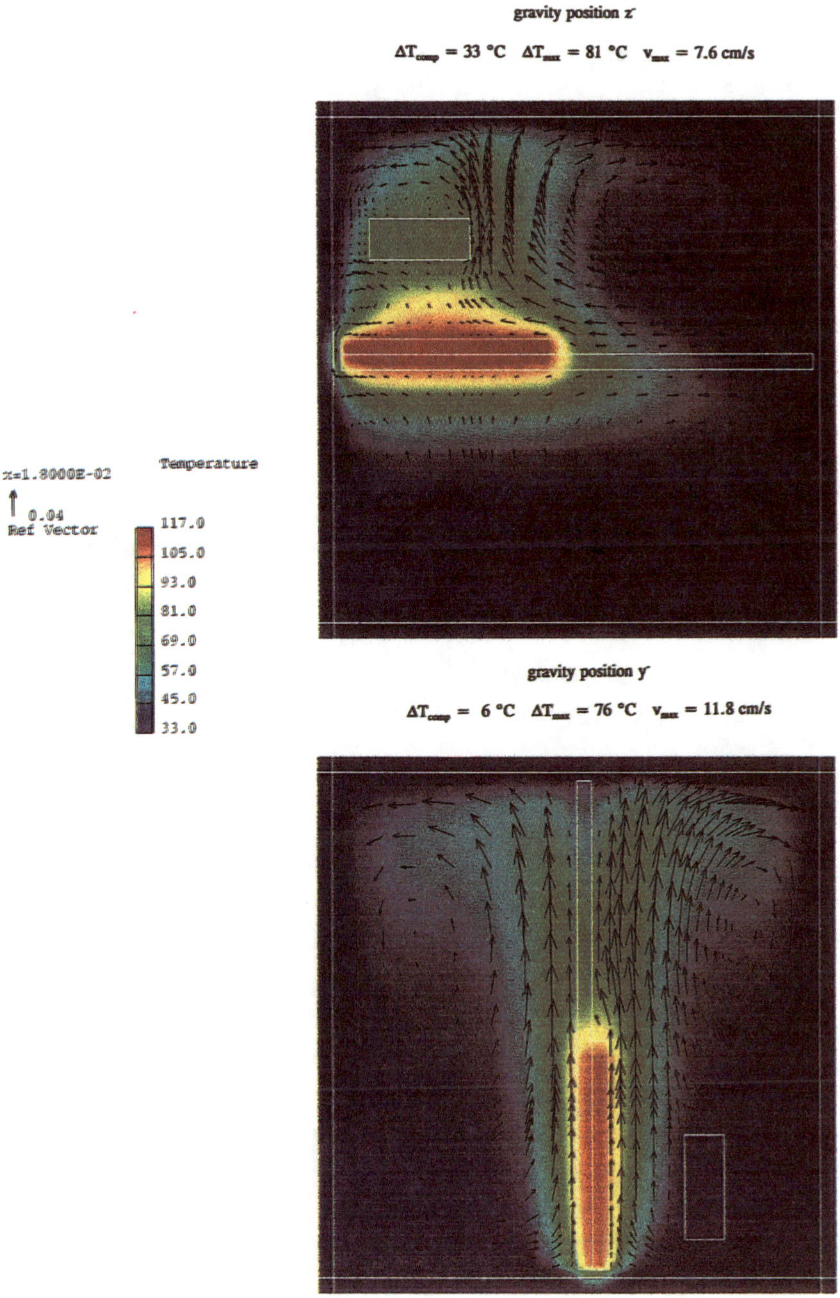

Figure 1.10. Temperature and velocity fields for a component in a small, enclosure, paper PCB, side.

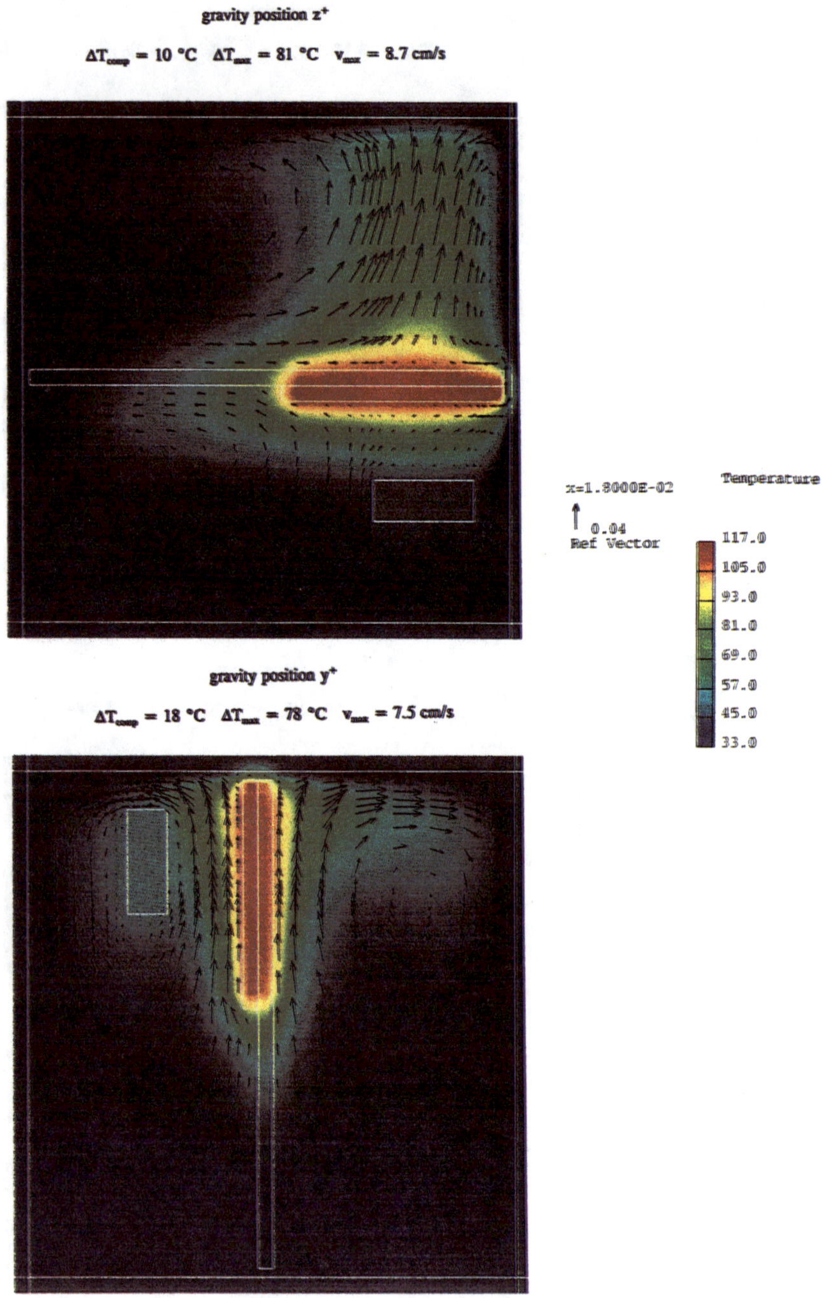

Figure 1.10 (Cont'd). Temperature and velocity fields for a component in a small, enclosure, paper PCB, side.

1.3.1.3 EXPERIMENTAL AND NUMERICAL RESULTS

The following set of parameters was chosen for both the experimental and numerical analysis:

PCB:	hard paper ($k = 0.4\,\text{W/m-K}$), aluminium ($k = 200\,\text{W/m-K}$)
Component:	small ($10 \times 5 \times 2\,\text{mm}$), large (circular, $15\times$)
Emissivity:	shiny ($\epsilon = 0.10$), black ($\epsilon = 0.95$)
Dissipation:	PCB (maximum 1 W), component (maximum 15 mW)
Position on PCB:	center side
Position in space:	gravity vector in z^+, z^-, y^+ y^- and x^- directions (see Fig 1.9)

Some typical results are presented here. The total estimated power transferred from the components into the PCB for a typical application is of the order of 1 W. Hundreds of experiments were performed. Every new measurement started from zero dissipation and was followed by activation of the heater foil and then waiting until steady state was reached. After starting the experiments with a PCB dissipation of 1 W and zero component dissipation, the first results showed surprisingly large variations of the component "adiabatic" temperature, when varying the position of the enclosure with respect to gravity.[1]

Subsequent CFD analysis showed the reasons for this. The vector plots shown in Fig 1.10 show remarkable differences in the flow patterns for variations in orientation and heater location. These variations result in the significant temperature variations observed. Note the "high" velocities that occur with a maximum around 12 cm/s for the vertical orientation and the dissipating area at the bottom. For the purpose of verification, the aluminum housing was replaced by a polycarbonate and smoke was injected with a needle. In doing so, rolls were observed rotating with a frequency of about 3 times per

[1]The adiabatic temperature is the temperature that the component achieves with zero power dissipation.

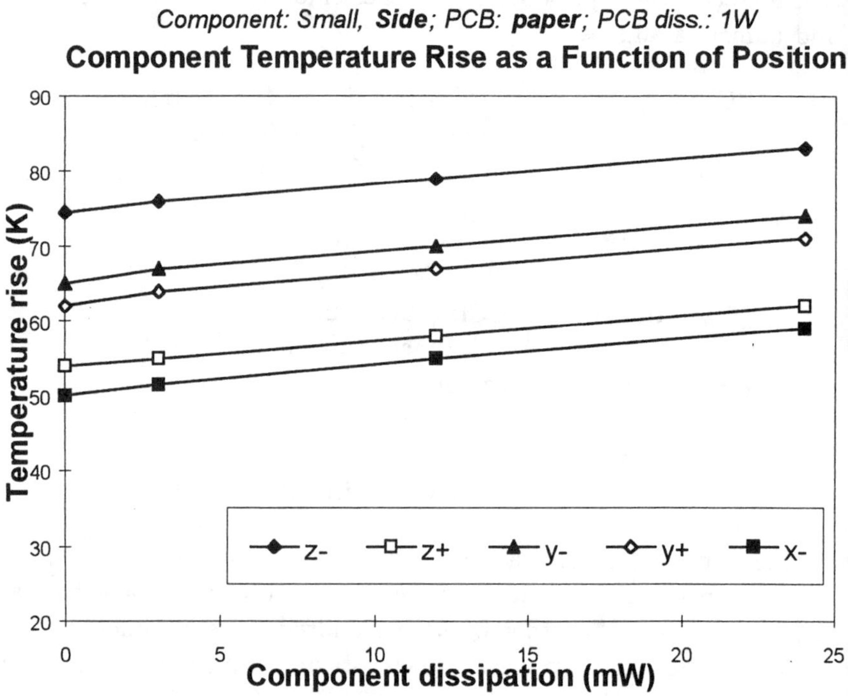

Figure 1.11: Component temperature rise as a function of board plus component dissipation. Note the dependency on board orientation.

second. Using a rough estimation of the roll length, the velocity lay somewhere between 10 and 15 cm/s thereby verifying the numerical results

Figure 1.11 shows what happens if additional power is added to the component. This power is in addition to the board dissipation of 1 W and the temperature rise is a function of the position of the PCB with respect to gravity. Observe that the slopes are virtually constant, indicating a purely local phenomenon caused by the component dissipation alone.

Finally, Figs 1.12 and 1.13 show the fairly good correlation between the experimental and numerical results; Fig 1.12 shows the data for a centered object on a hard paper PCB and Fig 1.13 those of a side-placed object on an aluminium PCB, both as a function of

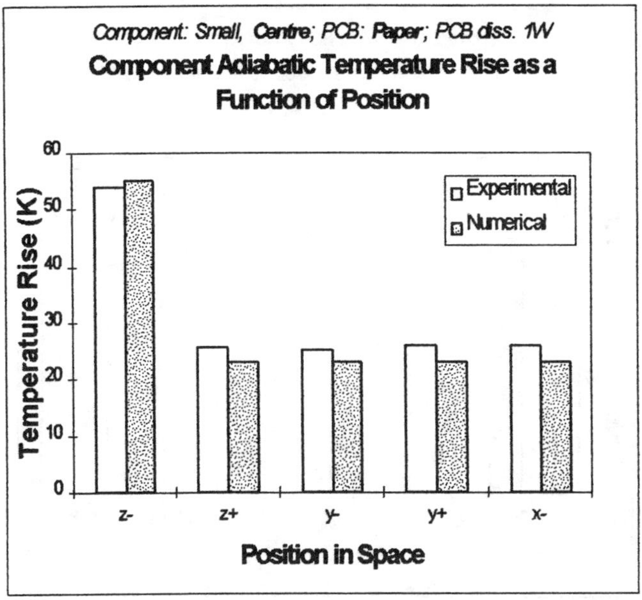

Figure 1.12: Example of experimental and numerical results for a small component located at the center of a paper PCB with a dissipation of 1 W.

the position with respect to gravity. Observe, in particular, the large differences in object temperature. No attempt was made to improve the correlation further because the discrepancies can be explained by taking into account the uncertainties in variables such as emissivity, thermal conductivity and positional accuracy in the experiments. Note that in almost all cases the measured temperatures are somewhat higher than the numerical ones indicating a common source of error. The heat transfer coefficient calculated from the slope is about 15 W/m^2-K, including convection and radiation to different temperatures.

One of the most important conclusions from this study is that it is not necessary to know (or to measure) the local heat transfer coefficient which, as shown in Fig 1.11, appears to be virtually independent of the orientation or the variations in the other parameters. Of considerably greater importance are the local air temperatures which are affected by the orientation, thermal conductivity of the PCB and dissipation. If this happens to be the case, a CFD anal-

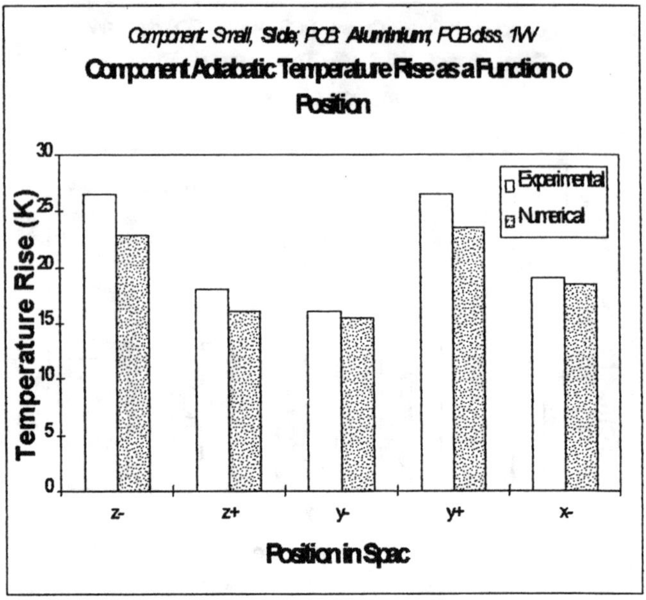

Figure 1.13: Example of experimental and numerical results for a small component located at the side of an aluminum PCB with a dissipation of 1 W.

ysis is almost the only way to arrive at an accurate temperature prediction, even in small enclosures such as the one considered here.

One of the objectives of this research was to look for a methodology to derive design rules. Figure Fig 1.11 seems to show that a promising method could be one based on the superposition principle. The underlying assumption is that the temperature of the component is the result of several heat transfer processes which can be treated independently. More details are given by Lasance (1994b) where suggestions are also given regarding the correction of the reference temperature used in the adiabatic heat transfer coefficient for the influence of the enclosure wall temperature. The conclusion of this part of the study is that under certain circumstances design rules can be formulated but that with increasing coupling between component and board, and certainly when more components are involved, design rules become so complex that one may wonder whether the name "rule" is still appropriate. On the other hand, a proper presentation of the combined results of CFD analyses may also be called a "design rule."

1.3.1.4 CONCLUSIONS

This subsection has presented an experimental and numerical analysis of a small enclosure consisting of an Aluminum housing including one component and one board. Emphasis was placed on gaining insight into the global and local phenomena which cause the temperature rise in the component. Various parameters have been explored, such as the influence of the position of the housing with respect to gravity, the size and location of the component, the thermal conductivity of the board and separate dissipation of the component and board. The most important conclusion is that CFD analysis is a "must" for predicting the thermal behavior of this small enclosure.

1.3.2 ANALYSIS AND ENHANCEMENT OF HEAT TRANSFER INSIDE OF A TV SET

This subsection focusses on the analysis of the natural convection heat transfer in a television set. A sketch of the set and the location of the most important elements are given in Fig 1.14.

In the earlier stages of the Product Creation Process the designer would often like to have answers to the following questions:

- What is the mean air temperature rise in the set? This question is related to the fact that most temperature specifications refer to a maximum allowable internal "ambient" temperature.
- How do the local flow patterns influence the local temperature?
- Are there any recirculation zones which give rise to any unwanted temperature rise?
- Are hot spots to be expected, and what is their location?
- What is the influence of the size and position of vents on local air temperatures throughout the set? This is also an important issue from a purely mechanical point of view. The molds for the plastic covers are very expensive, and a First-Time-Right design would be very valuable.
- What happens when the position of critical components is changed?

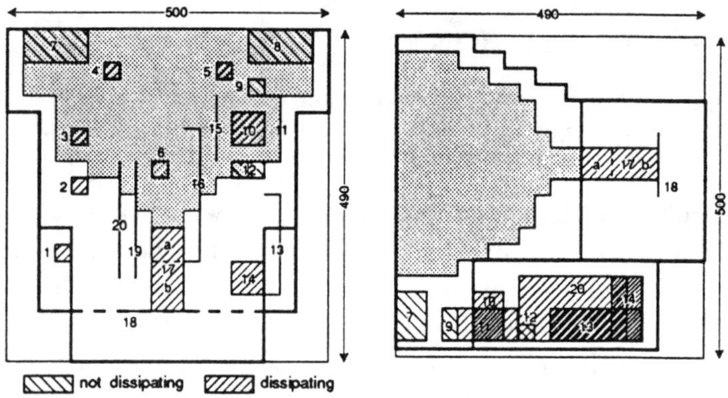

Position of elements in the TV set

1–6	resistors	14	LOT
7–8	speakers	15	sopspanel
9	elco1	16	heatsink3
10	SOPST	17	deflection
11	heatsink1	18	ptpanel
12	elco2	19	TXTpanel
13	heatsink3	20	PIPpanel

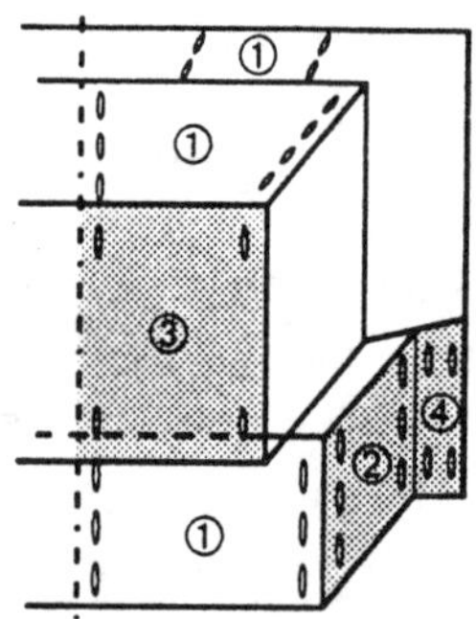

Figure 1.14: Location of the most important elements of a TV set.

- What is the optimal size of the heat sinks?
- What is the influence of the thermal conductivity of the board?

Note that prediction of accurate component temperatures is not included in this list because it is still rather difficult to use a CFD code for this purpose. This is due to the difficulties associated with the use of the very fine meshes required to determine the boundary layers around every object of interest and the influence of the ever-present instabilities of the flow upon the local heat transfer which are not covered by the CFD code. As will be shown in what follows, some pragmatic methods are used instead to calibrate the code. The philosophy underlying the proposed approach is based on CFD prediction of the local air temperatures and velocity fields in combination with experiments on a prototype to extract required information on several usually not well-known input parameters. Provided that the

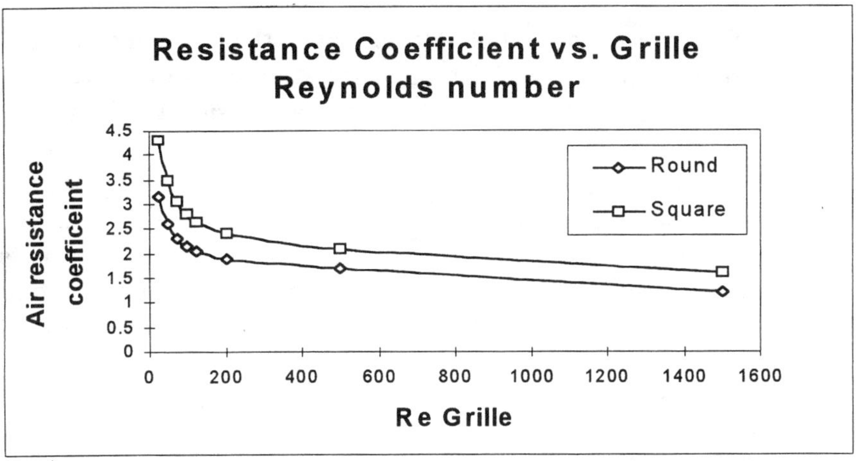

Figure 1.15: Air resistance coefficient as a function of the Reynolds number for commonly used grilles.

calibrated CFD analysis generates sufficiently accurate values, the code could be used to perform "numerical" experiments, which in many cases are more flexible, easier, cheaper and faster to perform than experiments with a prototype which is usually not available until the final stages of the Product Creation Process.

The local velocity and temperature fields were studied by means of a commercially available code (FLOTHERM) based on finite volume techniques. To simplify the analysis of the various conjugate heat transfer effects, the temperature fields caused by power transistors on flat heat sinks were separately calculated with software based on classical Fourier analysis techniques. These calculations were made in order to check the correctness of the assumption of a uniformly dissipating flat plate and to optimize the location of the transistors on their respective heat sinks.

For an accurate analysis of the air flow in the television set, values were needed for the grille air resistance coefficients. These coefficients have been measured for a variety of grilles, the parameters studied being the porosity and the shape and dimensions of the grille bars. Full 3-D numerical results provided clear evidence that the usual 1-D analytical equations failed to describe reality in this case.

To avoid the problem of excessive computer costs, measured values of the heat transfer coefficient were imposed on the faces of all dissipating components. Because in this study, performed in 1992, the mesh lines were at least 10-20 mm apart, the location of the reference temperature needed for the calculation of the heat transfer coefficient by using the imposed values lies outside the boundary layers. Note that in this case the problems become worse when the mesh size is reduced because the reference temperature will gradually increase. In the limit of a very fine mesh, the reference temperature becomes equal to the surface temperature. The numerical results were validated by comparison with numerous experiments on a full-scale model from which we obtained the temperatures of the critical objects and the "cell-averaged" air temperatures. Use was also made of smoke generators to visualize the flow inside the model which was fitted with a perspex back cover.

Observe, thanks to the remarkable increase in computer power, that after completion of the basic study, significant progress has been made in the simulation of various details. Mesh sizes can now be reduced to such an extent that 500,000 and more cells are no longer an exception. Conjugate and radiation heat transfer analyses have become standard. Still, the technology is far from the situation where it is possible to capture all of the details of the complex fluid flow that are required to perform an accurate analysis without experimental back-up.

In what follows, only a summary of this research is presented. More information can be found in Lasance (1991). From a purely experimental point of view, many details can be found in the paper by Yuki et al. (1990), which provides a wealth of information about the local effects on the temperature of several critical electronic parts of varying the position and area of inlet and outlet vents.

1.3.2.1 A PRAGMATIC APPROACH

The air resistance coefficient, the heat transfer coefficient, the optimization of flat plate heat sinks, and the experimental setup will now be discussed in some detail.

The Air Resistance Coefficient

Every opening, every grille and every object offers a resistance to air flow (Fig 1.16). When dealing with natural convection, lowering this air resistance is often the only way to increase the induced flow rate and thereby limit the air temperature rise. This requires

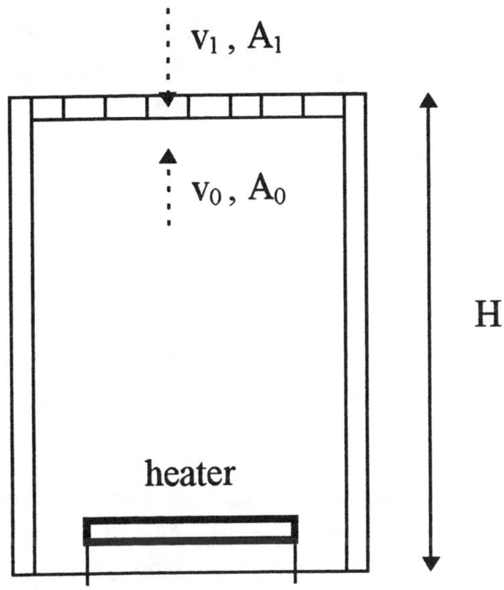

Figure 1.16: One-dimensional natural convection mass flow.

a knowledge of the ambient temperature and the location and the power dissipation of the components. However, the layout of many grilles in electronics applications are subject to several constraints such as safety regulations, electro magnetic compatibility requirements, styling aspects and manufacturability. In particular, industrial design aspects tend to become increasingly important and the ideas of the industrial stylists often clash with the ideas of the packaging engineer who is responsible for thermal management.

In addition, the mechanical engineer has his own opinion about the manufacturability of the parts which constitute a system such as, for example, the plastic back-cover of TV sets containing the required inlets and outlets. The complex 3D flow hampers the formulation of design rules [Lasance, (1991)]. For example, it has been observed that by closing parts of the grille, lower temperatures can be obtained locally. Hence, grille optimization, either by its location, area or resistance coefficient, is an important outcome of a 3-D CFD analysis. This optimization can be done as soon as the major aspects of the layout are known. One of the parameters required as input for the pre-processor is the air resistance coefficient of the inlet and outlet grilles.

First, some theoretical aspects will be discussed. The interested reader is referred to Idelchik (1986) where numerous data are presented for almost any imaginable grille configuration, together with the theoretical background.

The pressure loss due to the movement of air can be split into two major parts:

- Losses due to friction (frictional drag) near solid objects and
- Losses due to inertia; every change in the direction and velocity of the air causes an energy loss.

In electronics cooling, the first type of these losses can usually be neglected. This means that the air resistance is almost exclusively determined by the grilles and by the character of the flow near the grilles; laminar, turbulent or transitional. Because the pressure loss is a more or less quadratic function of the air velocity in most turbulent and transitional regimes, it is standard practice to define the resistance coefficient, ξ, as the ratio between the pressure and the product of the velocity squared and half of the air density. This ratio is also called the "velocity head," a well-known relationship emanating from theory based on first principles due to Bernoulli. Because in almost all practical cases the flow vs. pressure relationship can be described by a second order polynomial, the pressure drop can be expressed as [Idelchik (1986)]

$$\Delta p = \xi^{1/2} \rho v^2 = K_1 v + K_2 v_2 \tag{1.6}$$

where

$$\xi = \frac{K_3}{\mathrm{Re_g}} + K_4 \xi_{\mathrm{qu}} \tag{1.7}$$

in which K_1 through K_4 are proportionality factors, ξ_{qu} is the resistance coefficient for the square-law region, and $\mathrm{Re_g}$ is the Reynolds number, scaled with the hydraulic diameter of the grille. For $\mathrm{Re_g} > 10^5, \xi = \xi_{\mathrm{qu}}$ and hence, $K_4 = 1$. Generally, for $\mathrm{Re_g} < 25$, the second term in eq (1.7) can be neglected. At the upper extreme of $\mathrm{Re_g} > 2000$, the first term in eq (1.7) can be neglected.

Figure 1.15 shows an example of the air resistance coefficient of a grille commonly in use in consumer products.

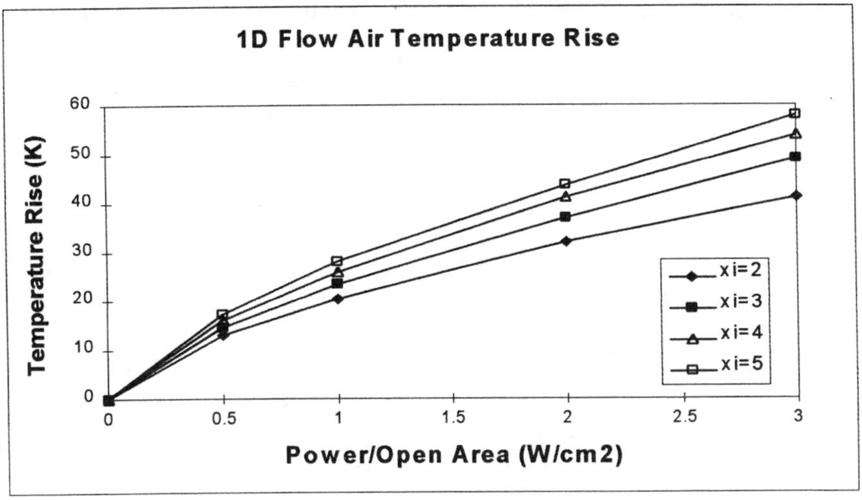

Figure 1.17: One-dimensional air temperature rise.

The horizontal axis presents Re_g and usually for natural convection, $\mathrm{Re}_g < 100$. As can be seen from the graph, ξ is heavily dependent on the air velocity in this region. It is this fact that introduces a serious problem for some numerical codes because in real systems 1-D flows are very rare. Consequently, in most cases of practical interest, ξ varies over the area of a grille. This is especially true for outlet grilles. To acquire some basic insight, the equations for a vented enclosure with a heat source generating a predominantly 1-D flow may be formulated. Typical values for a TV set are: $v = 30\,\mathrm{cm/s}$, $\mathrm{Re}_g = 100$, and $\xi = 3$. Then

$$\begin{aligned}
\Delta p &= \frac{1}{2}\xi \rho v^2 = (\rho_\infty - \rho) g H \\
\frac{\rho_\infty - \rho}{\rho} &= \frac{\delta T}{T_0} \\
\dot{V} &= v_0 A_0 = v_1 A_1 \qquad (1.8)\\
A_1 &= \beta A_0 \\
q &= \dot{V} \rho c_p \Delta T
\end{aligned}$$

These equations are valid when all of the power is concentrated in one plane. A more realistic situation is encountered when account is taken of the distributed sources of power in a real TV. In this case, the height, H, must be replaced by H_c which expresses the so-called "center of heat," which is analogous to the notion of the "center of mass." The center of heat is defined as [Noronha, (1965)]

$$q_{\text{tot}} H_c = \frac{\sum q_{\text{tot}} H_i}{n} \tag{1.9}$$

where n is equal to the number of heat sources and the reference plane for H is the top plane. The expression for ΔT can be derived from these equations

$$\Delta T = \left[\frac{\xi T_0}{2 g H_c} \left(\frac{q}{\rho c_p A} \right)^2 \right]^{1/3} \tag{1.10}$$

The air temperature difference is plotted as a function of the ratio of power to open area (q/A), with ξ taken as the parameter in Fig 1.17.

As an example, consider a TV with an open area of $150\,\text{cm}^2$, $q = 150\,\text{W}$, both inlet and outlet grilles with ξ of 2 (round) or 3 (square). Then the total resistance coefficients are 4 and 6, respectively. From Fig 1.17, ΔT's of 25°C and 30°C are found which could be a significant difference from the viewpoint of a designer. Observe, however, that eq (1.10) shows that the temperature rise varies as $\xi^{1/3}$ and A^2 so that an increase in A has much more effect than a decrease in ξ.

Unfortunately, the 3-D reality is quite different. Table 1.4 shows the FLOTHERM results for an air resistance coefficient varying between 1 and 50. The table presents the mass flow, the average air temperature rise of the outlet air flow and the maximum velocity. The 1-D approximation would predict a temperature rise factor of $50^{1/3} = 3.7$. From the full 3-D data, it is evident that the influence of the air resistance coefficient is far less than that predicted by the 1-D equation. This can best be explained by observing the flow patterns leaving the set. Usually the outlet grille is oversized and only part of it is really used by the main part of the flow. Increasing the air resistance results effectively in a larger area used by the outlet flow thereby compensating for the negative effect. On the other hand, for

Table 1.4: The influence of the air resistance coefficient on the temperature and velocity in the set.

ξ	**mass flow** kg/s	T_{out} °C	v_{max} m/s
1	1410	6.5	0.36
3	1390	6.6	0.34
10	1270	7.2	0.32
20	1180	7.8	0.30
50	1060	8.7	0.28

a completely optimized design operating near its limits, the effect described might again become important.

The Heat Transfer Coefficient

After opting to impose average heat transfer coefficients on all of the surfaces in contact with air, the problem that no data are available in the published literature must still be faced. Handbook correlations are not accurate enough because these correlations are almost exclusively measured under ideal wind tunnel conditions with no free stream turbulence and, for all but the highest dissipating elements, there is a buoyancy-induced forced convection situation with no natural boundary conditions. Moreover, general correlations are based upon numerous geometrical parameters and cannot easily be transferred to other geometries in 3-D. In fact, according to Moffat and Ortega (1988), no universal length-scale description has yet been found. In this case, the only solution seems to be to measure the heat transfer coefficient *in situ*. The basic idea is to use geometrically identical models made of polished copper fitted with a very small ceramic Pt100 element. These act both as a heat dissipator and a temperature sensor. Details are discussed in the previous section. In this case, the greatest problem is caused by the heat losses via the support. To be sure of a certain robustness, this support must have a diameter of at least one millimeter. For a common plastic, the heat losses are of the order between 10 and 15 percent estimated by repeating the experiments with different diameters and extrapolating to zero diameter.

The Optimization of Flat Plate Heat Sinks

Because of their essentially 2-D nature, low weight and favorable price, flat plate heat sinks are used in many consumer electronics applications to cool low to medium power transistors and IC's. Usually, more than one transistor is used on the same heat sink and odd shapes are the rule rather than the exception. Often, in the very early design stages, designers would like to have a fairly accurate estimation of the necessary size and shape. The optimization of these electronic parts is a good example of suboptimization because it is unnecessary to perform a parametric study on the entire system with the sole objective of optimizing some local structures. This point of view is only valid if the optimization of the subsystem is not dependent on the thermal behavior of the whole system. In other words, the arguments leading to an optimal design of the heat sink should be equally valid for a TV set with another layout or even other dimensions. Essentially, it requires independence of temperature and velocity fields. This assumption seems valid, because of the high thermal conductivity of aluminium and because the heat transfer coefficient is only a weak function of temperature. Briefly, the optimal layout of a certain heat sink with a given heat transfer coefficient and ambient temperature will also be optimal for other boundary conditions.

Design rules and analytical techniques useful for the optimization of flat plate heat sinks in TV sets are discussed by Lasance (1991, 1994b) and Krane et al (1993). Krane et al go one step further in the sense that the local heat transfer coefficient for a vertical heat sink on a horizontal board is also calculated while Lasance presumes a known average heat transfer coefficient.

1.3.2.2 THE EXPERIMENTAL SETUP

Because the high voltages in a real TV are a nuisance for the experimenter, most dissipating parts of the TV were simulated with resistors except for the power transistors on the heat sinks and the two transformers which are designated in Fig 1.14 by "SOPST" and "LOT." The transformers, operating with direct current, were esspecially designed for the experiments. Twenty thermocouples were positioned at characteristic locations in the set. Eleven thermocouples were fixed on solid objects while the rest measured air temperatures. Because large temperature gradients exist next to dissipating objects, the temperature measured by a thin thermocouple is very sensitive to its exact position. To fix this position and to average the temperatures around the thermocouple, the couple was fastened in a small copper cylinder which in turn was placed in a teflon holder. In this way the measured temperature can be more easily compared with the computed value, which is an average over a cell. The cover of the model of the TV set was made of perspex in order to facilitate viewing during smoke tests and was provided with numerous slots which could be easily closed with tape. This made it possible to investigate many different configurations.

After applying the power, temperatures were recorded at regular intervals by the data acquisition system ensuring that steady state was reached before the final data were recorded. The reference temperature in the final read-out was measured by a thermocouple inside a copper block outside the set thereby damping the temporal changes in the ambient temperature.

1.3.2.3 DISCUSSION OF SOME OF THE RESULTS

Some thirty five experiments were performed, the majority of them with a different layout of the grilles such as changes in porosity, position, size and area.

A few interesting conclusions could be drawn. For one thing, the influence of these various changes in layout on the temperature of solid objects appeared to be fairly minimal, of the order of 10 percent maximum. Substantial improvements could only be achieved locally by increasing the emissivity of the heat sinks (20 percent) and by reshaping or relocating these large flow-obstructing parts thereby redirecting the local flow (30 percent temperature reduction for the LOT). However, these results might easily change for TV sets with different power dissipations or layout.

The most important observation was that the agreement between experimental and numerical results was quite satisfactory. In particular, the flow patterns observed by the smoke visualization were correctly captured by the CFD analysis. Object temperature differences between runs were also correctly predicted within 10 percent or less. Air temperatures, as measured in the manner indicated in the previous section, are in the same inaccuracy range. It does not make much sense to compare absolute values because everything can be made accurate by simply tuning the unknowns such as thermal conductivities or imposed heat transfer coefficients. Recall that the grid size used was too large to analyze the boundary layers. Hence, the final conclusion of this research was that a CFD code, backed by a few experiments, could advantageously be used in the early stages of a design to study the various phenomena mentioned in the introduction.

1.3.2.4 CONCLUSIONS

It has been demonstrated that a CFD analysis with a relatively rough mesh applied to a natural convection-driven television set, can be satisfactorily used for the simulation of the local velocity and temperature field as a function of several important design parameters. This requires that some pragmatic measures are taken to estimate the unknown input parameters and the local heat transfer behavior around electronic parts such as flat plate heat sinks.

1.3.3 NATURAL CONVECTION HEAT TRANSFER IN AN ENCLOSURE FILLED WITH DIELECTRIC LIQUID

Characteristics of natural convection liquid immersion cooling of a three-by-three array of rectangular protrusions in an enclosure based on the recent studies by Joshi et al (1994) and Mukutmoni et al (1995) are presented in this section. The physical and computational challenges in liquid cooled electronic systems are illustrated by this model system. The high Rayleigh numbers encountered in dielectric liquids such as the Fluorinerts pose unique challenges to the numerical modeling of such systems. To reduce the complexity associated with the internal heat flow paths in the package, aluminum blocks equipped with thermo-foil heaters were employed in these studies. The elements geometrically simulated 20 pin dual-inline electronic packages and were mounted on a plexiglass substrate which formed one vertical wall of a dielectric liquid-filled rectangular enclosure.

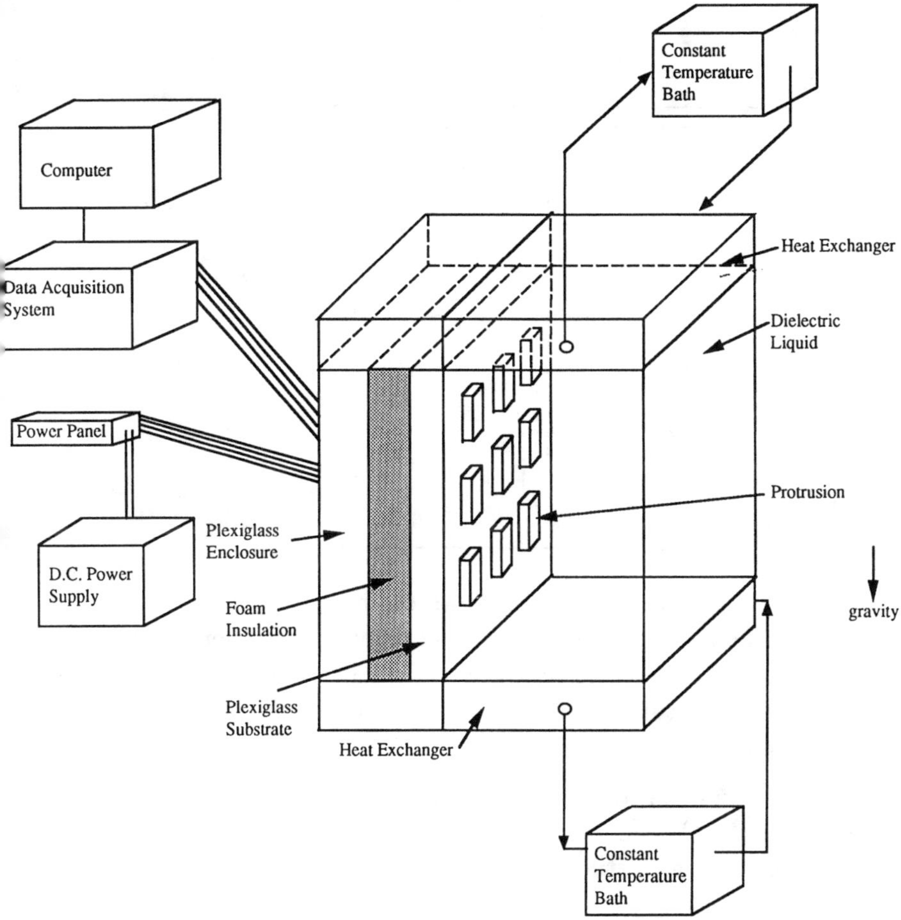

Figure 1.18: Schematic sketch of the experimental arrangement.

The remaining vertical boundaries of the enclosure were insulated while the top and bottom were maintained at prescribed temperatures using individual heat exchanger plates.

Protrusion surface temperatures in steady state were measured for a range of power dissipation levels for three Fluorinert liquids spanning a Prandtl number range from about 20 to 1400. The influence on element temperatures of the top and bottom surface boundary conditions in the enclosure and the width of the enclosure was investigated. Non-dimensional heat transfer results were empirically correlated. Changes in component temperatures due to partial powering of the array were also measured. Computations were carried out

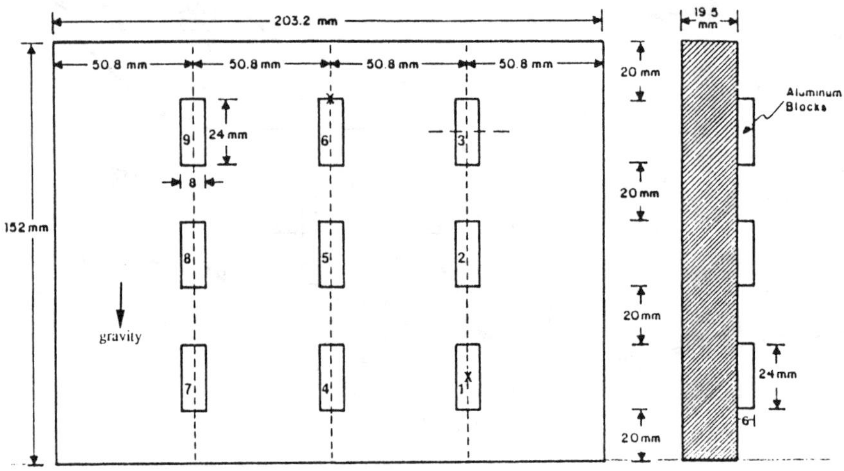

Figure 1.19: Front and side views of the plexiglass substrate and the aluminum protrusions. The component numbering scheme is identified. The crosses at the two locations indicate the positions where temperature fluctuations under nominally steady conditions were measured.

for FC75 and the code was first validated with experimental results. The effect of the substrate conductivity, k_s, on the heat transfer and fluid flow was also studied parametrically.

1.3.3.1 EXPERIMENTAL ARRANGEMENT AND PROCEDURE

The experimental apparatus is illustrated in Fig 1.18. The rectangular plexiglass enclosure was 203 mm long and 152 mm high. The chamber width, w, could be varied from 6 mm to 38 mm by repositioning the substrate containing the protrusions. The region behind this wall was filled with tightly packed foam insulation. The top and bottom boundaries of the enclosure were formed by removable water-cooled aluminum heat exchanger plates chosen to provide an almost isothermal surface condition. Once these plates were in place, the chamber could be filled with dielectric liquid through a capillary tube near the bottom. An additional partially fluid-filled capillary tube was connected near the top to allow for air escape and fluid expansion during component heating. All outer surfaces of the enclosure were insulated.

The geometric arrangement of the aluminum protrusions is shown in Fig 1.19. Thermofoil heaters were attached to the protrusion bases using a high thermal conductivity epoxy. The protrusions were then bonded in place using a fast-setting low thermal conductivity cement. Each heater consisted of an Inconel foil network on a Kapton backing. Protrusion surface temperatures were measured using embedded copper constantan thermocouples of 0.13 mm diameter. One thermocouple was attached at the center of each liquid-wetted face.

Electric power to the nine elements was provided with a DC power supply. The current through each heater was determined by measuring the voltage drop across a precision resistor connected in series. The nine heater and precision resistor pairs were connected in parallel to the power supply. An additional measurement of the voltage drop across the power supply allowed the determination of the heater power dissipation.

1.3.4 PROTRUSION TEMPERATURES IN FC75

The sensitivity of the heat transfer from the protrusions to changes in the enclosure boundary conditions was examined by considering three different cases. For the Type I enclosure condition, both the top and bottom boundaries were maintained nominally at 19°C, which was close to the ambient environment temperature. Two combinations of unequal, but constant, upper and lower boundary temperatures were also investigated. For the Type II condition the top surface of the enclosure was maintained at a nominal temperature of 10°C, while the bottom surface was kept at 15°C. This combination was reversed in the Type III runs where the top and bottom surfaces were at 15°C and 10°C respectively. In all of these runs the chamber width, w, was kept fixed at 13 mm because the objective was to study the effects of various parameters one at a time.

Average protrusion surface temperatures were determined from the measurements at various liquid exposed faces as

$$\theta_{\text{avg}} = \frac{\sum_{i=1}^{5} A_i \theta_i}{\sum_{i=1}^{5} A_i} \qquad (1.11)$$

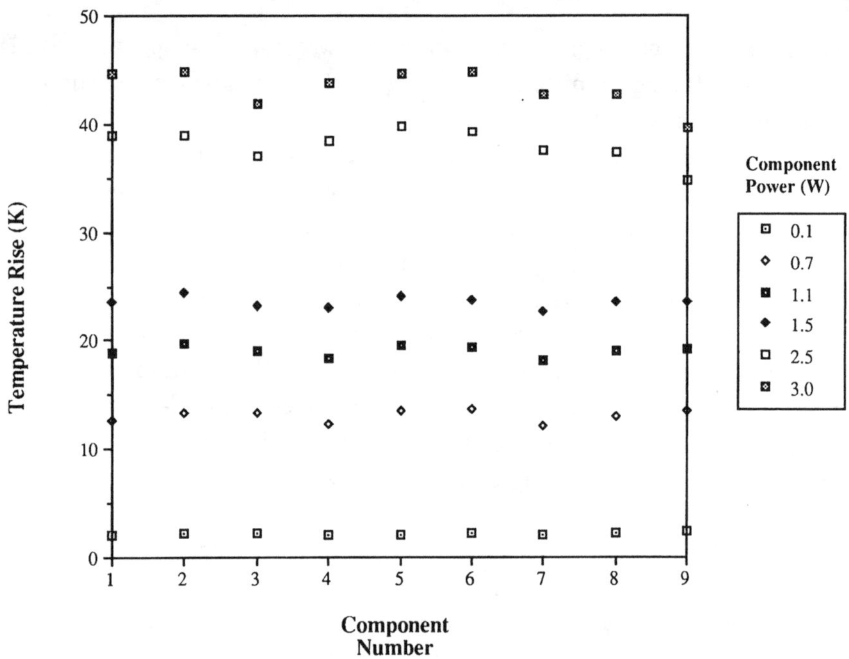

Figure 1.20: Average protrusion temperature rise in FC 75 for several power levels. Type I enclosure boundary conditions have been used with $w = 13\,\text{mm}$.

Measurements for each protrusion were made with all functioning thermocouples ($n = 5$ in most cases and $n = 4$ in some cases). The resulting average protrusion temperatures for a range of input power levels are shown in Fig 1.20 for the Type I condition. In Fig 1.20, at the lowest power level, the temperatures along each column show a gradual increase of about 10 percent from bottom to top. Each column shows a similar variation. Similar trends are seen for the power level of 0.7 W. These temperature profiles indicate a stably thermally stratified condition within the enclosure. Components along each horizontal row have almost the same average temperatures. Each column of protrusions appears to act almost independently of the others.

For higher input powers there are some changes in these characteristics. The bottom row of components still displays a thermally stratified pattern, evidenced by very close temperatures of protrusions 1, 4 and 7. Along the central column the middle component is the warmest followed by the top component except at the highest

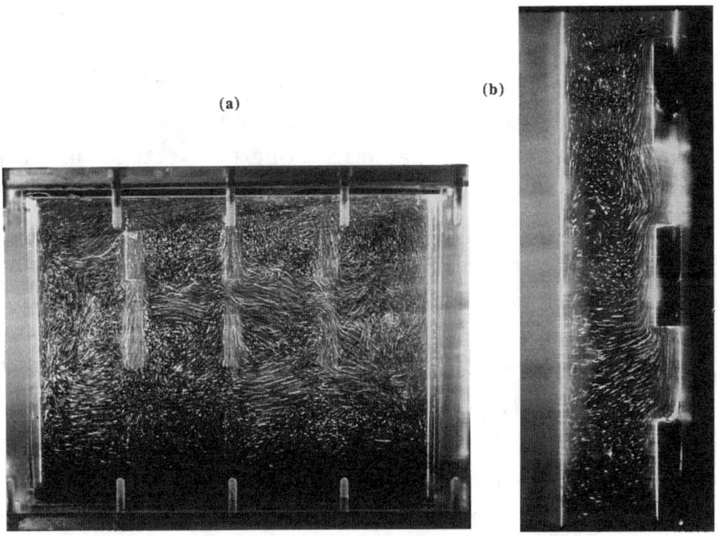

Figure 1.21: Flow patterns in FC 75 for input power of 1.5 W per protrusion. Exposure times are 10 seconds. In (a) the front view is seen in a plane adjacent to the protrusion front faces. In (b) the side view through the middle of the central protrusion column is seen.

power level. Moreover, in the flanking columns, the middle components are the warmest. However, at higher power levels the cooler upper boundary results in a drop in the temperatures of the top components compared to the bottom ones. The component temperatures within the array at a given power level do not vary by more than 15 percent.

The observed trends of temperature variations in Fig 1.20 are also supported qualitatively by the flow patterns shown in Fig 1.21 for an enclosure width of $w = 30\,\text{mm}$. These flow patterns were obtained by taking extended time exposure photographs of magnesium particles illuminated by a sheet of light from a helium-neon laser. Flow visualizations for smaller w were not successful due to the strong light reflections from the enclosure walls. Below the bottom row of protrusions almost no flow is seen in either the front or side views in Figs 1.21a and 1.21b. This confirms the existence of a stably thermally stratified region. A buoyant upflow is seen adjacent to the upper rows of protrusions. The side view through the middle of the central column of the array in Fig 1.21b shows that the upflow is largely confined to the region near the protrusions. The warm flow shed above the component columns is thermally unstable due to the presence of the colder liquid in contact with the upper boundary. This upflow spreads along the upper enclosure surface and returns downward, forming boundary layer type patterns along the vertical boundaries as shown in Fig 1.21a. The entrainment feeding the upflow along the central column of protrusions is seen to come primarily from the flanking columns. Because this fluid is warmer than the returning fluid along the vertical boundaries, higher temperatures along the central column, seen in Fig 1.20 are expected.

Figs 1.21a and 1.21b also show that the particle traces often follow irregular paths indicating the presence of timewise flow variations. It is to be noted that FC 75 has thermophysical properties which result in a value of the flow vigor parameter $g\beta/v\alpha$ of about an order of magnitude higher than that for water at a given temperature. Computational modeling of these flows at power levels of practical interest will require a three-dimensional transient formulation, as will be seen in a subsequent section. At power levels above 1.5 W in the present experiments, the flows appeared to be highly time-dependent and perhaps transitional or turbulent in nature. This is supported later by the timewise surface temperature responses.

Protrusion temperature measurements were also made with the upper and lower boundaries maintained at another equal and uniform temperature of 10°C. The general trends were still the same as in Fig 1.20. Except at the lowest power level, average protrusion temperatures were somewhat higher but within approximately 10 percent of the corresponding values in Fig 1.20 at the various power levels. These differences and their observed trend are to be expected due to the variations in the fluid thermophysical properties with temperature which cause a reduced flow vigor at lower fluid temperatures for the same input power. At the smallest power level, larger differences existed due to the greater influence of the enclosure boundaries. Even though the vertical boundaries were insulated, there was some heat transfer into the enclosure due to the lower heat sink surface temperatures compared to the ambient environment temperature of about 20°C. This is expected to have the greatest influence on component temperatures at the lowest power levels.

Protrusion temperatures for the type II and III boundary conditions are seen in Fig 1.22 as rises over the average of the top and bottom boundary temperature levels. In Fig 1.22a with the bottom boundary maintained at a higher temperature than the top, the temperatures are slightly lower than in Fig 1.20 for comparable power levels. The warmer bottom boundary temperatures presumably result in stronger flows and better heat transfer. This improvement is greatest at lower power levels and the temperatures are only about 10 percent lower compared to the Type I condition values at the highest power input. A stable thermally stratified condition in the absence of any input power to components is maintained in the enclosure by reversing the top and bottom boundary temperatures in Fig 1.22b. This is expected to be the worst of the three thermal conditions from a heat transfer standpoint because the tendency for fluid circulation by buoyancy adjacent to the protrusions is suppressed at low power levels. This is confirmed by the somewhat higher protrusion temperatures at all power levels except the highest two. Except for the two lowest power levels these differences are within about 10 percent of the Type I condition measurements.

In a comparison of Figs 1.20 and 1.22, it is clear that the protrusion temperatures are most sensitive to the enclosure top and bottom temperatures at the lower power levels. For power levels of 0.7 W and above, the deviations in average protrusion temperatures resulting from changes in the enclosure conditions are within 15 percent. Heat transfer from the protrusions is then dominated by the strong buoyant flows adjacent to them, with a relatively weak influence of the enclosure boundaries, for the three sets of conditions examined.

Measurements reported in Figs 1.20 and 1.22 are for $w = 13\,\mathrm{mm}$. This spacing was chosen to provide a reasonably small gap of 7 mm between the protrusion front faces and the opposite enclosure wall. This was done to realistically simulate electronic packaging applications where space is at a premium. Larger w values will only provide

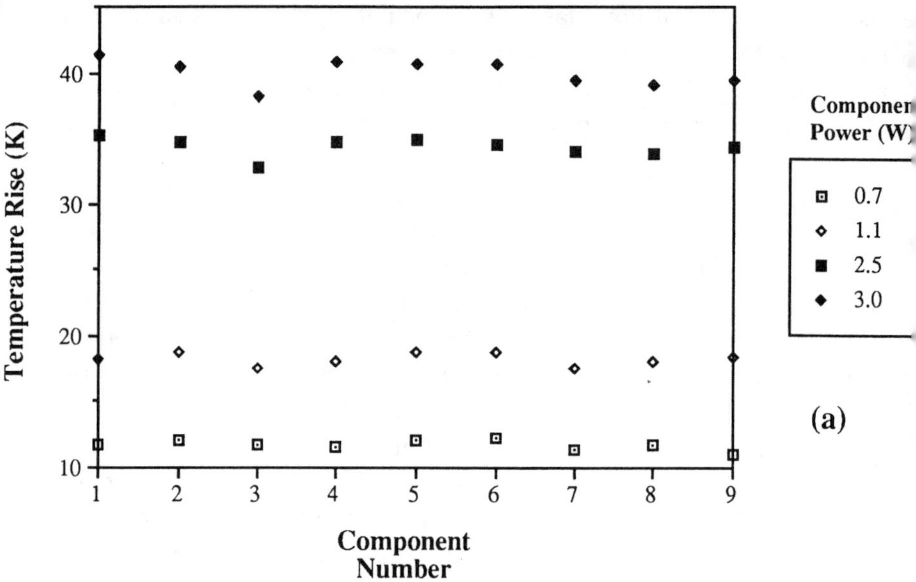

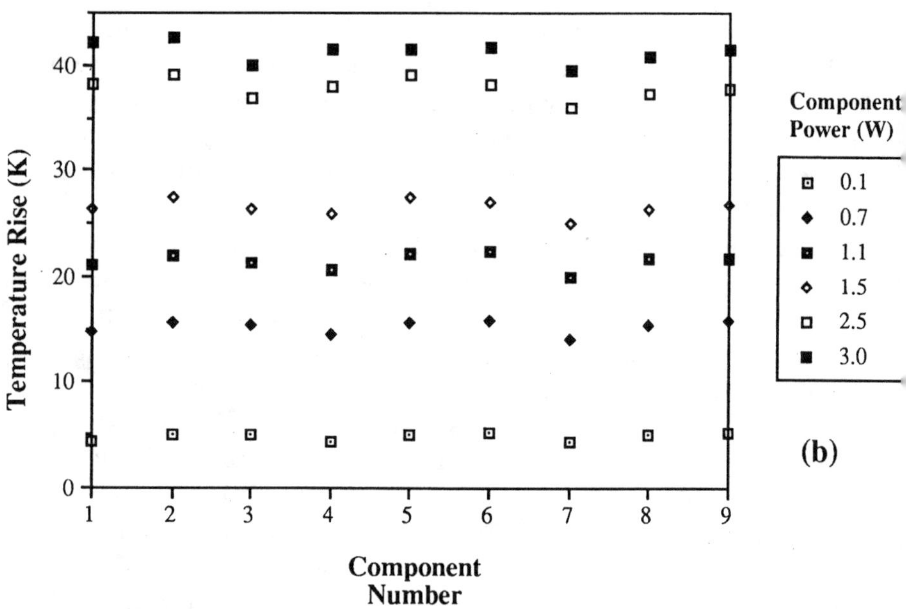

Figure 1.22. Average protrusion surface temperature rise in FC75 for several power levels with (a) Type II and (b) Type III enclosure boundary conditions for $w = 13\,\mathrm{mm}$.

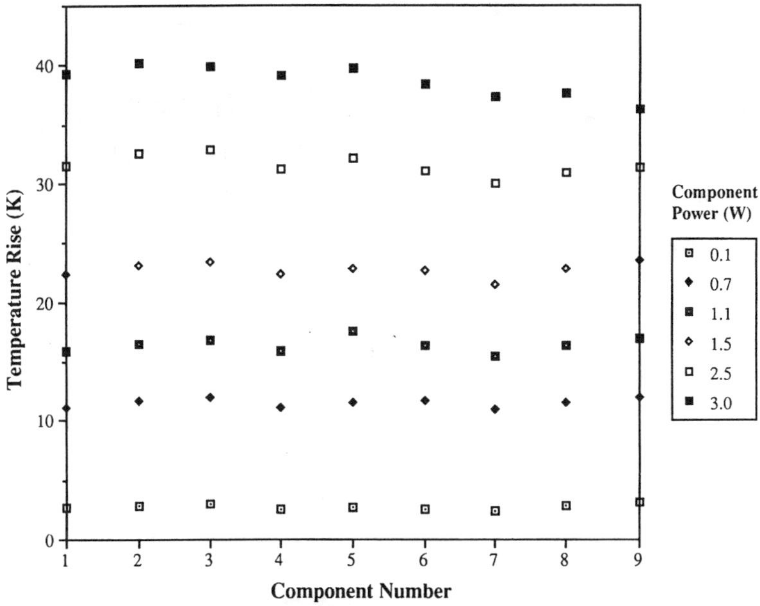

Figure 1.23: Protrusion temperatures in FC75 for a larger enclosure width of $w = 30\,\mathrm{mm}$ for Type I enclosure boundarty conditions.

improvements in heat transfer performance. To investigate the effects of enclosure width on the protrusion temperatures, a larger width of $w = 30\,\mathrm{mm}$ was selected and the boundary conditions of type I were used. The resulting steady protrusion temperature rises at several power levels are shown in Fig 1.23. As expected, these are lower than the corresponding levels for $w = 13\,\mathrm{mm}$ in Fig 1.20a. The differences are within about 10 to 15 percent at each power level indicating a relatively weak effect of w.

1.3.5 TEMPERATURE FLUCTUATIONS UNDER NOMINALLY STEADY CONDITIONS

Because the flow patterns in FC75 appeared to be time-dependent, the effect of these variations on the surface temperatures was determined. Several thermocouples were scanned for a period of 200 seconds at a rate of approximately three times per second. These measurements for the Type I enclosure boundary condition with top and bottom wall temperatures of 10°C are presented at several input power levels in Figs 1.24 and 1.25 for $w = 13\,\mathrm{mm}$ and $30\,\mathrm{mm}$ respectively. The two measurement locations, one in the top row and the

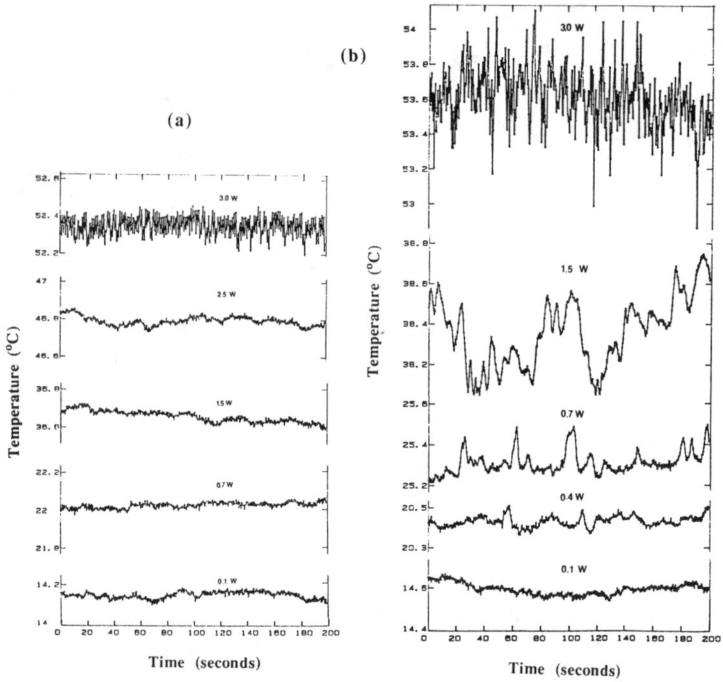

Figure 1.24: Temperature fluctuations measured by surface thermocouples under nominally steady conditions in FC75 at several protrusion power levels. Type I enclosure boundary conditions have been emloyed with a chamber width of $w = 13\,\text{mm}$. The fluctuations are displayed for protrusion locations in the (a) bottom and (b) top rows as identified in Figure 1.19.

other in the bottom row of protrusions, are identified in Fig 1.22a. Observe that that because these outputs are from surface-embedded thermocouples the temperature oscillations in the fluid are actually of larger amplitude and may also contain higher frequency components.

In Fig 1.24a, the temperature responses show no timewise fluctuations in the bottom row protrusion except at the two highest input powers. At the highest power of 3 W, the fluctuation amplitudes are about 0.2°C. At the upper location in Fig 1.24b at 0.4 W, fluctuation amplitudes of about 0.2°C are observed. These amplitudes become larger and higher frequency components are observed in the traces as the input power is increased. At the highest input power of 3 W the fluctuation amplitudes are about 1.4°C. The timewise variations at a larger enclosure width of $w = 13\,\text{mm}$ in Fig 1.25a show a steady pattern in the bottom row for all power levels examined. At the upper location in Fig 1.25b, the fluctuation amplitudes are comparable to those in Fig 1.24b. However, some of the higher frequencies are absent. It is interesting to point out that in an earlier study by Joshi et al (1990)) where the protrusions were arranged with their largest dimension aligned horizontally, the oscillation amplitudes in the top row were only about half of the values in Fig 1.25b.

The oscillation amplitudes in Figs 1.24 and 1.25 are well within 2 percent of the mean temperature increase at any location. The use of the instantaneous measured values of surface temperatures in the calculations of heat transfer characteristics was therefore considered appropriate once nominally steady conditions had been achieved. The non-dimensional transport correlations based on these measurements are presented next. It is noted that for some electronic components such as small plastic transistors, light emitting diodes (LEDs) and newer packaging schemes such as direct chip attach, much larger temperature fluctuations may arise than were found here.

1.3.5.1 NON-DIMENSIONAL HEAT TRANSFER CHARACTERISTICS

The average protrusion surface temperatures were used to develop general non-dimensional heat transfer correlations. The Nusselt numbers were obtained from the relationship

$$\text{Nu} = \frac{Q_{\text{net}} L}{k A (T_{\text{avg}} - T_{\text{ref}})} \tag{1.12}$$

where Q_{net} is the net rate of energy transfer to the fluid per protrusion, estimated by subtracting the back loss through the substrate from the input power, Q. Estimates of the back conduction loss were made using a one-dimensional model. These were within 5 percent of the power input. The area used in eq (1.12) is the total fluid exposed surface area of the protrusion. It is noted that there is also some transport of energy into the fluid resulting from the conductive spreading within the substrate from the protrusion bases. This effective increase in the area is not included in the calculation of A in eq (1.12) because this generally cannot be easily determined by the user. The present results are applicable to thermally insulating

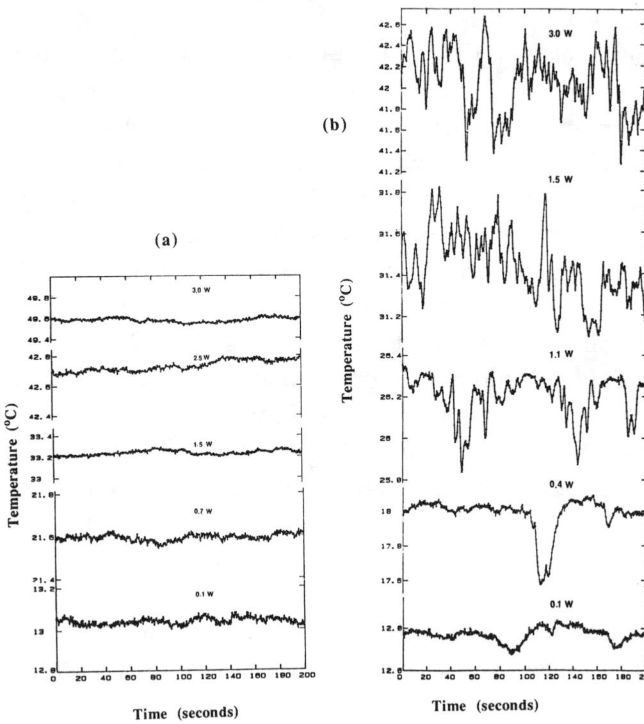

Figure 1.25: Temperature fluctuations in FC75 for a larger enclosure width of $w = 30\,\mathrm{mm}$. All other conditions are the same as in Figure 3.22. The fluctuations are displayed for protrusion locations in the (a) bottom and (b) top rows as identified in Figure 3.19.

substrates such as epoxy-fiberglass, which have about the same ratio of the liquid to substrate thermal conductivity as in the experiments. Protrusion temperatures for the same input powers will be lower for more highly-conductive substrates such as ceramics.

Several choices of the length scale are possible. The vertical extent of each protrusion was selected for the present calculations because it characterizes the local flow vigor. As will be seen subsequently, it also allows comparisons with previous experiments. The choice of a reference temperature for the calculation of temperature differences is also not unique when the upper and lower enclosure surfaces are maintained at different temperatures as for the Type II and III conditions in the present study. The average of the top and bottom enclosure surface temperatures was used as T_{ref} in the present calculations.

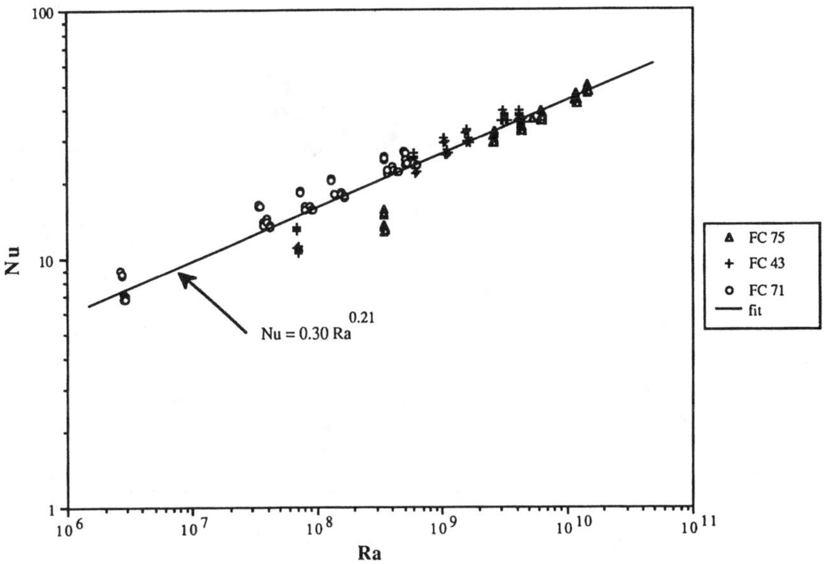

Figure 1.26: Non-dimensional heat transfer measurements for Type III boundary conditions for $w = 13\,\text{mm}$. Data for all protrusions in the three liquids are plotted, along with the linear best fit.

A modified Rayleigh number was defined by

$$\hat{\text{Ra}} = \frac{g\beta L^4 Q_{\text{net}}}{lAv\alpha} \tag{1.13}$$

This definition is based on the net rate of heat loss from the protrusion to the liquid and not on its temperature. In this form, a correlation developed for the Nusselt number as a function of the Rayleigh number will allow a direct estimation of component temperature based on power input. The fluid properties needed in the estimation of the Nusselt number and the Rayleigh number were evaluated at the film temperature

$$T_{\text{film}} = \frac{T_{\text{av}} + T_{\text{ref}}}{2}$$

The calculations of the Nusselt and Rayleigh numbers were done within a data reduction program which used curve fits in the range

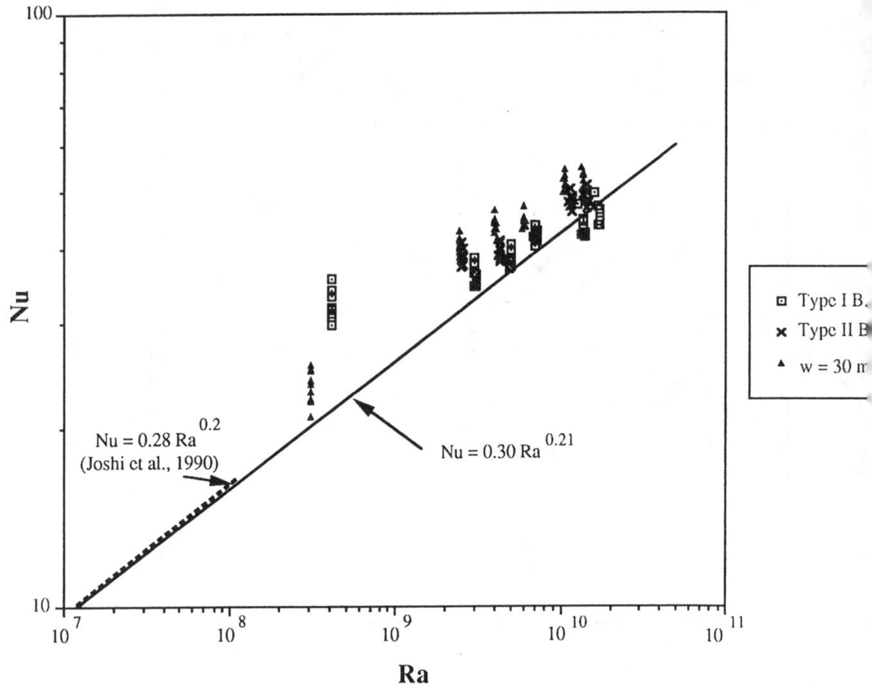

Figure 1.27: Effects of enclosure boundary conditions, enclosure width and protrusion orientation on the non-dimensional heat transfer measurements.

of interest for the relevant thermophysical properties of the three Fluorocarbon liquids.

The resulting data for the Nusselt number as a function of the Rayleigh number for the nine protrusions are shown in Fig 1.24 for the three dielectric liquids for the type III boundary condition with $w = 13\,\text{mm}$. The band at each Rayleigh number is the result of the variation in average protrusion temperatures within the array at a given power level. The data show approximately a six to sevenfold increase in the Nusselt number as the Rayleigh number increases by about four orders of magnitude for each protrusion. An empirical best fit to all the data over the range of $2 \times 10^6 \leq \hat{\text{Ra}} \leq 2 \times 10^{10}$ and $25 \leq \text{Pr} \leq 400$ is given by the equation

$$\text{Nu} = 0.30\hat{\text{Ra}}^{0.21} \tag{1.14}$$

The maximum deviation from eq (1.14) takes place at the lowest power level for each of the three liquids. For FC75 for $Q = 0.1\,\text{W}$, the data are as much as 32 percent below the correlation. As discussed earlier, the influence of enclosure boundaries and the ambient conditions is strongest for these conditions. This large deviation at the lowest power level may not be relevant in actual applications since the interest is usually in examining the transport at higher power levels. At power levels above $Q = 0.1\,\mathrm{W}$, 95 percent of the present measurements are within $\pm$ 15 percent of the correlation.

The correlation obtained in eq (1.14) was compared with the data obtained for the enclosure Type I and Type II boundary conditions with $w = 13\,\mathrm{mm}$. As shown in Fig 1.27, good agreement is seen except at the lowest power input.

Also seen in Fig 1.27 are the present data for $w = 30\,\mathrm{mm}$ and the empirical curve fit obtained for a different orientation of protrusions of the same size in FC75 [Joshi et al (1990)]. The protrusions in the earlier study were mounted with their largest geometrical dimension oriented horizontally. The enclosure dimensions were smaller than in the present study. In Fig 1.27 the earlier correlation obtained for $w = 30\,\mathrm{mm}$ is about 20 percent lower compared to the present data for the same w. This difference shows the expected trend since the vertical protrusion orientation is likely to result in more vigorous flows at a given input power.

It must be emphasized here that the applicability of heat transfer correlations is limited to the specific parametric ranges over which they are developed. In electronics cooling applications, geometrical complexity, discrete heating and the presence of conjugate heat transfer effects require extensive parametric studies to achieve some generality in results. The greatest potential for the use of such correlations is in conduction-type board and component level solvers. Another promising area is their use with thermally-optimized component placement schemes based on genetic algorithms. The interested reader is encouraged to consult the works of Queipo et al (1994) and Vikram (1996).

1.3.6 NUMERICAL FORMULATION

The experimental configuration in Fig 1.20 was also numerically simulated. Detailed geometric parameters are identified in Fig 1.28. Table 1.4 gives a complete numerical listing of the geometric and thermo-physical parameters in the present study. The thermal con-

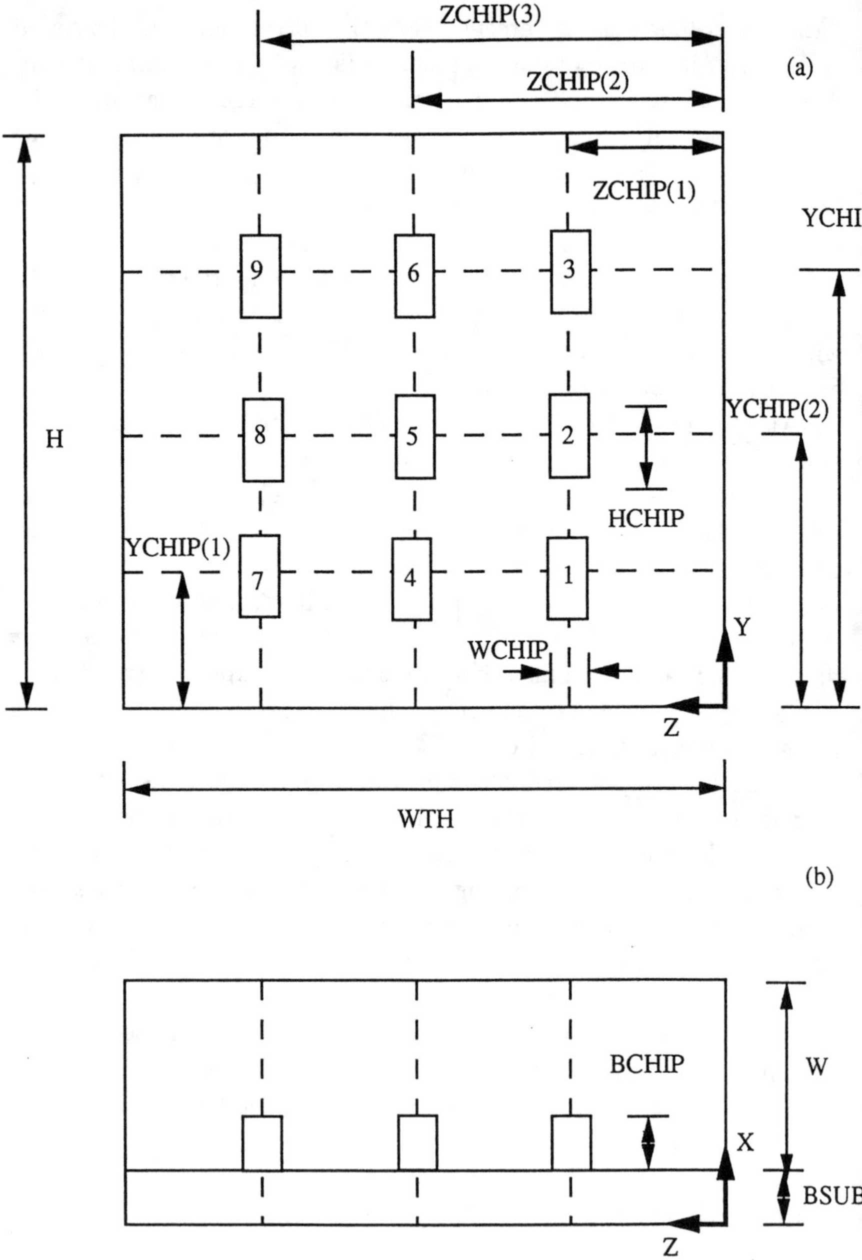

Figure 1.28. Dimensions of the enclosure and protrusions, projected on (a) y-z plane and (b) x-z plane.

ductivity ratio of the substrate and fluid was varied between 1 and 100 with all other parameters fixed. This range covers most of the substrates commonly used in applications.

The non-dimensionalized governing equations are

$$\nabla \cdot \vec{U} = 0 \tag{1.15}$$

$$\frac{\partial u}{\partial t} + \nabla \cdot (u\vec{U}) = -\frac{\partial p}{\partial x} + \nabla \cdot (\mathrm{Pr}\nabla u) \tag{1.16}$$

$$\frac{\partial v}{\partial t} + \nabla \cdot (v\vec{U}) = -\frac{\partial p}{\partial y} + \nabla \cdot (\mathrm{Pr}\nabla v) \tag{1.17}$$

$$\frac{\partial w}{\partial t} + \nabla \cdot (w\vec{U}) = -\frac{\partial p}{\partial z} + \nabla \cdot (\mathrm{Pr}\nabla w) \tag{1.18}$$

$$\frac{\partial T}{\partial t} + \nabla \cdot (T\vec{U}) = \nabla^2 T \tag{1.19}$$

In these equations, the standard Boussinesq approximations are used with one exception. The viscosity of FC 75 is treated as temperature dependent. The substrate temperature is governed by

$$\frac{R_s}{R}\frac{\partial T}{\partial t} = \frac{k_s}{k}\nabla^2 T \tag{1.20}$$

and the energy equation for each protrusion for most of its volume is

$$\frac{R_c}{R}\frac{\partial T}{\partial t} = \frac{k_c}{k}\nabla^2 T \tag{1.21}$$

The volumetric heat source is restricted to a layer in the protrusion that is one control volume thick in the x-direction immediately next to the substrate. This simulates the heating of the protrusions at their bases by thermofoil heaters in the experiments. The thickness of this layer for the present computations is 0.5 mm. The energy equation for this layer is

$$\frac{R_c}{R}\frac{\partial T}{\partial t} = \frac{k_c}{k}(\nabla^2 T + 1) \tag{1.22}$$

In the foregoing equations, the non-dimensional spatial coordinates x, y and z were obtained from the dimensional coordinates X, Y and Z by scaling with H, the enclosure height. The velocities were scaled by α/H, the time by H^2/α and the difference between the total pressure and the equilibrium hydrostatic pressure in the absence of a temperature gradient was scaled by $\rho\alpha^2/H$. The temperature was non-dimensionalized by first subtracting T_{wall} and then dividing by $q'''H^2/k_c$, where T_{wall} is the temperature of the top and bottom walls. In addition, k_c and k_s are the thermal conductivities of the chip and substrate, respectively, and R_c and R_s are the respective products of the density and specific heat for the protrusion and substrate. Note that Pr, the Prandtl number, is a function of the temperature and hence the spatial coordinates. Bo, the Boussinesq number, is the product of the Rayleigh number and the Prandtl number.

The boundary conditions consistent with the adiabatic side walls (including the back of the substrate) and isothermal top and bottom walls in a non-dimensional form are

$$x = 0, A_x, \quad 0 \le z \le A_x, \quad 0 \le y \le 1$$

$$u = v = w = 0 \quad \text{and} \quad \frac{\partial T}{\partial x} = 0 \tag{1.23}$$

$$x = 0, A_z, \quad 0 \le x \le A_x, \quad 0 \le y \le 1$$

$$u = v = w = 0 \quad \text{and} \quad \frac{\partial T}{\partial z} = 0 \tag{1.24}$$

$$z = 0, 1, \quad 0 \le x \le A_x, \quad 0 \le z \le A_z$$

$$u = v = w = 0 \quad \text{and} \quad T = 0 \tag{1.25}$$

From the governing equations and boundary conditions, four additional parameters are found, namely the thermal conductivity ratios and density-specific heat product ratios of the protrusion-fluid and the substrate-fluid combinations. The property parameter values used in this study, provided in Table 1.5, are chosen to match the experiments.

The temperature-dependent viscosity (Prandtl number in the non-dimensional formulation) for FC75 has been taken from a curve fit to the properties data

$$\begin{aligned}\Pr(T_d) = 29.088(1.4074 - 2.96 \times 10^{-2}T_d + 3.8018 \times 10^{-4}T_d^2 \\ - 2.7316 \times 10^{-6}T_d^3 + 8.168 \times 10^{-9}T_d^4)\end{aligned} \tag{1.26}$$

Table 1.5: List of the prescribed parameters. Units are given in the nomenclature.

Parameter	Value	**Parameter**	Value
W_e	0.2032	H	0.152
W	0.030	B_s	0.05
H_c	0.024	W_c	0.008
b_c	0.006	$Y_{c,1}$	0.038
$Y_{c,2}$	0.076	$Y_{c,3}$	0.114
$z_{c,1}$	0.0508	$z_{c,2}$	0.1016
$z_{c,3}$	0.1524	Q	0.1, 0.7
T_{wall}	10.0		
k_c/k	3732.0	k_s/k	A Variable
R_c/R	1.2	R_s/R	0.262
k	0.064		

The temperature T_d in eq (1.26) is dimensional and in degrees Celsius. For high power levels when the temperature rise in the protrusions is in the order of tens of degrees, the Prandtl number can vary by a factor of 2.

The governing equations were solved in primitive variables on a three-dimensional staggered grid, based on the finite volume method [Patankar (1980)]. The QUICK scheme [Leonard (1988)] and the SIMPLEX algorithm [van Doormaal and Raithby (1984)] were used for their solution. The time step was chosen such that the maximum Courant number $u\delta t/\delta x$ where δt and δx are the time step and grid size, respectively, was maintained at less than 1.0. The time-stepping was achieved by a first order backward Euler scheme. Non-uniform grids are required for this problem because the flow field has a boundary layer type of behavior near the wall along the substrate and protrusions. A $30 \times 36 \times 66$ non-uniform grid was used in the calculation, with extra resolution provided near the boundary layer regions. Further details of the numerical computations are provided by Mukutmoni et al (1993).

(a)

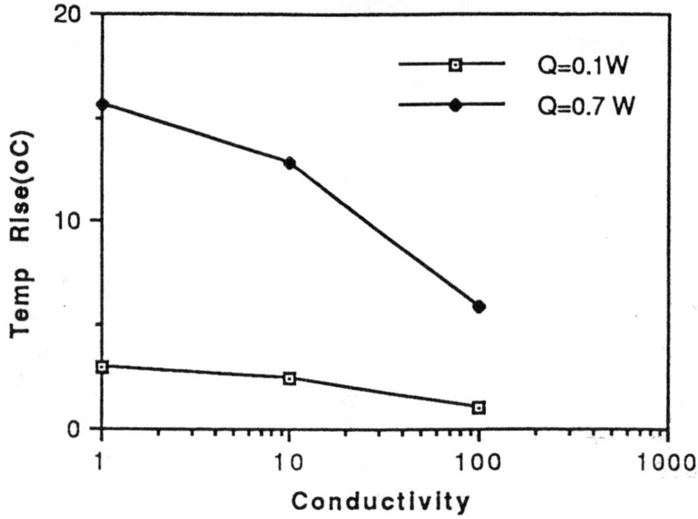

(b)

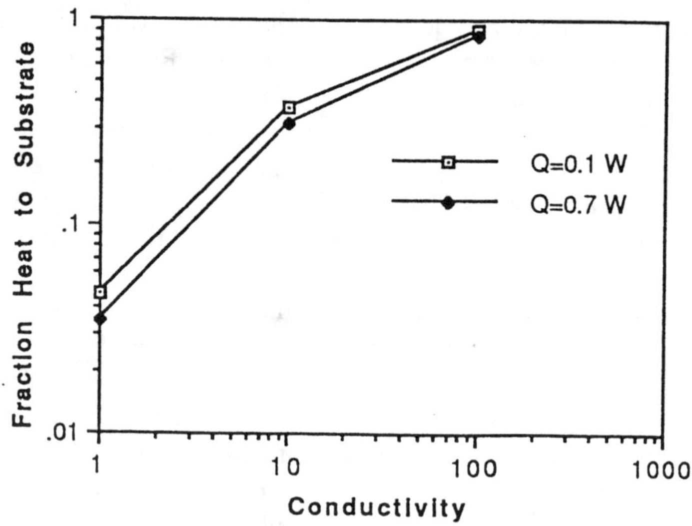

Figure 1.29. The effect of substrate-to-fluid thermal conductivity on (a) the average protrusion temperature rise and (b) the fraction of net heat generated that is conducted to the substrate.

Table 1.6: Comparison between protrusion temperatures (°C) of experiments of Joshi et al (1994) [referred to as Exp] and the computations [referred to as Num] for power levels of 0.1 W and 0.7 W. The conductivity ratio between the substrate and the fluid is unity.

Component	Q = 0.1 W		Q = 0.7 W	
Number	Num	Exp	Num	Exp
Top Row Average	13.1	13.1	26.2	N/A
Bottom Row Average	12.9	12.8	24.8	21.6

1.3.7 VALIDATION AND REFINEMENT STUDIES

The results of the calculations were first compared with experimental results for $Q = 0.1\,\mathrm{W}$ and $Q = 0.7\,\mathrm{W}$ per protrusion. The protrusion temperatures were calculated by taking a volume average of the control volume temperatures contained in the protrusions. The average protrusion temperatures of the bottom and top rows are compared in Table 1.5 and show good agreement.

The same qualitative features are also maintained within the protrusion array. In both the computations and experiments, the bottom row of protrusions is relatively cooler than the top row. The better agreement for the lower power level (Table 1.6) is due to the fact that in the numerical model the side walls were assumed to be adiabatic. In reality, as the protrusion temperatures rise, the heat loss through the side walls is no longer negligible. These heat losses were not monitored in the experiments, making it difficult to incorporate them in the numerical model.

The results given in Table 1.6 are the end result of a large volume of computations with various non-uniform grid sizes. Basically, the grid sizes were decreased until the protrusion temperatures were sufficiently close to those of the experiments. Various other parameters were also adjusted such as the resolution near the protrusions and the sizes of the control volumes next to the wall. The comparison with experiments also indicates the essential correctness and accuracy of the code as well as the adequacy of the grid resolution.

(a)

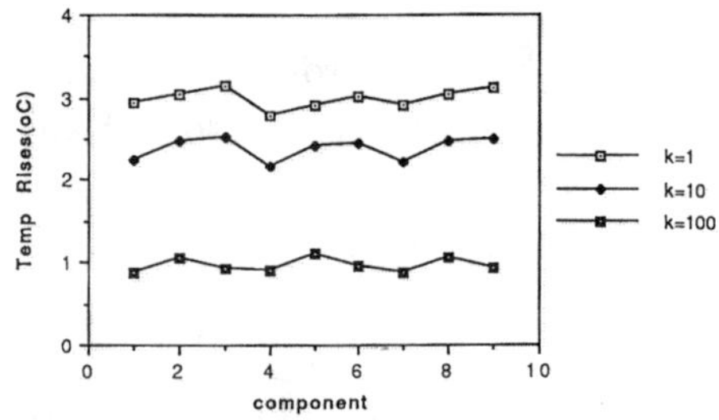

(b)

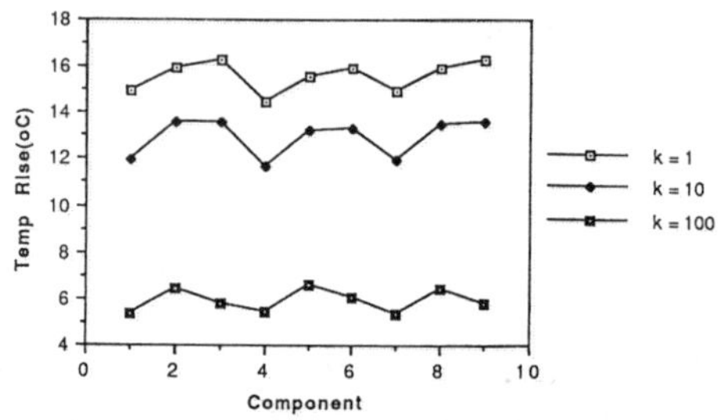

Figure 1.30: The temperature rise of individual protrusions for different substrate-to-fluid thermal conductivity ratios: (a) Q =0.1 W and (b) Q = 0.7 W. The component numbers are as shown in Figure 3.12.

1.3.7.1 THE EFFECT OF SUBSTRATE CONDUCTIVITY

With the validated code, the effect of varying k_s was investigated with all geometric dimensions and thermo-physical parameters fixed. Two different power levels were investigated, namely 0.1 W and 0.7 W per protrusion. The resulting velocity and temperature fields for all substrate thermal conductivities were unsteady. However, the temperatures within the protrusions and the substrate were found to be relatively steady due to the disturbance-dampening effect of the solid regions. Overall, it was found that the substrate thermal conductivity significantly affects the heat transfer from the protrusions. The flow field is affected less profoundly. These features will now be examined in more detail.

The Temperature Field

The average protrusion temperature rise as a function of k_s/k is shown in Fig 1.29a. As k_s/k is increased from 1 to 100, the average

(a)

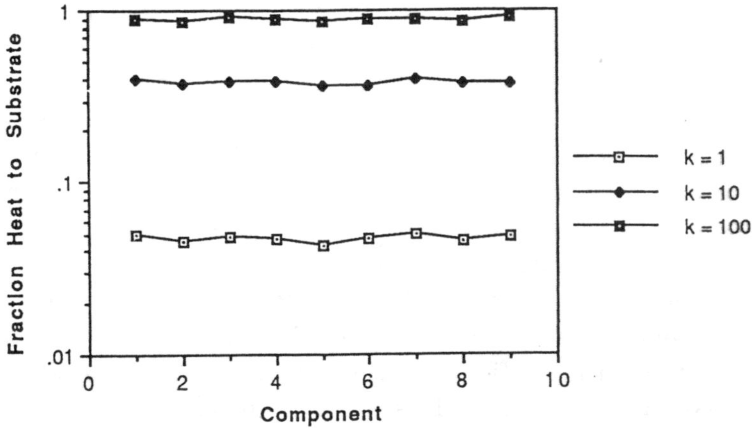

(b)

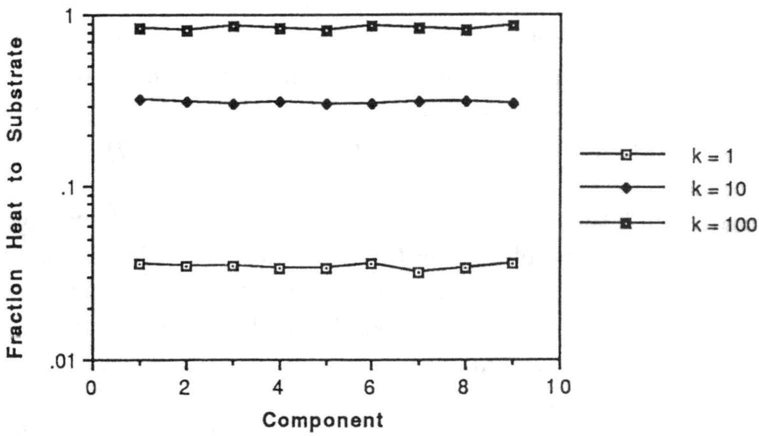

Figure 1.31. The fraction of net generated power conducted to the substrate for individual protrusions for different substrate-to-fluid thermal conductivity ratios: (a) $Q = 0.1\,\mathrm{W}$ and (b) $Q = 0.7\,\mathrm{W}$.

protrusion temperatures are reduced by almost 50 percent and for the same conductivity ratio range, the percentage of heat passing directly into the substrate from the protrusions is increased significantly. At a conductivity ratio of unity, as shown in Fig 1.29b, barely 5 percent of the generated power is conducted directly into the substrate. This demonstrates the relative dominance of the convection process. At a conductivity ratio of 100 which is typical of a ceramic substrate in a fluorocarbon liquid, almost 90 percent of the power generated, prior to being convected into the fluid, passes directly into the substrate.

A high thermal conductivity substrate acts as a heat sink and reduces the protrusion temperatures. Heat conducted from the protrusions to the substrate is also efficiently transferred to the top and bottom enclosure walls. In Fig 1.29b, for the same substrate conductivity, the fraction of heat transferred to the substrate is slightly lower at the higher Q. This is a reflection of the fact that convection is relatively stronger for higher Q's.

The temperatures of the individual protrusions are shown in Fig 1.30 for Q's of 0.1 W and 0.7 W. The component numbers are indicated in Fig 1.28a. The temperatures of the components for different thermal conductivity ratios for $Q = 0.1\,\text{W}$ are shown in Fig 1.30. It can be seen that a rough vertical symmetry is maintained. The component temperatures in the left column (7,8 and 9) are roughly the same as the right column temperatures (1,2 and 3).

For all cases, the coolest temperatures are located at the bottom row and the coolest component is always 4 (bottom row, middle column). It is interesting to note that, for the lower conductivities, the hottest row is at the top. For the highest conductivity, Fig 1.30 shows that it is the middle row that is the hottest and that the results for $Q = 0.7\,\text{W}$ are qualitatively similar (Fig 1.30b). The component temperatures are higher, as expected.

Figure 1.31 shows the fraction of heat loss to the substrate for individual components. The fraction of heat loss to the substrate was found to be a maximum for the components in the bottom row. The only exception is for a conductivity ratio of 100 where it is the top row of components whose fraction of heat to the substrate is highest. This is consistent with the protrusion temperature results. It was found that, as k_s is increased, the top row components get relatively

(a)

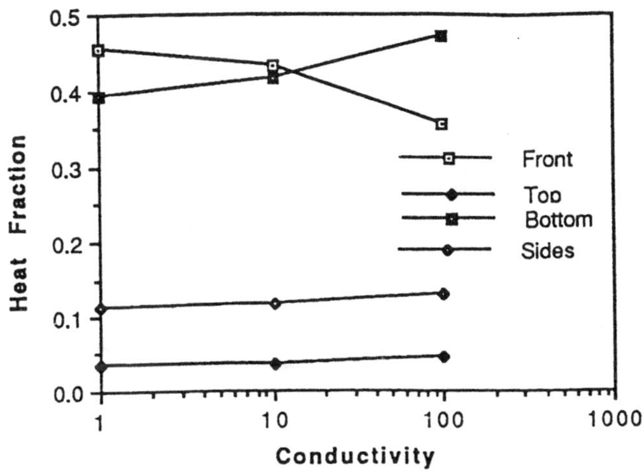

(b)

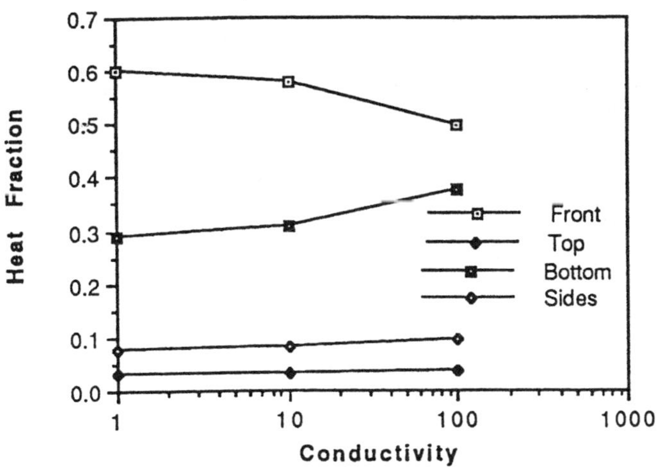

Figure 1.32. Effect of k_s/k on the fraction of net generated power convected to the fluid from the various faces for (a) $Q = 0.1\,\mathrm{W}$ and (b) $Q = 0.7\,\mathrm{W}$.

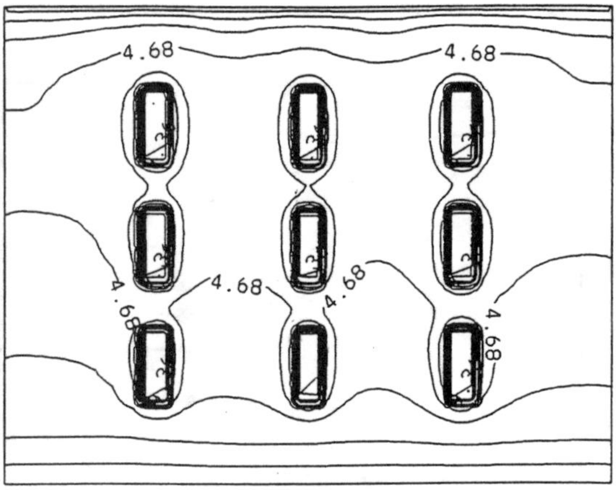

(b)

CONTOUR FROM 0.00000E+00 TO 16.381 CONTOUR INTERVAL OF 1.1701

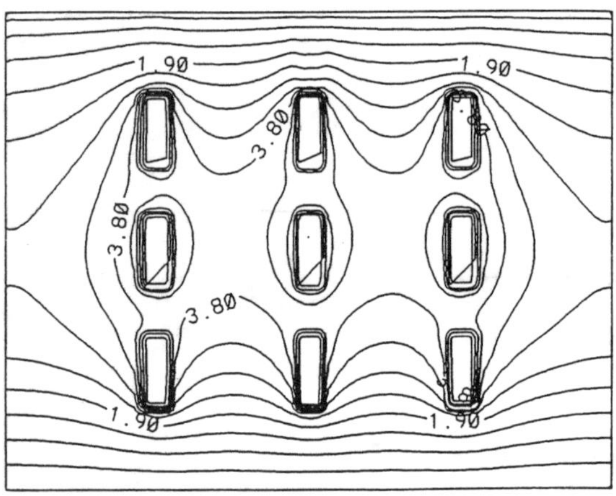

CONTOUR FROM 0.00000E+00 TO 6.6414 CONTOUR INTERVAL OF 0.47439

Figure 1.33. Dimensional temperature contours in the Z-Y plane for $X = 0.0195\,\text{m}$ and $Q = 0.7\,\text{W}$ for (a) $k_s/k = 1.0$ and (b) $k_s/k = 100$.

cooler. This trend is observed both for $Q = 0.1\,\mathrm{W}$ (Fig 1.31a) and for $Q = 0.7\,\mathrm{W}$ (Fig 1.31b).In Fig 1.32a, the fraction of the generated heat leaving each surface orientation is shown as a function of the substrate thermal conductivity ratio for a power level of 0.1 W. This fraction is the ratio of heat transferred from a particular face orientation to the total heat transferred directly from the protrusions to the fluid. One surprising result is that the relatively high fraction of heat that leaves the bottom face is of the same order as the front face loss. Together, the front and bottom faces account for more than 80 percent of the net heat loss. It is seen that, as the substrate conductivity is increased, the fraction of the heat leaving the front face decreases and the fraction from the bottom face increases with relatively small changes in the top and side faces. The corresponding fractions at a power level of 0.7 W, in Fig 1.32b, show numerical values different from Fig 1.32a, but the general trends are the same. One important exception is that the front face yields the highest fraction over the entire range of thermal conductivities of the substrate.

Figure 1.33a plots the dimensional temperature contours in the Z-Y plane at the fluid-substrate and protrusion-substrate interface. The power level is 0.7 W per protrusion and the substrate conductivity ratio is unity. The contour intervals are uniform. A crowding of isotherms implies sharper gradients as seen, for example, at the substrate-protrusion boundaries and at the top and bottom walls (Fig 1.33a). At the same power level, but at higher conductivity ratio (Fig 1.33b), more evenly spaced isotherms are observed in the substrate. This is to be expected because as the conductivity increases, a lower temperature gradient is required to transport the heat by conduction.

The Velocity Field

The fluid velocity field in the enclosure is extremely complicated. While the temperature field within the solid regions was nominally steady, the fluid velocity field was highly unsteady in certain regions. This field was investigated with the help of particle path lines, as shown in Fig 1.12a for $k_s/k = 100$ and $Q = 0.7\,\mathrm{W}$. The particle path lines are in the Z-Y plane at a distance of 0.5 mm from the front faces of the protrusions. Assuming steady state consitions, the path lines were calculated by integrating the velocity field. The results match experimental flow visualization techniques if the flow field is steady or changes slowly in the exposure period (using aluminum particles in an illuminated sheet, for instance)

Figure 1.34b shows the path lines for the same section at different instants. It can be seen that the main features of the flow, such as quiescence at the bottom and the vertical plumes along the front faces, are relatively unchanged. However, the region between the vertical plumes in Figs 1.34a and 1.34b is highly vortical, changing significantly with time. The overall stability of the main features of the flow may be one reason why the component temperatures are effectively time-independent.

Figures 1.34c and 1.34d show the path lines in the Z-Y plane near the wall opposite to the front faces. The flow descends along the plane and no plume-like structures are observed. However, rapidly changing vortices (Figs 1.34c and 1.34d) are seen, even in this plane which is only 0.5 mm from the wall. The unsteady nature of the flow is further emphasized in Fig 1.34e, where the u-velocity at (0.190, 0.700, 0.894) is plotted as a function of the non-dimensional time. It is clear that an unsteady formulation is necessary to capture the details of the flow patterns.

1.3.7.2 SUMMARY AND CONCLUSIONS

The following significant conclusions can be drawn from the experiments:

1. The effects of enclosure boundary conditions are most significant at lower power levels. Because the interest is usually in the maximum power dissipation rates, their influence appears to be secondary for the three choices studied. A simple heat transfer correlation for all protrusions is obtained which fits 95 percent of the data above $Q = 0.1\,\mathrm{W}$ within 15 percent.

2. Because of its low viscosity and a boiling temperature of 102°C at atmospheric pressure, for single phase liquid cooling, FC75 is the most likely candidate of the three liquids considered. The present results show that, of the three liquids, it also yields the lowest protrusion temperatures at a given input power level.

3. The flow patterns in FC75 under typical expected operating conditions are strongly time dependent. Temperature traces, however, show relatively weak fluctuations in surface temperatures. The computations of these strongly buoyant flows will require three-dimensional transient models. The present measurements provide a data base for validating these computations.

4. An increase in enclosure width from $w = 13\,\mathrm{mm}$ to $w = 30\,\mathrm{mm}$ improves the heat transfer coefficients by only about 15 to 20 percent in FC75.

The computations for the same configuration are within a maximum difference of 15 percent from measured protrusion temperature rises for power dissipation levels of 0.1 W and 0.7 W. The effect of varying the substrate thermal conductivity was investigated in detail. The results indicate that component temperatures decrease significantly as the conductivity is increased. The percentage of heat flowing directly into the substrate from the protrusions also increased. It was found that most of the heat rejected from the components came from the front and bottom faces.

The temperature field in both the substrate and the protrusions was nominally steady for both power levels considered. The components in the bottom row were found to be the coolest. The temperature field had sharp gradients at the top of the substrate-fluid interface for substrates of low thermal conductivity. Although the main features such as the vertical plumes along the protrusions changed very little, the velocity field showed marked unsteadiness. The bottom region of the enclosure was found to be relatively stagnant.

This study clearly demonstrates the need for a combined experimental and computational approach for the thermal design of electronic systems. The complexities in geometry and heat dissipation patterns, along with the unique fluid thermophysical properties of the fluorocarbon liquids, result in great unsteadiness in the flow field. In order to achieve a predictive capability for design, it is essential that the numerical parameters such as grid sizes and time steps be refined by comparisons against experiments.

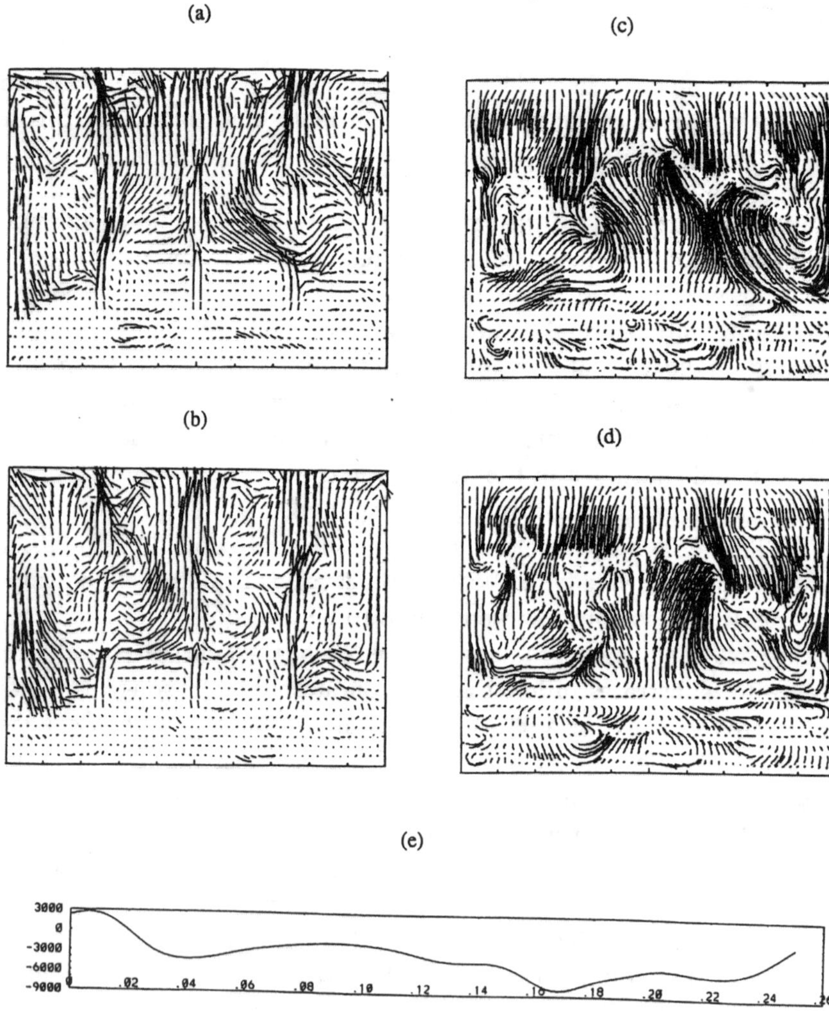

Figure 1.34. Particle pathlines for $Q = 0.7\,\mathrm{W}$ and $k_s/k = 100$ in the Z-Y plane at $X = 0.026\,\mathrm{m}$ at (a) $t = 0$, (b) $t = 0.25$ and $X = 0.049\,\mathrm{m}$, (c) $t = 0$ and (d) $t = 0.25$. Also shown is (e), the timewise variation in u-velocity at the non-dimensional location of $x = 0.190$, $y = 0.700$ and $z = 0.894$.

1.3.8 RECOMMENDATIONS

It is worthwhile to provide a summary of the features in common that pertain to the three examples.

1. When the objective of the analysis is to obtain a high accuracy in solid-object temperatures, the investigation will be time-consuming and very costly (see examples 1 and 3). This is caused mainly by the fairly sophisticated experiments that must be designed.

2. When parametric studies are the main goal, simple experiments are well suited to enhance the temperature prediction by elimination of one or more "fit" parameters. The presence of a prototype is, of course, very welcome. Even if a real prototype is not available, a mockup incorporating the essentials from a thermal viewpoint can be fairly easily constructed especially when dealing with completely new products for which no experience has been obtained.

The following approach is recommended if thermal characterization of a complex system is required in the earlier stages of a design.

- Get as much information as possible about local heat dissipations from electrical engineers, about the location of all the objects that influence the air flow from the mechanical engineer (especially about the allowable location of the vents) and about the constraints imposed by various design objectives.

- Perform some back-of-an-envelope calculations based on engineering judgment (see examples 1 and 2) to get some idea about the contribution of the various heat paths (such as the importance of radiation and the local fluid temperature rise caused by upstream heating). These exercises provide justification for possible simplifications to be used in subsequent CFD analyses. Some guide lines can be found in the chapter by Azar (1996).

- Make a CFD model. Don't bother about details of the individual parts but model these as metal blocks. Use a mesh according to the available computer power but at least with cells on the order of component dimensions. Remember that a finer grid does not necessarily result in better accuracy because, in most cases, it will not be possible to simulate all boundary layers around all objects of interest.

- For some parts, a refinement in the analysis is needed. Consider, for example, a flat heat sink. To optimize its dimensions or thickness, full system analysis is an overkill. Analytical approaches or conduction-only analyses, using the boundary conditions from the relatively rough system analysis, will provide sufficient information at a fraction of the cost. For extruded heat sinks, analytical approaches usually fail, and for their optimization, the fluid side of the problem must also be included. Moreover, in this case, only the analysis of a subdomain is sufficient provided the boundary conditions are taken from the full system analysis.

- Measure the unknown input parameters such as air resistance coefficients, emissivities, thermal conductivities, interface resistances and the like with proper experimental facilities. Keep in mind that the goal is not to measure the real physical properties but the effective values important for the system. Measurements *in situ* are therefore preferred over measurements in certified laboratories. This is probably the most expensive and time-consuming activity when people start to work in this field.

- Perform sensitivity analyses to gain an insight into the influence of the unknowns on the information that is required. It will probably be apparent that an exact knowledge of some of these unknowns is not needed.

- Make extensive use of past information and keep a library.

- It is not usually advisable to try to get accurate temperature information on solid objects out of a system level thermal analysis. If this information is nonetheless required, use imposed heat transfer coefficients instead and set the parameters in such a way that reliable temperatures are calculated. The temperature changes that are the result of varying some parameter can be trusted with some confidence.

- Don't spend too much time in trying to derive correlations. It is usually a waste of time. Why? Because basically, correlations are only valid if congruency can be realized and this really is never the case in a system with a complex geometry. Therefore, the rule to bear in mind is that correlations cannot be trusted beyond the parameter space investigated. Instead, common practice should be to use a CFD tool to grasp the influence of a changing parameter and display the results in a tabular or graphical form which can be understood by a designer.

- Make it a principle never to trust a 2-D approximation of a 3-D system unless there is clear evidence to do so. The fluid flow around 2-D objects is very different from the flow around 3-D objects and the visualizations, shown in Fig 1.35, taken from a paper by Azar and Russell (1990), confirm this. From results like these, it is evident that correlations developed for 2-D objects certainly fail for 3-D objects.

- Don't spend too much time in formulating simple design rules or tools. In a very competitive business, there is a tendency to cut expenses for personnel training. The consequence is that management is increasingly asking for simple design tools which can be used by inexperienced designers. As an example, consider the problem of the right choice of a heat sink when there is a series of commercially available extruded heat sinks. This is essentially an optimization problem with cost, weight and volume as constraints. It is an illusion to think that this problem can be satisfactorily solved by a simple design tool. For example, the design cannot be concluded based on analytical relationships describing the performance of a heat sink in terms of well-established correlations for fully-developed flow and fin theory, unless the heat sink is completely shrouded. Indeed, a CFD tool is the preferred choice to solve such a problem. Alternatively, when a CFD tool, or the expertise, is lacking, an appropriate experimental setup simulating the application is a much better way of tackling the problem than using fast, user unfriendly, but highly innacurate design rules. A useful design rule is that the number of pull-down menus and colors is inversely proportional to accuracy.

1.4 CHALLENGES IN ACCURATE THERMAL CHARACTERIZATION OF NATURAL-CONVECTION COOLED ELECTRONIC SYSTEMS

As has been argued in the preceding section, CFD tools are extremely useful in the early design phases for exploring alternative designs and realizing First-Time-Right designs thereby shortening Time-To-Money. One step further is to predict various reliability metrics such as Field Call Rates and Life Cycle Costs. It is generally believed that accurate estimates of absolute and relative temperatures are needed to meet future reliability-assessment requirements. The question that will be addressed in this section is, given the state-of-the-art, can this goal be realized, and if not, what are the critical issues that are waiting to be resolved?

In general, it is tacitly assumed that by increasing the number of computational cells, the accuracy will improve accordingly. It will be shown that only a small fraction of the problem can be solved in this way. In Section 1.4.1 an idealized and rather simple system will be used to demonstrate the various problem areas associated with the use of CFD codes and their input data, while section 4.2 deals with the added complexity when real electronic systems are the subject of the analysis.

1.4.1 THERMAL CHARACTERIZATION OF AN IDEALIZED ELECTRONIC SYSTEM

The analysis of canonical cases such as channel flow with uniform temperature or heat flux boundary conditions, or of the more complex fictitious cases such as thermal wakes behind discrete, but uniformly distributed, components on an adiabatic wall, is aimed at understanding one, usually rather complex, physical phenomenon. In general, researchers go to great lengths in trying to avoid the simultaneous occurrence of other "disturbing" physical phenomena and to use materials with well-known physical properties. For these cases, an abundance of information is available in the published literature in the form of correlations, equations or design rules which, unfortunately possess a fairly limited applicability. Characteristic terms are channel flow, developed, laminar or turbulent, known view factors, adiabatic, uniform and constant temperature. Unfortunately, real-life systems are characterized by the coupling of various complex physical phenomena and difficult-to-measure physical properties and other input parameters. This section will discuss the consequences of these complications. Some observations are summarized in the following list in greater or lesser detail:

- Complex physical phenomena
    - no purely natural or forced convection (mostly mixed convection).
    - no purely laminar or turbulent flow but mostly transitional or unstable.
    - no fully developed flow but mostly developing.
    - no adiabatic boundaries.
    - non-uniform dissipation, source distribution and component dimensions.
    - both vertical and horizontal boards.
    - complex radiation effects.
- Not sufficiently well-known physical properties
    - thermal conductivities, especially of ceramics and plastics.
    - strongly anisotropic board thermal conductivity.
    - emissivities, often changing in time, due to oxidation and dust settling.
    - temperature dependency of many physical properties.
- Not sufficiently well-known input parameters
    - thermal data of components.
    - air resistance coefficients.
    - contact and interface resistances.
    - surface roughness.

To illustrate the various problem areas, consider the performance of a thought-experiment and look at the thermal characterization of

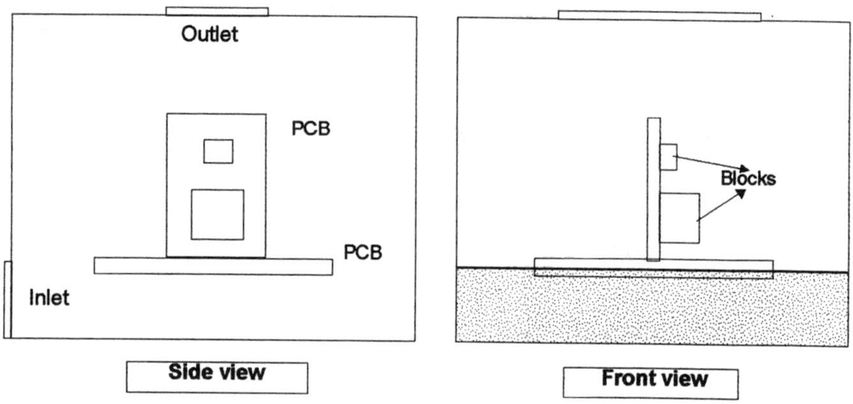

Figure 1.35: A simple electronic system.

an idealized electronic system consisting of one uniformly dissipating horizontal board and a vertical board with two metal blocks of unequal size placed inside a vented enclosure provided with inlet and outlet grilles. The two blocks are necessary for studying the coupling phenomena caused by conjugate heat transfer and by wake effects. In Fig 1.35, the position of the boards and the blocks can be changed.

The most important idealization is the replacement of electronic components by solid structures, henceforth denoted as "metal blocks", with uniform properties, especially a high thermal conductivity. This simplifies the experimental and numerical analysis considerably because temperature gradients in the metal blocks are virtually absent. The second idealization concerns the neglect of the interface resistances between the blocks and the board. The final simplification disregards the complex layout of a real electronic system with its effective roughness of the boards caused by many small components and the presence of wires and cables. The entire setup is located in a "normal" laboratory with central heating, lighting, sun irradiation and other people. The objective of the thought-experiment is to investigate which factors are responsible for possible discrepancies between the CFD analysis and suitable experiments. If the differences observed are smaller than a predefined uncertainty level, the CFD code could be used to perform parametric studies much more quickly and cheaply while at the same time providing information about the complete temperature, flux and velocity fields. In this case, the designer might want to acquire more insight relating to the sensitivity of the simulated component temperatures with respect to many parameters. These include:

- air resistance coefficients,
- distance between components,
- power dissipation of the horizontal board,
- position with respect to the grilles and
- thermal conductivity of the vertical board.

Consider the situation which occurs when substantial differences are found between the results of the CFD analysis and the experiments. It is important to make a clear distinction between three different sources of uncertainty; experimental uncertainties, uncertainties in the numerical model and finally uncertainties in the input parameters. Assume further that the experiment is well-designed such that the results are correct to within 5 percent based on 20:1 odds. The realization of this accuracy, by the way, could be the subject of an entire separate chapter.

The next two sections attempt to formulate an answer to questions of how accurately are the real physics simulated by the numerical analysis, including the initial and boundary conditions, and how accurately are the physical properties and input parameters known?

1.4.1.1 ABOUT THE ACCURACY OF THE NUMERICAL MODEL

Despite the incredible computer power available today, and promised teraflop rates in the near future, the question remains whether numerical results describe reality to a sufficient degree of accuracy. Bear in mind that computational models comprise many levels of modeling and a model is inherently a simplification.

The following arguments are taken from an editorial by Worlton (1996). A scientific model of reality, a mathematical model of the scientific model, a discrete numerical model of the continuous mathematical model, an application-program model of the numerical model and a computational model of the application model may all be distinguished. Each model in this list is an approximation to its predecessor. In addition, all computers do their operations with limited precision and the number of significant digits degenerates as operations are performed, especially in subtraction operations. Further discussion of these topics and how to verify the underlying assumption in computational physics that the results correspond to reality are beyond the scope of this chapter. However, the following items are, to a certain extent, under control of the scientist/designer.

The influence of the algorithms used.

For example, what is the influence of the method (FEM, FDM, FVM or spectral), false diffusion, temperature-dependent physical properties, the Boussinesq approximation and the convergence criteria? Ideally, CFD vendors should be able to answer these questions. It can safely be assumed that these types of uncertainty will decrease in the future given the fact that all approximations can be relaxed by using a more powerful computer.

The influence of the mesh size.

This type of error can be substantially reduced by increasing the number of mesh lines until no further improvement is obtained. However, for complex geometries in 3-D the most favorable mesh size could exceed time requirements. For some time to come, it will be almost impossible to model every boundary layer around the many components, printed circuit boards and system boundaries present in, for example, a television set. One is forced to use multigrid technique, local zooming or other refinement algorithms, if available.

The radiation heat transfer approximations.

When looking at the three heat transfer mechanisms, the theoretical methods of addressing the radiation effects are quite different from the convective and conductive modes. Contrary to convection and conduction, radiation is not restricted to local effects. Its global behaviour leads to integro-differential equations that are inherently more difficult to solve.

Almost all engineering radiation calculations for electronic systems are based on the following assumptions:

- All surfaces are diffuse (or, in some cases, specular), gray, opaque, isothermal and uniformly irradiated.
- The medium between surfaces is transparent for the wavelengths ranges of interest.

It is felt that the usual generalization of diffuse surfaces causes unrealistic oversimplifications in a few applications in which specular surfaces are present such as flat heat sinks and copper planes. It is not realistic to assume that even future codes will be used to model all the combined radiation heat transfer mechanisms occurring in a real system, not because of theoretical limitations, but because of excessive computing costs. Thus, in most CFD benchmarks full radiation analysis is not part of the test protocol and, instead, correction factors are introduced. The best way to proceed is to reduce the influence of radiation to negligible levels by covering all test surfaces with a low emissivity material. After the code has been validated in this way, the influence of radiation effects can be studied by blackening the surfaces one by one. In practice, usually only a few critical objects should be given the *full radiation treatment.*

The modeling of some complex physical phenomena.

While in the past research was almost exclusively directed towards finding Nusselt-Grashof and Nusselt-Reynolds correlations for uniform boundary conditions and smooth surfaces, the need to have data available for boards populated with electronic components stimulated researchers to focus on the mutual influence of discrete elements on each other. Because research in this field has mainly been driven by the computer and telecommunication industries, the majority of the papers deal with channel flow. Excellent reviews on this subject, focusing on the cooling of electronics, can be found in Kraus and Bar-Cohen (1983), Peterson and Ortega (1990), Anderson (1990), Aung (1991) and Pecht (1991).

Conjugate heat transfer.

Relatively few studies have incorporated the conjugate heat transfer problem or the non-uniform heat release [Zinnes (1970), Culham (1988), Graham and Witzman (1988), Barozzi and Nobile (1990), Bertin and Saulnier (1990), Kishinami et al (1990), Culham et al (1991), Wroblewski and Joshi (1992, 1993)]. It is generally assumed that conjugate heat transfer can be dealt with by the current state-of-the-art in CFD and that the use of Nusselt number correlations for convective heat transfer will eventually become obsolete.

The ultimate challenge for a correct numerical analysis, however, is formed by the modeling of some complex physical phenomena that are inherently present in a real system. The modeling of the environment, the grilles, horizontal plates, and transition and turbulence are discussed in greater or lesser detail in what follows.

Grilles and environment.

Boundary conditions on the openings at the inlets and outlets of electronic systems need to be prescribed. Standard engineering practice is to prescribe pressure losses caused by inlets, outlets and vents. The accuracy of such approximation under various conditions has not yet been examined. A more serious problem emerges when considering the influence of the instabilities and "mini-jets" generated by the grille itself. These effects may persist long enough to cause a measurable influence on downstream local heat transfer coefficients.

Other largely unknown parameters are the local heat transfer coefficients at the outside envelope of the system. It is to be expected that the temperature of a component is far less sensitive to this type of uncertainty than to those discussed before. Usually, values taken from any textbook would serve these goals.

Horizontal plates.

A more serious problem relates to the phenomena emerging from the introduction of horizontal plates within electronic systems. In all cases where miniaturization with respect to volume is not of paramount importance, the packaging engineer favors a vertical position of the PCB's because of their greater natural convective heat transfer capabilities.[2] This is often the case in the mainframe and telecommunication industries. The situation is entirely different in the consumer electronics sector. Think of car radios, mini-audio sets, LCD TV, notebook computers, cellular phones and electric light bulbs. In all these cases, it is very difficult, if not impossible, to use

[2]It is interesting to conjecture that this simple fact could be one of the reasons for the evolution of man some two million years ago. Bipeds have a cooling advantage over quadripeds in the hot savannahs that supposedly were the environments favoring such an evolutionary breakthrough.

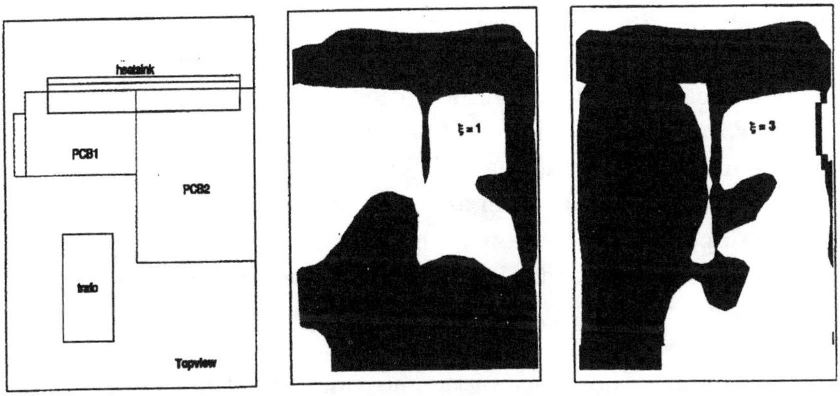

Figure 1.36: Flow reversal in an electronic system (a) The configuration, (b) $\xi = 1$ and (c) $\xi = 3$.

vertical boards. A literature survey reveals that much more attention has been paid in the past to the analysis of natural convection heat transfer in vertical channels and isolated plates as compared to horizontal and inclined plates. Only a few papers focus on the cooling of electronics and almost none discuss discrete boundary conditions. Appendix B briefly reviews the existing literature on horizontal-plate heat transfer problems in general.

An interesting observation on the effects of horizontal plates in a real system concerns unexpected flow reversals. In a combined experimental/numerical analysis of some consumer product, a remarkable and worrying effect appeared. Figure 1.36a presents a schematic view of the most important objects. In the top view, two horizontal boards, the large transformer and the audio heat sink can be recognized. As an input for the code, the air resistance coefficient, ξ, of the outlet grid was imposed. By performing a sensitivity analysis on this parameter, it turned out that by changing ξ from 1 to 3, the flow field changed significantly. Figures 1.36b and 1.36c show the sign of the z-velocity in a plane just above the top of the set. Black is the negative direction, that is, the airflow out of the system, and white represents the flow into the set.

Observe that the dominant heat sources (transformer and heat sink) generate an upward velocity, as expected, but that the flow in the other regions changes its direction completely. If this effect is indeed apparent in real applications, a serious problem is encountered because a change in the average resistance coefficient of the grilles brought about by styling considerations, or worse, by the customer partly covering the outlet grille can be the cause of an unexpected change in the overall thermal behavior of the system. Another interesting observation is that, in any case, the computer analysis shows that an appreciable part of the inlet air enters the set via the top grille despite the fact that it was thought that ample room had been provided at the bottom of the set. This effect was indeed demonstrated by flow visualization using the Helium bubble technique. This example clearly shows one of the most important advantages of using CFD codes; the acquisition of physical insight by numerical flow visualization. Moreover, in this case, much more experimental and numerical research is needed to clarify the basic phenomena underlying the observed change in the flow field, and the consequences on the critical temperatures, in order to design systems in such a way that the flow field is invariant with respect to influences that can be imposed by the customer.

The most important conclusion emerging from the literature review presented in Appendix B is that natural convection above a hot plate is not very well understood, either from a physical or a computational point of view. Another thing that becomes evident is that in many cases the heat transfer behaviour is not a steady phenomenon. Apparently more than one solution exists as is often the case in situations far from equilibrium. From an engineering viewpoint, this behavior could introduce many problems which can be divided into two different categories:

- A minor disturbance in an otherwise steady state operating system could cause a transition to a completely different flow field in an unpredictable way.

- If more than one power source contributes to the heat transfer, the resulting flow field could depend on the sequence of powering up. For example, take an Application Specific Integrated Circuit (ASIC) on a horizontal board with some other components dissipating as well. There could be a difference in the final flow field, depending on whether the ASIC is turned on after a steady state has been established for the rest of the board, or whether the ASIC was powered first and then the rest of the board. In other words, superposition principles cannot be applied in such cases.

If these phenomena do indeed affect the heat transfer behavior in electronic systems with predominantly horizontally placed boards, it could be envisaged that such behavior, triggering multiple solutions, adversely influences the predictability of the temperature in the field and hence the reliability. However, it is not known to what extent the effects arise, and if so, what the influence is on the local temperatures. Of course, flow fields could look very different but what really matters is the consequence with respect to the local temperatures. As regards the cooling of electronics, a few interesting research fields suggest themselves.

- Stratification influences: could they occur, and how serious could the influence be?
- Interaction of plumes arising above localized heat sources could generate a much stronger plume. What are the optimum locations as a function of the interaction?
- The influence of the experimentally observed random and periodic oscillations on the maximum component temperatures attained.
- What is the influence of unstable plumes on downstream components?
- Mixed convection will be increasingly important, especially in the consumer electronics industry, because the introduction of small fans could gradually take over the purely natural convection driven systems. The question is to what degree superposition principles still apply when the natural and forced convection flows are comparable.

- Is it unrealistic to assume that it is essentially the geometrical complexity of electronic systems that prevents the occurrence of multiple solutions?

- Another topic that remains to be explored in a much wider area is 3-D internal flows in an enclosure, either completely closed or (partly) covered with vents.

One example is given by Lasance (1994c) in a discussion of a steady-state analysis of a small 2-D enclosure ($36 \times 18\,\mathrm{mm}$) with a horizontal dissipating plate in the center and a positive temperature difference between bottom and top. For example, it makes a difference when the power is applied to the plate before or after the temperature difference is set. From a practical point of view, there could be a difference in the heat transfer behavior of a system depending on whether the system is switched on starting from ambient conditions, or whether some feature is turned on when starting from an already powered system. The results make it clear that the initial conditions should be taken into account. At a later stage, transient studies were also performed, basically supporting these conclusions [Adam (1993)].

It is of interest to note that similar behavior has been observed on a much larger scale in the study of the ventilation in a concert hall. With a fully populated hall, it turns out to be impossible to overrule the natural convection flow field raised by the audience by using low-velocity ventilation fans, while no problems arise when the fans were switched on before the people filled the hall [TNO (1992)].

Transition and Turbulence.

The last complex phenomenon that must be mentioned is discussed in some detail in Appendix A and deals with transitional and turbulent natural convection heat transfer in geometrically simple enclosures.

Two observations are worth mentioning here are:

- Sophisticated turbulence models (any other than standard k-ϵ) are not expected to be implemented in system-level analysis tools for quite some time. Even for this fictituous, relatively simple system, the computer requirements are way beyond what is practically useful. While "real" transition and turbulence are not expected for the system under consideration because low Reynolds number-turbulence effects triggered by plumes and affecting downstream components require a very fine grid and a transient approach. A few weeks of CPU time on workstations may be tolerable for a PhD thesis, but certainly not in a designer's environment.

- On the other hand, there is no idea whatsoever about the errors originating from the use of laminar or zeroth-order turbulence models in a complex system level analysis. This is simply because there are no numerical data available. There are, of course, lots of experimental data but possible differences might well be attributed to variations in all the other unknowns.

The question that remains to be answered is how do all these more or less intractable problems influence the objective of the analysis which is the determination of the temperature of the two metal blocks presented in Fig 1.35? For example, the horizontal, uniformly dissipating board generates a plume. Depending on the location of the vertical board, the air resistance coefficients, and the ratio in power dissipation, the plume will either be "attracted" by the vertical board or generate instabilities downstream affecting the local heat transfer conditions around the metal blocks. To what extent, however, is not known.

1.4.1.2 ABOUT THE ACCURACY OF PHYSICAL PROPERTIES AND OTHER INPUT PARAMETERS

For a fair comparison between experimental and numerical results (at least to answer the question of how well can a CFD code predict the behavior of real systems), ideally all input parameters should be known to, say, within a few percent of the final uncertainty, especially in the case of many components and a strong coupling between them. The following survey shows which input parameters are needed in the subject simple system.

Air Resistance Coefficients for the Inlet and Outlet Grilles.

Although these coefficients can be measured with relative ease, it is still a matter of dispute as to whether the results acquired with low velocity forced convection flow are equally applicable to natural convection flow. A more serious practical problem has to do with the fact that the air resistance coefficient is not a constant but is strongly dependent on the velocity, especially for low velocities. Particularly, in the case of the outlet grille, a very strong non-uniform flow is observed and one may ask whether it is permissible to impose a constant air resistance coefficient for the entire grille which is common practice in most commercially available CFD codes. Again, there are no studies that address the influence of velocity-dependent air resistance coefficients on the temperature of the component under study. To reduce this type of error, the CFD code must be capable of handling velocity-dependent air resistance coefficients.

In this light, it is rather difficult to estimate the uncertainty caused by the imposed (measured) air resistance coefficient. The only solution is to perform a sensitivity analysis but the acquisition of reliable data will be seriously hampered if this data is confounded with other unknown input parameters, especially when non-linear interactions are to be expected, as in a buoyancy-driven system like this. On the basis of previous research, discussed in section 1.3.2, it appears that changes in air resistance coefficients have only a minor influence on the temperatures of solid objects, and can be neglected compared to other sources of error. Note that this conclusion applies to only the <u>value</u> of the coefficient and <u>not</u> to the effects of the instabilities generated by the grille.

Physical properties.

The uncertainty caused by variations in material properties is much easier to assess than the errors caused by the modeling of complex fluid flow phenomena. As a rule of thumb, the material properties of the electronic parts that are closest to the component whose temperature is to be studied have the strongest impact. So, in this case, by far the greatest problems are caused by the uncertainty in the thermal conductivity of the vertical board and the emissivity of the metal blocks.

Emissivity. As already discussed, apart from the mathematical difficulties, there is another problem that impedes an accurate radiation analysis. Oxide layers, roughness, surface treatment, dust and grease all influence the emissivity to a significant extent. This is precisely why it is often useless to look up the values in the tables that can be found in many textbooks unless, of course, rough estimates are the objective of the calculation. Furthermore, emissivity is a function of temperature. For metals, this influence can often be described by a linear function, and it should therefore be recommended that the Stefan-Boltzmann law describing the radiation heat loss from a grey and diffuse surface should be rewritten as $q \sim T^5$. However, for the cooling of electronics, the absolute temperature and the temperature differences are usually not large enough to take this effect into account. A final point to make is that technical surfaces tend to deteriorate with time due to the settling of dust and oxidation.

When radiation heat transfer accounts for 20 percent of the total heat transfer, which is a commonly encountered situation in natural convection cooling, and the emission coefficient has an error of 25 percent, a 5 percent error in the calculated temperature results.

Thermal conductivity. For many engineering materials, such as plastics, ceramics and alloys, no accurate data are available, much less as a function of temperature. An especially intriguing problem shows up when printed circuit boards are part of the analysis. The values for the thermal conductivities of PCB's cited in literature are mostly conductivities of the epoxy in the direction normal to the PCB. However, these values are useless in practice because

- the thermal conductivity of the reinforced epoxy can be highly anisotropic,
- the in-plane thermal conductivity is much more important from a practical point of view and
- the very complex patterns of metal tracks and layers will influence the thermal conduction behavior considerably.

For an analysis at system level, and to a certain extent also at board level, average values for the thermal conductivity of the PCB's are needed. An experimental technique has therefore been devised for determining the "effective thermal conductivity" of the PCB's. It is based on a steady-state analysis of the temperature distribution over the PCB to be measured with contactless dissipation at the center and with very well defined boundary conditions. The numerically found temperature is fitted to the contactless measured center

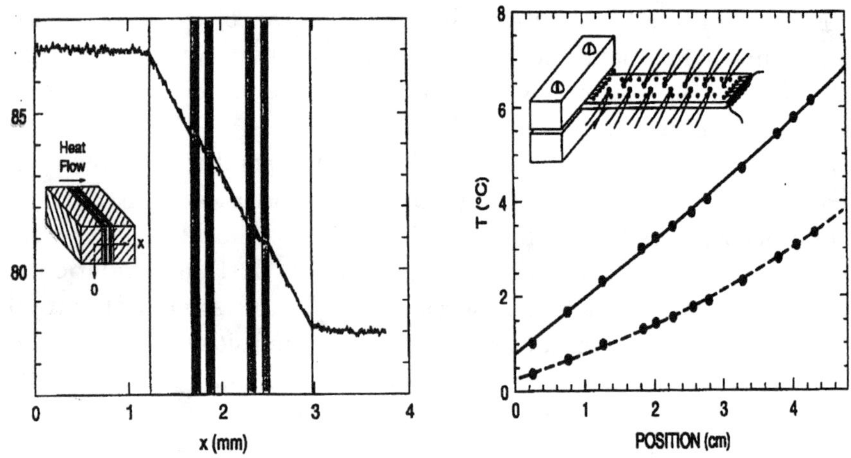

Figure 1.37: Measurements of in-plane and out-of-plane thermal conductivity of boards. (a) Measurement of through-plane thermal conductivity of a PCB with embedded copper layers by means of scanning IR thermography abd (b) Temperature distribution in a PCB to determine in-plane thermal conductivity using thermocouples. The upper curve was taken in vacuum and the sample covered with radiation insulation, the lower in air and no insulation. [Azar and Graebner (1996)] (Courtesy of IEEE Press).

temperature. The resulting thermal conductivity of the model can be interpreted as the "effective,' thermal conductivity. It should be noted, however, that this value does not represent a genuine material property because it depends on the dimensions and the boundary conditions [Vanlaer and Lasance (1994)]. Another approach, based on separate measurements of the in-plane and out-of-plane thermal conductivity of real boards (including vias and tracks) was recently proposed by Azar and Graebner (1996) (Fig 1.37).

A future way of tackling this problem could be to use the detailed board layout resulting from electrical computer aided design (ECAD) tools and to use the computer as the measurement facility to calculate an effective thermal conductivity based on the areas that can be allocated to various components of interest. It would be preferable

Table 1.7: Estimated contributions to temperature uncertainty in numerical analysis of an idealized system.

Numerical errors	Error in percent	Input Data	Error in in percent
Algorithms	5	Air Resistance Coefficient	<5
Mesh Size	5	Thermal Conductivity	10
Radiation	5	Emissivity	5
Complex Physics	10		

to observe the position of the symmetry lines in an isotherm plot starting with an educated guess for the thermal conductivity.

In any case, if standard printed circuit boards are used in the experiment, validation will be difficult if the exact value of the effective thermal conductivity is not known a priori. If it is assumed that 50 percent of the heat transfer of the metal block is by conduction via the board for surface mount components without a heat sink, this value can easily be 80 percent). A 20 percent error in the thermal conductivity results in a 10 percent error in temperature prediction.

Estimation of the total error. The overall uncertainty in the computation of temperatures inside the idealized electronic system can be determined by combining the individual estimates. As an example, Table 1.7 provides plausible uncertainty contributions to temperature for individual parameters.

Taking the root mean square of the individual percentages, the final uncertainty is of the order of 20 percent, notwithstanding the fact that the numerical analysis does meet current standards. As an example, suppose that the temperature rise of the metal block is about 60°C. With a 20 percent uncertainty, the real value can be anywhere between 54 and 66°C. As pointed out in the introduction to this section, this is definitely not the answer needed for reliability analyses.

To conclude, there is currently no way of testing whether or not the numerical analysis makes sense when comparing the results with an equally well performed experiment. For the case under study, given the imposed boundary conditions and material properties, the numerical analysis may or may not be correct and there is no way to validate this by the proposed experiment.

In closing this section, it should again be emphasized that the current state-of-the-art is very well capable of providing useful answers if the purpose of the analysis is to get accurate information on temperature differences as a consequence of changing parameters.

1.4.2 SEVERAL WAYS TO ENHANCE THE ACCURACY OF CFD ANALYSES

Looking into the future, it may be expected that, on the numerical side, some errors will decrease to more acceptable levels. Others, especially those relating to the treatment of complex fluid phenomena, will not be reduced to the same extent, unless direct numerical simulation or large eddy simulation techniques become available for 3-D complex systems. These, according to Fisher and Rhodes (1994), will not be the case for another ten years. The problems caused by the errors in the input data will eventually be solved by the development of more accurate test facilities.

For some time to come, a pragmatic approach to enhance the predictability of CFD analyses will be needed. Three possible alternatives will be discussed: to impose known heat transfer coefficients, to apply the concept of Integrated Thermal Analysis, and to use experimental benchmarks to generate correction factors.

1.4.2.1 IMPOSE KNOWN HEAT TRANSFER COEFFICIENTS

One way to improve the accuracy of system modeling is to use imposed heat transfer coefficients to replace the numerical calculation of the local heat transfer around electronic parts. In practice, heat transfer coefficients are often determined via appropriate correlations, or rather, via suitable experiments. In this way, a relatively rough mesh can be used, precluding excessive running times, but still able to resolve sufficiently fine details in the fluid flow in order to be sure of accurate local air temperatures. Appropriate correction factors can be incorporated in the imposed heat transfer coefficients to eventually guarantee a better accuracy than can be obtained by using a very fine mesh alone. Unfortunately, other problems show up, as will be discussed in what follows. First consider the implications caused by the definition of the heat transfer coefficient itself.

The fundamental equation for describing the heat flux at the interface between a solid surface at T_s and a fluid at T_{ref} is

$$q \equiv -k\frac{dT}{dy} \tag{1.27a}$$

or

$$q \equiv h(T_s - T_{\text{ref}}) \tag{1.27b}$$

In fact, the two definitions are fundamentally different. Equation (1.27a) is based on a solution of the temperature profile adjacent to the solid/fluid interface. As argued before, it is to be expected that it will take many years before it will be possible to calculate such a profile for every electronic part of interest at the system level. Equation (1.27b) defines the heat transfer coefficient.

Apart from the problem with the non-constant h, the second and much more challenging problem is how to define T_{ref}. It is interesting to note that as early as 1953, Jakob, in his book *Heat Transfer*, discussed this problem in a chapter titled *A Difficulty in Principle Concerning the Coefficient of Heat Transfer*. However, some canonical cases aimed at defining the reference temperature can be treated in a straightforward way.

The Far Field Temperature.

Application: an isolated object with uniform temperature or flux in an infinite reservoir of fluid at a constant temperature and at steady state.

Obviously, the reference temperature has to be taken as the temperature of the undisturbed fluid. The definition is well-suited for a boundary layer type of flow over plates subjected to uniform boundary conditions.

The Cup-Mixed-Mean Temperature.

Application: channel or tube flow, uniform boundary conditions, "well-behaved" flow.

A logical choice would be the mean temperature over a defined cross-section. But how is it defined? As an average of the temperature in the cross-section? The integral of the heat energy transport over the area divided by the total mass flow? The temperature at the axis of the channel? The last certainly has the great advantage of relative ease of measurement, but what if a channel with asymmetric heating is the purpose of the analysis? The first, the arithmetic average temperature, suffers from the problem that with uniform flux boundary conditions, regions with a very low velocity, and consequently a high temperature, contribute unrealistically to the mean value. The second option seems the most appropriate choice from a theoretical point of view. But then the question arises: how can this reference temperature be measured in a real-life system? Traversing the cross-section with laser doppler techniques and tiny thermocouples is perhaps a nice academic exercise, but it clearly fails as a prototype of an electronic system.

There is a way out, provided uniform boundary conditions and well-mixed flow prevail. This is usually the case with heat exchangers. Then, the reference temperature at any cross section can be calculated via $q_{\text{upstream}} = \rho c_p \Delta T$. However, this approach becomes questionable if the conditions of the flow are not so well-behaved, for example if the mixing is not perfect. Nakayama (1988) noted that the ratio of the actual temperature rise of the air in the immediate vicinity of a component to the rise of the mixed mean temperature clearly shows that differences of up to a factor of four could be expected.

The inlet Temperature.

Application: practical systems.

The possibility of defining the inlet temperature as the reference temperature is based purely on practical considerations because this temperature is, at least, easy to measure and is, at most, not subject to variations. The main flaw is that the definition has no physical significance and does not lend itself to a more accurate analysis because the whole system under test is involved. There is no way of separating local and global effects and optimization becomes a difficult task. In some cases, such a definition could offer an advantage as a kind of Figure of Merit to compare different designs. However, in that case, it is advisable to reformulate eqs (1.27) to abandon the use of the heat transfer coefficient as

$$q = \frac{T_{\text{comp}} - T_{\text{inlet}}}{R_{\text{Th,T}}} \tag{1.28}$$

An "Ambient" Temperature.

Application: temperature specifications of electronic parts inside systems.

Numerous official specifications for components state some reference temperature, generally called the "ambient" temperature, above which the component is no longer guaranteed. This "ambient" differs from the one discussed previously as the far field temperature in that the applciation is neither restricted to one dissipating object nor to the far field.

Very often, such a temperature is measured with a thermocouple, positioned at some undefined place inside a powered system somewhere in the neighborhood of the component in question. It is often not realized that large temperature gradients frequently exist around these objects apart from the fact that radiation and conduction effects adversely affect the accuracy of the thermocouple reading. Readers who have been working in a research environment for years might find this shocking but the method described is "standardized" in many factories where testing is a costly and time-consuming activity.

Apart from these observations, it is obvious that, inside a real system, it will be very difficult to define a reference temperature in a proper and manageable manner because the exact position of the thermocouple should be prescribed for every practical situation.

Considering the foregoing items, one can conclude that there is a problem, not only because a choice should be made between various alternatives, but also because the non-uniform boundary conditions and the poorly behaved flow patterns in most of the practical systems do not allow for a logical and sensible way of dealing with local air temperatures.

The "Adiabatic" Temperature.

Application: complex systems.

A potentially more attractive approach to defining the reference temperature in complex systems is offered by the so-called "adiabatic" temperature, proposed by Arvizu and Moffat (1982) for forced convection and Ortega and Moffat (1986) for buoyancy-induced flows. The adiabatic temperature is defined as the temperature a component or object achieves when its power is switched off.

All experiments, which are based on heating one element at a time, are the basis for the determination of the adiabatic heat transfer coefficient. If the literature search is restricted to those papers that have some relationship with the cooling of electronics, it becomes evident that the majority of the results obtained are based on this method. Perhaps the most important property of the definition is that it makes sense from a physical viewpoint for certain problems. Heat transfer, especially in electronic systems with their many discrete boundary conditions is a very localized phenomenon, and with a reference temperature assigned to the component itself, it stays that way.

Finally, such an approach paves the way to superposition principles, which is simply a very efficient way of separating important effects that can be treated subsequently. For purely forced convection-dominated systems, in which temperature-dependent properties may be neglected, it can be demonstrated that superposition has a firm scientific basis. For pure natural convection, such a base is lacking. Fortunately, the local fluid flow around most electronic parts in a natural convection-cooled electronic system is again dominated by forced convection, albeit buoyancy-induced. From a designer's viewpoint, the adiabatic-temperature concept can provide very valuable information. For example, by switching off the power of one particular component, one can study the effects that cause the "unpowered" temperature rise, by observing the various paths of the heat entering or leaving the component. In doing so, possibly alternative measures can be studied to lower its temperature. More information can be found in Moffat and Ortega (1986), Maciejewski and Moffat (1990,

1992a, 1992b), Moffat and Anderson (1990), Anderson (1990), Anderson and Moffat (1992) and Meinders et al (1996).

Only a few studies have been devoted to conjugate heat transfer based on the adiabatic heat transfer coefficient concept but their number is rising [Witzman et al (1989), Ortega and Kabir (1991), Ortega and Lall (1992), Azar and Moffat (1991), Manno and Azar (1991), Lasance (1994b) and Anderson (1994)]. Unfortunately, when dealing with conjugate heat transfer, a practical problem is being faced because an adiabatic boundary condition cannot be realized. What is needed is a method to use the "unpowered" instead of the "adiabatic" temperature as a reference because this situation can be easily measured. The problem could be solved if means existed for the prediction of the ratio of convective/conductive heat transfer without using a very fine mesh. A problem that still remains to be solved is how to impose known heat transfer coefficients in CFD codes [Lasance, (1991)] because the calculation of an appropriate reference temperature poses a serious challenge. Research is currently under way at several research laboratories to address these fundamental problems.

However, whatever definition of the heat transfer coefficient is used, the concept of embedding such values in relatively simple CFD analyses of complex systems is certainly an alternative to using complex fluid flow models coupled to a very fine grid.

1.4.2.2 INTEGRATED THERMAL ANALYSIS

The concept of integrated thermal analysis of electronic systems is relatively recent and the relevant literature is limited. Przekwas et al (1994) proposed a Virtual Prototyping Environment for multi-scale thermal packaging of electronic systems. To avoid using an unnecessarily fine grid for the entire computational domain, they proposed several techniques such as multi-domain matching grids, patched/overset grids, Chimera arbitrary grids, unstructured and hybrid structured/unstructured grids to be employed for modeling complex components. Both overset grid and unstructured grid techniques were applied to their study of flow over two single chip modules between parallel boards. They also used a multi-dimension, multi-scale grid technique to study the board scale electronic cooling.

A finite-element-based multi-level thermal model for electronic systems was presented by Klein (1994). The capability of automatically establishing accurate boundary conditions by transferring the results of computations at the system level was mentioned. However, only a convection-type boundary condition obtained from system level analysis was considered for incorporation into board level analysis and the details of integration were not presented. The use of suitable simplified component models in system level analysis was considered as another way to reduce the computer storage load and computational time. Linton and Agonafer (1994) presented a coarse finned heat sink model that can be used in system models. As few as $3 \times 4 \times 3$ cells were used to represent a finned heat sink. Agreement among the experimental data, the detailed model and the coarse numerical model was reported to be quite good. Ewes (1995) reported on building compact models for IC-packages. Knowledge of the thermal resistance value of the IC-packages played a key role to building up a proper compact model. By applying the compact model to system analysis, a relatively coarse grid was required to define a IC-package and crucial information, such as the junction temperature of the components, could be calculated reasonably accurately. Rosten and Lasance (1995) deal with more general methods of generating compact models for electronic parts.

Tang and Joshi (1996) describe a technique for the integrated thermal analysis of a laminar natural-convection-air-cooled electronic system. This approach, summarized in Fig 1.38, is illustrated by modeling an enclosure with electronic components of different sizes mounted on a printed wiring board. First, a global model for the entire enclosure was developed using a computational fluid dynamics/computational heat transfer (CFD/CHT) approach on a coarse grid. Component representations were simplified in this model in order to save memory and computational time. To ensure the accuracy of the final results, a non-uniform grid was used with finer grids placed near solid surfaces to resolve the hydrodynamic and thermal boundary layers. The solution to the global model was obtained by using a finite volume method [Patankar, (1980)]. Thermal information from the global model, in the form of board and component surface temperatures, local heat transfer coefficients and reference temperatures, and heat fluxes, was extracted. These quantities were

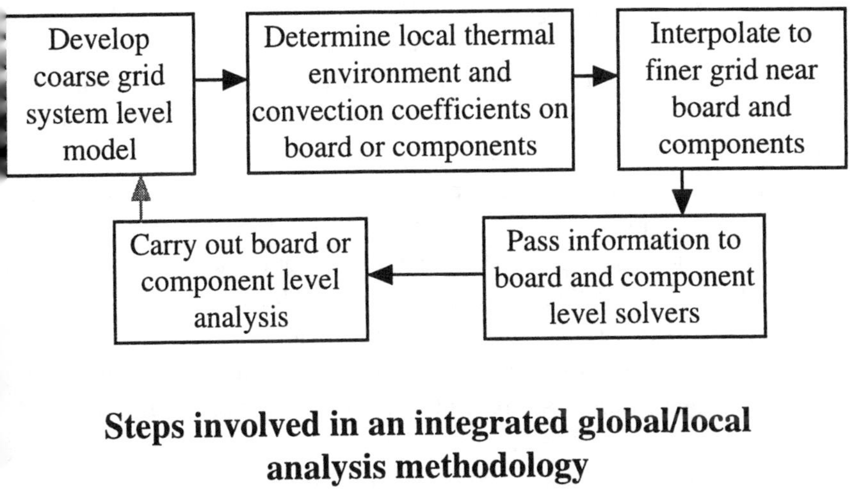

Figure 1.38: A flow chart for integrated thermal analysis.

interpolated on a finer grid using Lagrangian polynomials and further employed in board and component level thermal analyses as various boundary conditions.

In component level analysis, various parts of the component were modeled with a fine grid to predict detailed thermal information such as junction temperature and temperature gradients. The component investigated was a leadless ceramic chip carrier (LCCC). The use of various boundary condition choices extracted from the global model was assessed for component or package level simulations. Because the component is the foundation for building the package hierarchy, proper modeling of it at the system level becomes very important. The desirable criteria for a simplified component model, when used in system level analysis, are:

- inclusion of the relevant influences attributed to the air flow around components and conduction phenomena,
- avoidance of an unnecessarily fine grid. Widely accepted standards for component simplification are not yet available, and the selection of simplification methods is largely user and component dependent.

Nigen and Amon (1992) presented a methodology for concurrent thermal design of electronic products. An evolutionary five stage approach is employed, beginning with rough "back of the envelope" calculations and culminating in a conjugate conduction/convection simulation for a portion of the sample board. The level of approximation included at each stage of the design is selected based on both time and accuracy constraints. The application of the approach is demonstrated for a simple example of a channel containing an array of in-line protrusions.

The need for integrated system and component level modeling has also been articulated by Reynell et al (1991). They illustrated the application of the CFD code FLOTHERM to the calculation of the airflow and convection near two components on a board and also for an entire electronic system. The need for using fluid temperatures more representative of the local conditions, as part of the component level calculations was identified. The overall conclusion was that a two stage approach be followed for the analysis of convection cooled equipment. The macro (or system level) approach models the entire system and determines local conditions that are then to be employed in a micro (or component level) calculation. The details of the integration methodology between the two levels were not addressed.

1.4.2.3 THE USE OF EXPERIMENTAL BENCHMARKS FOR VALIDATION AND CALIBRATION

A promising way to measure correction factors is by designing "ideal" experimental benchmarks for complex geometries where all boundary conditions and material properties are under control and well known to within a few percent. First of all, a reference situation should be created, serving as a baseline experiment. For example, the walls of the experimental enclosure could be covered with copper and the measured temperature can be imposed in the numerical analysis, thereby circumventing all problems related to the heat transfer to the outside world. No grilles should be present and especially the inflow of air should be exceptionally well controlled. All relevant surfaces should be covered with a non-deteriorating low-emissivity material. PCB's made of materials with well-known properties should be used. Heat dissipators should be flush-mounted in order to exclude backward- and forward-facing step effects. The dimensions should be

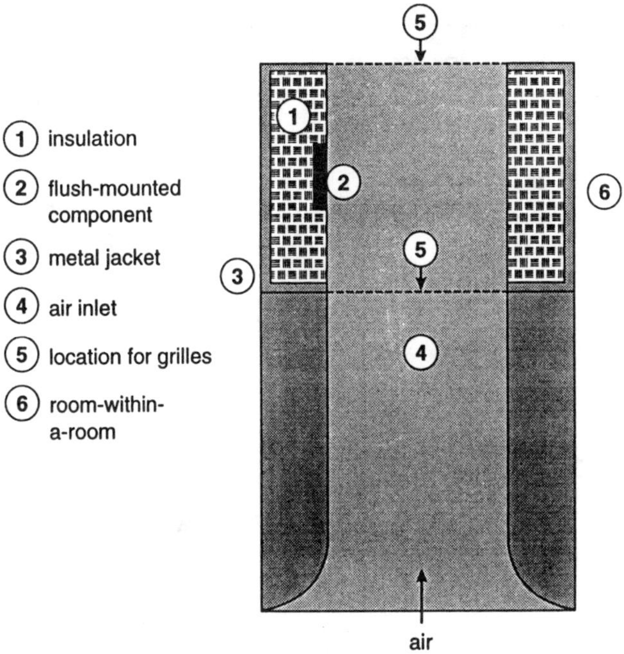

Figure 1.39: Experimental benchmark for base-line study of laminar flow.

such as to ensure that the flow is in the laminar regime everywhere. A sketch of an experimental setup based on these principles is shown in Fig 1.39. Such a benchmark apparatus can be used to validate the numerical analysis at a more complex level resulting in the answer to the question: how accurate is the basic numerical model, without taking into account the effects of the complex real-world inlet and boundary conditions? In reality, a somewhat more complex benchmark has been built, comprising several flush-mounted components in a vertical array, to permit the study the influence of wake effects on the adiabatic temperature of the downstream metal blocks.

The most difficult part of the problem was found to be the thermal conductivity of the insulation material. In almost all heat transfer experiments, the fact that no real "insulation" material is available seriously hampers accurate validation. For this particular case, there is an interplay between flush-mounted heat dissipators and natural convection. Figure 1.40 shows the convective/conductive heat transfer ratio as a function of the thermal conductivity of the insu-

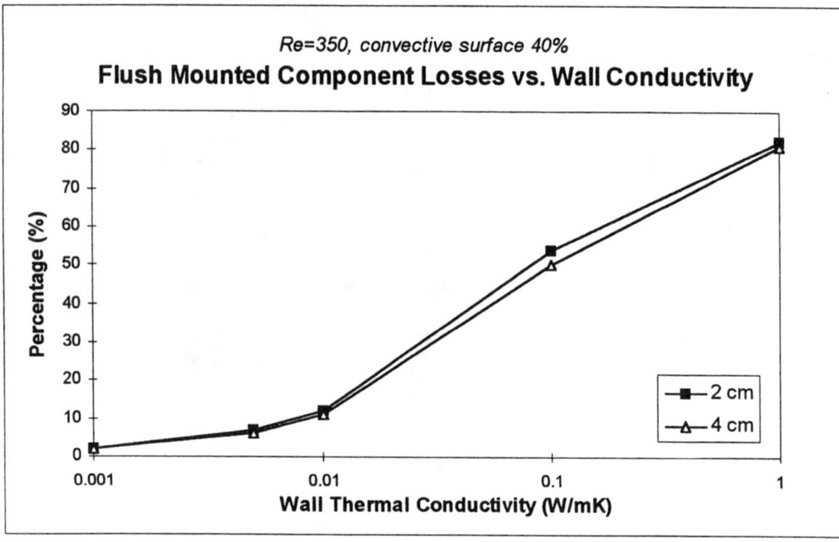

Figure 1.40: Percentage heat loss as a function of insulation conductivity.

lation material for two different thicknesses. Note that even for a thermal conductivity equivalent to air, 25 percent of the heat still flows into the insulation.

As yet unpublished research shows clearly that after this problem has been solved by using dedicated experiments to retrieve the thermal conductivity *in situ*, it was indeed possible to show the experimental and numerical results to within a few percent thereby establishing the base-line situation.

The next step is to investigate the many unknown influences one by one. The influence of recirculating zones just behind larger electronic parts can be studied by comparing flush-mounted and protruding components when all other parameters are known. This is the *validation* part. To the *calibration* domain belong the analysis of the effects of the ever-present wires which, of course, are never part of the model but might induce instabilities. Local adiabatic heat transfer coefficients of the metal component, influenced by instabilities caused by the inlet air and the grille, could be compared to detailed numerical studies and correction factors, if necessary, could be deduced. By applying different grilles and porosities, the sensitivity of the component temperature to these input parameters could be assessed. Preliminary experiments provided evidence that the instabilities mentioned have a substantial impact on the local heat transfer coefficients (of the order of 10 percent). Evidently, this "enhancement" factor is a function of the distance between the component and the grille.

After this, in the numerical analysis of a *real* system with real PCB's, the emissivities, conductivities and wall conditions can be tuned to their experimental counterparts by applying appropriate correction factors for real world material properties.

Finally, the combination of fitted input parameters and correction factors found in this way can be used with confidence to perform the accurate parametrical studies required for reliability prediction. Observe that it has been tacitly assumed that no fundamental problem exists in extrapolating the results from a relatively simple to a real complex system. Not all experts agree and it remains for the future to settle this point.

1.4.3 THERMAL CHARACTERIZATION OF REAL ELECTRONIC SYSTEMS

A real system mainly differs from the idealized system discussed in that there is an enormous increase in small details, which are virtually impossible to simulate in the CFD model but nevertheless influence the final result. By far the most important aspect here is the increased complexity of a real electronic part as compared to the metal blocks. It is worth noting here that in prototyping, it is often tacitly assumed that a real component can be appoximated by a metal block or more specifically, that the heat transfer coefficient derived for the block can be translated to an average heat transfer coefficient for the real component. Very recently, however, results became available indicating that the errors introduced by such an assumption can be of the order of 10 percent, but are usually less [Vinke and Lasance (1995)] By replacing the blocks with real electronic parts, two additional sources of uncertainty are introduced. These are the thermal data of electronic parts and the ever-present interface and contact resistances. These two items will be briefly discussed in the following section.

1.4.3.1 THE THERMAL RESISTANCE OF ELECTRONIC PARTS

Sufficient details of the layout in the final stages of a design are usually available to provide the designer with answers to two important questions:

- does the system meet the temperature specifications?
- What is the expected reliability?

To formulate answers to these questions, temperature information is required on a very small scale. It is not to be expected that detailed models of electronic parts will be used at the system level for a long time to come. At least 10,000 nodes are needed to model a Plastic Quad Flat Package (PQFP) to sufficient detail. An additional difficulty is that detailed models will probably not enter the public domain for proprietary reasons. For the time being, reliance must be placed on the thermal data made available by the component manufacturers.

For decades, the manufacturers of electronic components have represented their thermal data by using some expression for a thermal resistance defined as

$$R_{\mathrm{Th}} = \frac{T - T_{\mathrm{ref}}}{Q} \tag{1.29}$$

temperature, and Q is usually the steady state dissipation of the component. There is abundant evidence that these data cannot be used for accurate temperature prediction in a board or system level analysis. Many of the apparent paradoxes presented in the literature on this subject are attributed to the application and measurement of thermal resistances to situations in which their physical meaning is not well defined [Rosten and Lasance (1994, 1995), Rosten (1995) and Lasance (1996)].

Several papers have recently been published dealing with boundary condition independent compact models aimed at eventually replacing the existing thermal data provided by the component manufacturers. These models, preferably, but not exclusively, in the form of a simple resistance network, can be embedded in board and system level analysis software. For members of the PQFP and Ceramic Pin Grid Array Package (CPGA) families, it has been demonstrated that their accuracy is better than 95 percent compared to the detailed models [Lasance et al (1994) and Vinke and Lasance (1995, 1996)].

The accuracy of the detailed model depends largely on its input parameters. Apart from the difficulty of getting hold of the material properties such as the thermal conductivity of the molding compound and an appropriate value for the die attach, the variance inherent in the manufacturing process should be included as well. Particularly when the die is not soldered but glued to the die pad as is, for example, the case with consumer power devices, the variance in the thermal resistance as published in the data sheets can be as large as 25 percent or more [Childress and Peterson, 1989)]. The only way to deal with this problem is to measure the thermal resistance of a particular component used in the validation by performing a separate experiment.

Another important issue concerns the experimental validation of the detailed numerical models. Novel methods had to be developed because existing standards could not meet the requirements with respect to accuracy, mainly because of the lack of well-defined and reproducible boundary conditions. The interested reader is referred to a paper by Temmerman et al (1995).

In conclusion, it is expected that boundary-condition-independent compact models will greatly improve the predictability of junction temperatures at the board and system level, provided the detailed models from which they are derived are experimentally validated by proper measurement protocols.

1.4.3.2 INTERFACE AND CONTACT RESISTANCES

Component manufacturers have succeeded in achieving a continuous decrease in the overall thermal resistance of the package by optimal choice of materials, thinner packages and the use of heat spreaders. Consequently, the influence of the resistance formed by the interface between device and board, heat sink or cold plate is increasing. Because the theoretical treatment of contact heat transfer is still very difficult, heavy reliance must still be placed on experiments. The literature reveals that most methods currently in use are based on steady-state experiments. These are difficult to perform if a high accuracy is required and they also take a fairly long time, typically

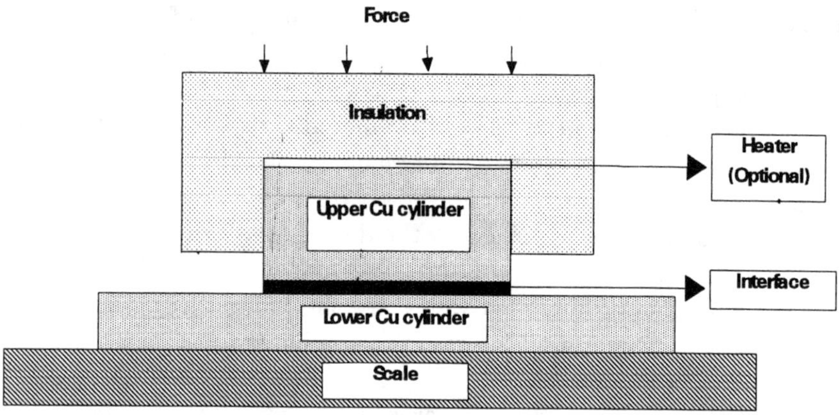

Figure 1.41: Sketch of the interface resistance measurement facility.

hours. Recently, a description has been given of a transient method which overcomes these problems [Lasance and Lacaze (1996)]. Here, the term "interface resistance" is used to define the thermal resistance of the layer between one surface of an electronic device and its mounting surface. The layer may consist of only contact resistances, heat sink compounds, metallic coatings, insulation sheets or combinations of these. The total thermal resistance of the interface is called the "interface resistance."

Fig 1.41 presents an overall view of the basic features. It consists of an upper cylinder of copper, insulated with an appropriate insulation material, a lower copper cylinder and the interface to be measured between the two cylinders. The entire setup is placed on a scale in order to measure the force. Some characteristic dimensions are the diameter of copper block, 4 cm and its height, 2 cm. After studying several alternatives, a temperature pulse applied to the lower block was considered to be the most optimal solution. The easiest way is to use a single cold plate with two water inlets or one inlet plus a valve with two water supplies at different temperatures.

At the start of the experiment, the cold water is suddenly changed to hot water and the heating curve is monitored. After about 20 seconds, the hot water is replaced by cold water and the cooling curve is then taken. Both curves are well suited for easy data acquisition. Numerical analysis, combined with some preliminary experiments, showed that a resolution of 0.02 °C/W could be realized.

1.4.4 SUMMARY

It has been shown that it is currently impossible to test the hypothesis that a numerical model really does represent a real electronic system at an acceptable confidence level, at least when the objective of the analysis is the accurate prediction of the temperature of critical electronic parts.

Part of the problem arises from the fact that in the current state-of-the-art, no existing code is capable of solving the physics of a complex system in sufficient detail. It is the lack of sufficiently accurate input parameters and boundary conditions that constitutes some of the other obstacles.

A possible solution is proposed by developing suitable experimental benchmarks which are able to provide, on the one hand, correction factors for the intractable small-scale physical phenomena and, on the other hand, useful values for the unknown input parameters and boundary conditions.

In this way, it is believed that the wealth of relative information that can already be extracted from CFD analyses, supported by proper experiments, can be transferred to the domain of absolute information.

1.5 CONCLUSIONS

The continually shrinking time-to-market for electronic products will require an ever-increasing reliance on computational simulations. The timely availability of three-dimensional computational fluid dynamics capabilities is offering unprecented opportunities for the thermal characterization of natural convection cooled electronic systems. A number of complexities, however, still impede a greater reliance on predictive modeling capabilities. Some of these issues, including computational resources for handling large, realistic problems and databases on thermophysical properties of electronic packaging materials will be addressed as part of the advances in computing hardware and the ongoing research attention from the thermal community.

On the other hand, many challenging barriers to better predictability of temperatures in electronic systems will remain unless attention is specifically focused on them. Some of these require a deeper physical understanding of the fundamental thermofluid phenomena involved such as transport from horizontal heated surfaces within enclosures, natural convection in partially open and vented enclosures and characterization of transitional and turbulent flows in geometries of interest in electronic systems.

Significant attention needs to be simultaneously devoted to developing efficient global-local modeling strategies that make simulation of realistic systems possible. Appropriate coupling approaches between the global and local models need to be identified. Efficient simulation techniques for combined conduction, convection and radiation processes in realistic enclosures need to be developed. Indeed, the success of the foregoing advances can only be assessed by detailed experimental validation. Currently, a number of barriers exist which make such validation difficult. Significant advances in characterization of anisotropic thermal behavior of packaging materials, accurate determination of physical properties and assessment of interface thermal resistances should result in benchmark-experimental studies for realistic enclosures. Such studies will form the foundation for the advancement of current computational codes to truly provide a predictive thermal characterization capability.

1.6 ACKNOWLEDGEMENTS

The authors would like to acknowledge the very valuable comments and suggestions by Prof A. Bar-Cohen of the University of Minnesota. Dr. Azar's (Lucent Technologies) review of section 4 is also highly appreciated. One of the authors (CL) is greatly indebted to Prof Hanjalic of Delft University for reviewing the section on geometrically simple enclosures.

The reader should be aware that **Certain commercial equipment, instruments, materials or software are identified in this chapter in order to specify the procedure adequately. Such identification does not imply recommendation or en-**

dorsement by the authors nor does it imply that the items identified are necessarily the best available for the purpose.

1.7 NOMENCLATURE

Roman Letter Symbols[3]

A	area, m^2
A	total liquid wetted surface area of each protrusion, m^2
A_i	area of the ith liquid wetted protrusion face, m^2
A_x	aspect ratio in the x-direction
A_z	aspect ratio in the z-direction
B_c	size of protrusion in x-direction, m
Bo	Boussinesq number
B_s	thickness of substrate, m
c_p	specific heat, J/kg-K
d	thickness, m
g	gravitational acceleration, m/s^2
g	gravity constant, m/s^2
h	heat transfer coefficient, W/m^2-K
H	height of the cavity, m
	height, m
H_c	size of protrusion in y-direction, m
k	thermal conductivity, W/m-K
K_0, K_1, I_0, I_1	Bessel functions dimensionless
L	vertical dimension of each protrusion, m
Nu	Nusselt number, dimensionless
Nu_{avg}	average Nusselt number at the faces
p	non-dimensional pressure
P	pressure, Pa
Pr	Prandtl number, dimensionless
q	heat flux, W/m^2
Q	input power to each protrusion, W
q	volumetric heat generation rate, W/m^3
Q_{net}	net power loss from each protrusion to the fluid, W
R	product of density and Specific Heat, J/m^3-K
	thermal resistance, K/W or °C/W

[3]The nomenclature also applies to the appendices.

Roman Letter Symbols (Cont'd)

r	radius, m
$\hat{\text{Ra}}$	modified Rayleigh number dimensionless
t	non-dimensional time
T	non-dimensional temperature
ΔT	temperature rise, K
$\vec{U}$	non-dimensional velocity vector
v	velocity, m/s
V	volume flow, m^3/s
w	enclosure width; distance between the substrate and the parallel enclosure wall, mm
W	width of enclosure in x-direction minus the substrate, m
W_c	size of protrusion in z-direction, m
W_e	dimensions of enclosure in z-direction, m
X, Y, Z	dimensional spatial coordinates, m
x, y, z	non-dimensional spatial coordinates, scaled by H m
$y_{c,i}$	location of ith protrusion center in y-direction, m
$z_{c,i}$	location of ith protrusion center in z-direction, m

Greek Symbols

α	thermal diffusivity of liquid, m^2/s
β	coefficient of volumetric expansion of liquid, K^{-1} porosity dimensionless
γ	Euler's constant (0.57), dimensionless
ϵ	emissivity, dimensionless
ν	kinematic viscosity of liquid, m^2/s
θ	protrusion surface temperature
ρ	density, kg/m^3
σ	Stefan-Boltzmann constant, W/m^2-K^4
ξ	air resistance coefficient, dimensionless

Subscripts

avg	average protrusion surface value
c	protrusion
film	value at which fluid properties are evaluated
i	value at the center of the ith fluid exposed protrusion face
ref	reference taken as an average of top and bottom enclosure boundary values

Subscripts (Cont'd)

s	substrate
wall	value at horizontal walls
x	x-direction
z	z-direction

1.8 REFERENCES

Adam J (1993). Personal Communication.

Adams V, Joshi Y, and Blackburn D (1995). Natural Convection from an Array of Electronic Packages Mounted on a Horizontal Board in a Narrow Aspect Ratio Enclosure, presented at the *2nd ISHMT/ASME Conf*, Surathkal, India, 911-917.

Afzal N and Hussain T (1984). Mixed Convection over a Horizontal Plate, *ASME, J Heat Transfer*, 106, 240-241.

Ahmed I, Krane R, and Parsons J (1994). A Preliminary Investigation of the Cooling of Electronic Components with Flat Plate Heat Sinks, *ASME J Elec Packaging*, 116, 60-67.

Aihara T (1991). Air Cooling Techniques by Natural Convection, Chapter One in *Cooling Techniques for Computers,* W Aung ed., Hemisphere Publishing Company, New York, NY.

Amin M (1993). Natural Convection Heat Transfer in Enclosures fitted with a Periodic Array of Hot Roughness Elements at the Bottom, Int J Heat and Mass Transfer, 36, 755-763.

Anderson A (1990). Convective Heat Transfer from Arrays of Modules with Non-Uniform Heat Release, *PhD Thesis*, Stanford University, Stanford, CA.

Anderson A (1994). Decoupling Convective and Conductive Heat Transfer Using the Adiabatic Heat Transfer Coefficient, *ASME J Electric Packaging*, 116, 310-316.

Anderson A and Moffat R (1992). The Adiabatic Heat Transfer Coefficient and the Superposition Kernel Function: Part 2-Modeling Flatpack Data as a Function of Channel Turbulence, *ASME J Electric Packaging*, 114, 22-28.

Arvizu D and Moffat R (1982). The Use of Superposition in Calculating Cooling Requirements for Circuit Board Mounted Electronic Components, *32nd Proc Electr Comp Conf*, 133-144.

Aung W (ed) (1991). Cooling Techniques for Computers, Hemisphere Publishing Co, New York, NY,

Azar K (1996). Chapter Five in *Air Cooling Technology for Electronic Equipment,* Kim S and Lee, S (eds), Chemical Rubber Publishing Co, Fort Lauderdale, FL.

Azar K and Graebner J (1996). Experimental Determination of Thermal Conductivity of Printed Wiring Boards, *Proc 12th IEEE SEMI THERM*, Austin, TX, 169-182.

Azar K and Moffat R (1991). Heat Transfer Coefficient and its Estimation in Electronic Enclosures, *Proc NEPCON East*, Boston, MA, 361.

Azar K and Russell E (1991). Effect of Component Layout and Geometry on the Flow Distribution in Electronics Circuit Packs, *ASME J Electric Packaging*, 113, 50-57.

Baker, E (1972). Liquid Cooling of Microelectronic Devices by Free and Forced Convection,*Microelectronics and Reliability*, 11, 213-222.

Baker E (1973). Liquid Immersion Cooling of Small Electronic Devices, *Microelectronics and Reliability*, 11, 163-173.

Balaji C and Venkateshan S P (1993). Interaction of Surface Radiation With Free Convection in a Square Cavity, *Int J Heat and Fluid Flow*, 14 (3), 260-267.

Balaji C and Venkateshan S P (1994a). Correlation for Free Convection and Surface Radiation in a Square Cavity, Int J Heat and Fluid Flow, 15 (3), 249-251.

Balaji C and Venkateshan S P (1994b). Interaction of Radiation With Free Convection in an Open Cavity, Int J Heat and Fluid Flow, 15 (4), 317-324.

Bar-Cohen A (1993). Thermal Management of Electronic Components With Dielectric Liquids, *JSME International Journal*, Series B, 36 (1), 1-12.

Bar-Cohen A, and Schweitzer H (1985). Convective Immersion Cooling of Parallel Vertical Plates, *IEEE Trans Components, Hybrids and Manufacturing Technology*, CHMT-8, 343-351.

Barozzi G and Nobile E (1990). Conjugate Heat Transfer in Forced Convection Cooling of Chip Arrays, *in "Heat Transfer in Electronic and Microelectronic Equipment,"* A E Bergles ed., Hemisphere Publishing Company, New York, NY.

Beckermann C, Smith T F, and Pospichal B (1994). Use of a Two-dimensional Simulation Model in the Thermal Analysis of a Multi-Board Electronic Module, *ASME J Heat Transfer*, 116, 126-133.

Bejan A (1984). *Convection Heat Transfer*, John Wiley and Sons, New York, NY.

Bergles A E and Bar-Cohen A (1990). Direct Liquid Cooling of Microelectronic Components,, *in "Advances in Thermal Modeling of Electronic Components and Systems*, Bar-Cohen, A and Kraus, A D (eds), ASME Press, New York, NY, 2, 233-342.

Bertin Y and Saulnier J (1990). A Detailed Analysis of the Conductive/Convective Heat Transfer in a Single Electronic Board, in *"Heat Transfer in Electronic and Microelectronic Equipment"*, A E Bergles (ed), Hemisphere Publishing Company, New York, NY.

Betts P and Dafa'Alla A (1986). Turbulent Buoyant Air Flow in a Tall Rectangular Cavity, *Exp Heat Transfer*, 2, 59.

Buller M and Duclos T (1982). Thermal Characteristics of Horizontally Oriented Electronic Components in an Enclosed Environment, *IEEE Trans Components, Hybrids and Manufacturing Technology*, CHMT-5, 493-498.

Cadre M, Riault D, Buannec J, and Kuntz J (1989a). L'Architecture Mecanique des Equipments Electroniques, Optimisation par Simulation Numerique du Comportement Thermique, *Proc 3rd Conference on Heat Transfer, Energy and Environment Applied to Communication and Computer Fields*, Pierros-Guirec France.

Cadre M, Viault A, Pimont V, Bourg A, and Latrobe A (1989b). Modelisation des Phenomenes Thermiques dans des Enceintes Electroniques Fermees, *Proc 3rd Conference on Heat Transfer, Energy and Environment Applied to Communication and Computer Fields*, Pierros-Guirec, France.

Cadre M, Latrobe A, and Le Jannou J (1986). Simulation and Experimentation of Air Flow in Electronic Racks, *Proc Intelec '86*, Toronto, Canada.

Cadre M, Viault A, Pimont V, and Bourg A (1990). Modeling of PCB's in Enclosures, *in "Heat Transfer in Electronic and Microelectronic Equipment"*, A. Bergles, (ed), Hemisphere Publishing Co, New York, NY, 327-338.

Catton I (1978). Natural Convection in Enclosures, *Proc. 6th Int. Heat Transfer Conf*, Toronto, Ontario, 6, 13-30.

Chao P, Lior N, Churchill S, and Ozoe H (1989). The Effect of Partitions on Natural Convection in Enclosures, *in "Heat Transfer in Convective Flows", ASME HTD-107*, 315-322.

Cheesewright R and King K (1986), Experimental Data for the Validation of Computer Codes for the Prediction of Two Dimensional Buoyant Cavity Flows, *Proc. ASME HTD-60*, 75-81.

Chen L and Li L (1993). Numerical Analysis of Natural Convection in an Enclosure with Discrete Heat Sources and a Vertical Heat Sink, *in "General Papers in Heat Transfer-Natural and Forced Convection" ASME HTD-127*, 1-8.

Chen C C, Young, P L, and Cech J M (1988). Thermal Performance of a High Performance Multichip Mosule, *Proc* 38th *Elec Comp Conf*, Los Angeles, CA, 302-304.

Childress W and Peterson G (1989). Quantification of Thermal Contact Conductance in Electronic Packages, *IEEE Trans Components, Hybrids and Manufacturing Technology* CHMT-12, 717-723.

Chu T and Goldstein R (1973). Turbulent Convection in a Horizontal Layer of Water, *ASME J Fluid Mech*, 60, 141-159.

Clausing A and Berton J (1989). An Experimental Investigation of Natural Convection From an Isothermal Horizontal Plate, *ASME J Heat Transfer*, 111, 904-908.

Cohen C and Reshotko E (1956) The Compressible Laminar Boundary Layer with Heat Transfer and Arbitrary Pressure Gradient, *NACA Report 1294*.

Culham J (1988) Conjugate Heat Transfer From Surfaces with Discrete Thermal Sources", *Thesis University of Waterloo*, Waterloo, Ontario.

Culham J, Lemczyk T, Lee, S, and Yovanovich M (1991). META-A Conjugate Heat Transfer Model for Air Cooling of Circuit Boards with Arbitrarily Located Heat Sources, *ASME HTD-171, "Heat Transfer in Electronic Equipment"*, 117-126.

de Vahl Davis G (1983). Natural Convection of Air in a Square Cavity: a Benchmark Numerical Study, *Int J Num Meth Fluids*, 3, 227-264.

de Vahl Davis G (1986). Finite Difference Methods for Natural and Mixed Convection in Enclosures, *Proc 8th Int Heat Transfer Conf*, San Francisco, CA, 1, 101-109.

Dehghan, A A (1994). Natural Convection in Discretely Heated Open Cavities, *PhD Thesis*, The University of New South Wales, Sydney, Australia.

Dehghan A and Behnia M (1996), Combined Natural Convection-Conduction and Radiation Heat Transfer in a Discretely Heated Open Cavity, *ASME J Heat Transfer*, 118, 56-74

Desai C, Vafai K, and Keyhani M (1995). On the Natural Convection in a Cavity with a Cooled Top Wall and Multiple Protruding Heaters, *ASME J Electric Packaging*, 117, 43-45.

Dol H, Hanjalic K, and Kenjeres S (1996). A Comparative Assessment of the Second-Moment Differential and Algebraic Models in Turbulent Natural Convection, *presented at the Int Conf on Turbulent Heat Transfer* March, San Diego, CA.

Dol H, Peeters T, Henkes R, and Hoogendoorn C (1995). Application of a Differential Reynolds-stress Model to a Turbulent Natural Convection in a Cavity, *Proc. 10th Turbulent Shear Flows Conf*, Pittsburg, PA.

Drabbels J (1995). Natural Convection Heat Transfer of Metal Cuboids Flush-Mounted in a Horizontal Plate, *presented at 2nd EUROTHERM Seminar on "Thermal Management of Electronic Systems"*, Leuven, Belgium.

Drake S, White M, Felthouse A, Ganderton C, and Glanfield S (1994). Computational Analysis of Air and Heat Flow in Electronic Systems, in *"Thermal Management of Electronic Systems", Proc Eurotherm Seminar no. 29*, C Hoogendoorn, R. Henkes, C. Lasance (eds), Kluwer Academic Publishers, Dordrecht, the Netherlands, 45-54.

Du Z and Bilgen E (1990), Effects of Heat Intensity, Size and Position of the Components on Temperature Distribution within an Electronic PCB Enclosure, Trans ASME J Elec Pack, 112, 249-254.

Ewes I (1995). Modeling of IC-packages based on Thermal Characteristics, *EUROTHERM Seminar No 45, Proceedings Thermal Management of Electronic Systems*, Leuven, Belgium.

Facas G (1993), Natural Convection in a Cavity with Fins Attached to both Vertical Walls, *J Thermoph Heat Transfer*, 7, 555-560.

Fisher E and Rhodes N (1994). Uncertainty in Computational Fluid Dynamics, *EPSRC/I Mech E Experts Meeting*, Bournemouth, UK.

Fluorinert Liquids, Product Manual (1985). *3M Corporation*, Minneapolis, MN.

Frederick R (1992). Heat Transfer Enhancement in Natural Convection in Finned Cavities, *2nd Int Conf Adv Comp Meth Heat Transfer*, Milan, Italy. 157-169.

Free J, Russell R, and Louie J (1995). Recent Advances in Thermal/Flow Simulation: Integrated Thermal Analysis into the Mechanical Design Process, *Proceedings of 11th SEMITHERM Conf*, San Jose, CA, 136-145.

Fuchs G (1989). Rack Thermal Measurements and a New Rack Simulation Model, *Proc 3rd Conf on Heat Transfer, Energy and Environment Applied to Communication and Computer Fields*, Pierros-Guirec, France.

Fusegi T and Hyun J (1994). Laminar and Transitional Natural Convection in an Enclosure with Complex and Realistic Conditions, *Int J Heat and Fluid Flow*, 15, 258-268.

Garcia M, Cosley R, and Marongiu M (1995). Experimental and Computational Studies on the Thermal Management of Electronic Enclosures Using Natural Convection, *presented at 2nd EUROTHERM Seminar on "Thermal Management of Electronic Systems,"* Leuven, Belgium.

Gebhart B (1979). Buoyancy Induced Fluid Motions Characteristic of Applications in Technology, *J. Fluids Eng*, 101, 5-28.

Gebhart B, Jaluria Y, Mahajan R, and Sammakia B (1988), *Buoyancy Induced Flows and Transport*, Hemisphere Publishing Corp., New York, NY.

Goldstein R and Lau K (1983), Laminar Natural Convection from a Horizontal Plate and the Influence of Plate-edge Extensions, *J Fluid Mech*, 129, 55-75.

Goldstein R and Volino R (1995). Onset and Development of Natural Convection Above a Suddenly Heated Horizontal Surface, *ASME J Heat Transfer*, 117, 808-821.

Graham K and Witzman S (1987). Analytical Correlations of Thermal Design of Electronic Packages, in Cooling Technology for Electronic Equipment, W Aung (ed), 249-264.

Guglielmini G, Milano G, and Misale M (1985). Electronic Cooling by Natural Convection in Partially Confined Enclosures, *Heat and Technology*, 3, 43-57.

Guglielmini G, Milano G, and Misale M (1990). Some Factors Influencing the Optimum Free Air Cooling of Electronic Cabinets, in *Heat Transfer in Electronic and Microelectronic Equipment*, A Bergles (ed), Hemisphere Publishing Corp, New York, NY, 311-325

Hadim A and Ramot M (1993). Natural Convection in an Enclosure with Discrete Heat Sources on the Vertical Walls, *ASME J Elec Pack*, 115, 444-448.

Hanjalic K (1994). Achievements and Limitations in Modeling and Computation of Buoyant Turbulent Flows and Heat Transfer, *Proc Int Heat Transfer Conf*, Brighton, UK, 1, 1-18.

Hanjalic K, Kenjeres S, and Durst, F (1996). Natural Convection in partitioned Computation Two-Dimensional Enclosures at High Rayleigh Numbers, *Int J Heat and Mass Transfer*, 39, 1407-1427.

Hanjalic K and Vasic S (1993) Computation of Turbulent Natural Convection in Rectangular Enclosures with an Algebraic Flux Model, *Int J Heat and Mass Transfer*, 36, 3606-3624.

Hasnaoui M, Bilgen E, and Vasseur P (1992a). Natural Convection Heat Transfer in Rectangular Cavities Partially Heated from Below, *J Thermophys and Heat Transfer*, 6, 255-264.

Hasnaoui M, Vasseur P, and Bilgen E (1992b). Natural Convection in Rectangular Enclosures with Adiabatic Fins attached on the Heated Wall, *J. Warme und Stoffubertragung*, 27, 357-368.

Hawkins L, Khan J, and Yao G (1992). Numerical Solution of Buoyancy-Driven Flow through Small Openings between Two Enclosures, *ASME HTD-198*, 105-112

Heindel T J, Ramadhyani S, and Incropera F P (1995a). Laminar Natural Convection in a Discretely Heated Cavity: I Assessment of Three-Dimensional Effects, *ASME J Heat Transfer*, 117, 902-909

Heindel T J, Ramadhyani S, and Incropera F P (1995b). Laminar Natural Convection in a Discretely Heated Cavity: II Comparisons of Experimental and Theoretical Results, *ASME J Heat Transfer*, 117, 910-917

Henkes R (1990). Natural Convection Boundary Layers, *PhD Thesis*, Delft, Holland.

Henkes R, and Hoogendoorn C (1993). Turbulent Natural Convection in Enclosures: a Computational and Experimental Benchmark Study, *Proc Eurotherm Seminar*, 22, EETI, Paris, France

Henkes R, and Hoogendoorn C (1995). Comparison Exercise for Computations of Turbulent Natural Convection in Enclosures, Num. Heat Transfer A, vol. 28, pp. 59-78

Henkes R, and Le Quere P (1996). Three-dimensional Transition of Natural-convection Flows, J. Fluid Mechanics, vol. 319, pp. 281-303

Hetsroni G, Yarin L, and Kaftori D (1996). A Mechanistic Model for Heat Transfer from a Wall to a Fluid, Int. J. Heat Mass Transfer, vol. 39, pp.1475-1478

House J, Beckermann C, and Smith T (1989) Natural Convection Heat Transfer in a Square Enclosure Containing an Obstruction, in *Heat Transfer in Convective Flows, ASME HTD-107*, 289-297.

Huang Y, and Aggarwal S (1995). Effect of Wall Conduction on Natural Convection in an Enclosure with a Centered Heat Source, *ASME J Elec Pack*, 117, 301-306.

Husar R, and Sparrow E (1968). Patterns of Free Convection Flow Adjacent to Horizontal Heated Surfaces, *Int J Heat and Mass Trans*, 11, 1206-1208.

Idelchik I (1986). *Handbook of Hydraulic Resistance*, Hemisphere Publsihing Company, New York, NY.

Ince N and Launder B (1995). Three-dimensional and Heat-loss Effects on Turbulent Flow in a Nominally Two-dimensional Cavity, *Int J Heat and Fluid Flow*, 16, 171-177.

Incropera F (1988). Convection Heat Transfer in Electronic Equipment Cooling, *ASME J Heat Transfer* 110, 1097-1111.

Ishiguro R, Abe T, Nagase H, and Nakanishi S (1978). Heat Transfer and Flow Instability of Natural Convection over Upward-facing Horizontal Plates, *Proc 6th Int Heat Transfer Conf*, Toronto, Ontario, 229-234.

Ishizuka M, Miyazaki Y, and Sasaki T (1987). Air Resistance Coefficients for Perforated Plates in Free Convection, *ASME J Heat Transfer*, 109, 540-543.

Jakob M, (1953), *Heat Transfer*, 1, John Wiley and Sons, New York, NY.

Janssen R (1994). Instabilities in Natural Convection Flows in Cavities, PhD Thesis, Delft

Janssen R and Henkes R (1995a). Influence of Prandtl Number on Instability Mechanisms and Transition in a Differentially Heated Square Cavity, *J Fluid Mechanics*, 290, 319-344.

Janssen R and Henkes R (1995b). The First Instability Mechanism in Differentially Heated Cavities with Conducting Horizontal Walls, *ASME J Heat Transfer*, 117, 626-633.

Joshi, Y and Paje, R A (1991). Natural Convection Cooling of a Ceramic Substrate Mounted Leadless Chip Carrier in Dielectric Liquids, *Int Comm Heat Mass Transfer*, 18, 39-47.

Joshi, Y, Kelleher M D, and Benedict, T J (1990). Natural Convection Immersion Cooling of an Array of Simulated Electronic Components in an Enclosure Filled with Dielectric Fluid, in *Heat Transfer in Electronic and Microelectronic Equipment* A E Bergles (Ed), Hemisphere Publishing Corp, New York, NY, 445-468

Joshi, Y, Kelleher M D, Powell, M and Torres, E I (1994). Natural Convection Heat Transfer From an Array of Rectangular Protrusions in an Enclosure Filled With Dielectric Liquid, *ASME J Elec Pack*, 116, 138-147.

Joshi, Y, Willson T, and Hazard, S J III (1989a). An Experimental Study of Natural Convection From an Array of Heated Protrusions on a Vertical Surface in Water, *ASME J Elec Pack*, 111, 121-128.

Joshi, Y, Willson, T, and Hazard, S J III (1989b). An Experimental Study of Natural Convection Cooling of an Array of Heated Protrusions in a Vertical Channel in Water, *ASME J Elec Pack*, 111, 33-40.

Kang B and Jaluria Y (1990). Natural Convection Heat Transfer Characteristics of a Protruding Thermal Source Located on Horizontal and Vertical Surfaces, *Int J Heat and Mass Transfer*, 33, 1347-1357.

Kangni A, Vasseur P, and Bilgen E (1995). Natural Convection in Inclined Enclosures with Multiple Conducting Partitions, *J Thermoph and Heat Transfer*, 9, 270-277.

Kangni A, Yedder R, and Bilgen E (1991). Natural Convection and Conduction in Enclosures with Multiple Vertical Partitions, *Int J Heat and Mass Transfer*, 34, 2819-2825.

Kassemi M (1993). Analysis of Radiation-Natural Convection Interactions in 1-g and low-g Environments Using the Discrete Exchange Factor Method, *Int J of Heat and Mass Transfer*, 36(1), 4141-4149.

Kato K, Takarada T, Miyazaki H, Sato H, and Nakagawa N (1991). Heat Transfer in a Channel between Vertical Electronic Circuit Boards Cooled by Natural Air Convection, *J Chem Eng Japan*, 24, 568-574.

Kelkar K and Patankar S (1990). Numerical Prediction of Natural Convection in Square Partitioned Enclosures, *Num Heat Transfer A*, 17, 269-285.

Kelleher, M D and Knock R H (1987). Laminar Natural Convection in a Rectangular Enclosure Due to a Heated Protrusion on One Vertical Wall-Part I: Experimental Investigation, *Proc 2nd ASME/JSME Thermal Eng Joint Conf*, Honolulu, HA, 169-177.

Kenjeres S and Hanjalic K (1995a). On the Prediction of Thermal Convection in Horizontal Slender Enclosures, *presented at Tenth Symp on Turbulent Shear Flows*,

Kenjeres S and Hanjalic K (1995b). Prediction of Turbulent Thermal Convection in Concentric and Eccentric Horizontal Annuli, *Int J Heat and Fluid Flow*, 16, 429-339.

Kennedy K and Kanehi J (1983). Free Convection in Tilted Enclosures, in *Heat Transfer in Electronic Equipment*, ASME HTD-28, 43-47.

Kern D Q and Kraus A D (1972). *Extended Surface Heat Transfer*, McGraw-Hill Book Company, New York, NY.

Keyhani M, Chen L, and Pitts D R (1991). The Aspect Ratio Effect on Natural Convection in an Enclosure with Protruding Heat Sources, *ASME J Heat Transfer*, 113, 883-891.

Keyhani M, Prasad V, and Cox, R (1988). An Experimental Study of Natural Convection in a Vertical Cavity with Discrete Heat Sources, *ASME J Heat Transfer*, 110, 616-624.

Kim D and Viskanta R (1984). Study of the Effects of Wall Conduction on Natural Convection in Differentially Oriented Square Cavities, *J Fluid Mech*, 144, 153-176.

Kim S, Anand N, Fletcher L (1989). Free Convection Between Series of Vertical-Parallel Plates with Embedded Line Heat Sources, in *Numerical Simulation of Convection in Electronic Equipment Cooling*, ASME HTD-121, 49-57.

Kim S Y, Sung H J, and Hyun J M (1992). Mixed Convection from Multiple-layered Boards with Cross-streamwise Periodic Boundary Conductions, *Int J Heat and Mass Transfer*, 35, 2941-2951.

Kimura S and Bejan A (1984). The Boundary Layer Natural Convection in a Rectangular Cavity with Uniform Heat Flux from the Side, *ASME J Heat Transfer*, 106, 98-103.

Kishinami K, Saito H, and Tokura I (1990). A Fundamental Investigation of Laminar Natural Convection Heat Transfer from a Vertical Plate with Discontinuous Surface Heating, in *Heat Transfer in Electronic and Microelectronic Equipment*, A E Bergles (ed), Hemisphere Publishing Corp, New York, NY.

Kitamura K and Kimura F (1995). Heat Transfer and Fluid Flow of Natural Convection Adjacent to Upward-facing Horizontal Plates, *Int J Heat and Mass Transfer*, 38, 3149-3159.

Klein M K (1994). Multi-Level Thermal Analysis applied to the Design of Electronic Products. *ASME EEP-9 "CAE/CAD Application to Electronic Packaging*, 59-65.

Kordyban, T (1994). Estimating the Influence of PCB and Component Thermal Conductivity on Component Temperatures in Natural Convection, *Flowtherm Third International Users Conference.*, London, UK.

Krane R and Phillips T (1986). Natural Convection Cooling of a Horizontally-Oriented Component Board Mounted in a Low-Aspect-Ratio Enclosure, *ASME HTD-57, "Heat Transfer in Electronic Equipment"*, 113-122.

Krane R and Phillips T (1991). Natural Convection Cooling of a Horizontally-Oriented Component Board Mounted in a Low-Aspect Ratio Enclosure, *ASME J Elec Pack*, 113, 301-308.

Krane R, Ahmed, I, and Parsons, R (1993). The Cooling of Electronocs with Flat Plate Heat Sinks, in *"Cooling of Electronics,"* S Kakac et al (eds), Kluwer Academic Publishers, Dordrecht, the Netherlands, 825-858.

Kraus A (1982). *Analysis and Evaluation of Extended Surface Thermal Systems*, Hemisphere Publishing Corp, New York, NY.

Kraus A D and Bar-Cohen A (1983). *Thermal Analysis and Control of Electronic Equipment*, Hemisphere Publishing Corp, New York, NY.

Lacroix M and Joyeux A (1995). Natural Convection Heat Transfer around heated Cylinders inside a Cavity with Conducting Walls, *Num Heat Transfer A* 27, 335-350.

Lage J L, Lim J S, and Bejan A (1992), Natural Convection With Radiation in a Cavity With Open Top End, ASME J Heat Transfer, 114, 479-486.

Lall B, Ortega A, and Kabir H (1994). Thermal Design Rules for Electronic Components on Conducting Boards in Passively Cooled Enclosures, *Proc ITHERM*, Washington, DC, 50-61.

Lasance C (1986). Computer Analysis of Heat Transfer Problems to Check the Validity of Engineering Formulae, *Proc 8th Int Heat Transfer Conf*, San Francisco, CA, 325-330.

Lasance C (1991). Numerical Simulation of Natural Convection in a Television Set, *ASME HTD-171*, 101-108.

Lasance C (1994a). Accurate Temperature Prediction in Consumer Electronics: a Must but still a Myth, in *"Cooling of Electronics,"* S Kakac et al (eds), Kluwer Academic Publishers, Dordrecht, the Netherlands, 825-858.

Lasance C (1994b). Pragmatic Methods to Determine the Parameters Required for the Thermal Analysis of Electronic Systems", in "Cooling of Electronics," S Kakac et al (eds), Kluwer Academic Publishers, Dordrecht, the Netherlands, 859-898.

Lasance C (1994c). Thermal Management of Air-Cooled Electronic Systems: New Challenges for Research, in it "Thermal Management of Electronic Systems', Proc Eurotherm Seminar no 29, C Hoogendoorn, R Henkes, and C. Lasance (eds), Kluwer Academic Publishers, Dordrecht, The Netherlands.

Lasance C (1997) Thermal Characterization of Electronic Parts with Compact Models: Interpretation, Application, and the Need for a Paradigm Shift", submitted to IEEE SEMI THERM, January 1997.

Lasance C and Lacaze C (1996). A Transient Method for the Accurate Measurement of Interface Resistances, *Proc 12th IEEE SEMI THERM*, Austin, TX, 43-50.

Lasance C, Vinke H, and Rosten H (1995). Thermal Characterization of Electronic Devices by Means of Boundary Condition Independent Compact Models, *IEEE Trans CPMT-A*, 18, 1-9.

Le Jannou J, Cadre M, Latrobe A, and Viault A (1986). Thermal Field Prediction in Electronic Equipment, *Proc Int Heat Transfer Conf*, San Francisco, CA, 6, 2983-2987.

Le Quere P (1994). Onset of Unsteadiness, Routes to Chaos and Simulations of Chaotic Flows in Cavities heated from the Side: a Review of Present Status, *Proc Int Heat Transfer Conf*, Brighton, UK, 1, 281-296.

Le Quere P and Penot F (1987). Numerical and Experimental Investigation of the Transition to Unsteady Natural Convection of Air in a Vertical Differentially Heated Cavity, *ASME HTD-94*, 75-82.

Lee S and Yovanovich M (1989). Conjugate Heat Transfer from a Vertical Plate with Discrete Heat Sources Under Natural Convection, *ASME J Elec Pack*, 111, 261.

Lee T, Chambers B, and Mahalingam M (1994). Application of CFD Technology to Electronic Thermal Management, Proc ECTC, Washington, DC, 411-420.

Lee, J J, Liu K V, Yang, K T, and Kelleher, M D, (1987). Laminar Natural Convection in a Rectangular Enclosure due to a Heated Protrusion on One Vertical Wall - Part II: Numerical Simulation, *Proc Second ASME/JSME Therm Engr Joint Conf*, Honolulu, HA, 179-185.

Leonard, B P (1988). Elliptic Systems: Finite Difference Method IV, Ch 9 in Handbook of Numerical Heat Transfer, W J Minkowycz et al (eds), John Wiley and Sons, New York, NY, 347-378.

Lewandowski W and Kubski P (1983). Methodical Investigation of Free Convection from Vertical and Horizontal Plates, *Warme-und Stoffubertragung* 17, 147-154.

Linton R and Agonafer D (1994). Thermal Model of a PC, J Elec Pack, 116, 134-137.

Liu K V, Yang K T and Kelleher M D (1987a). Three-Dimensional Natural Convection Cooling of an Array of Heated Protrusions in an Enclosure Filled with a Dielectric Fluid, *Proc Int*

Symp on Cooling Technology for Electronic Equipment, Honolulu, Ha, 486-497.

Liu K V, Yang K T, Wu Y W, and Kelleher M D (1987b). Local Oscillatory Surface Temperature Responses in Immersion Cooling of a Chip Array by Natural Convection in an Enclosure, *Proc Symp on Heat and Mass Transfer in Honor of B. T. Chao*, University of Illinois, Urbana-Champaign, IL.

Maciejewski P and Moffat R (1990). A Correlation for Boundary Layer Heat Transfer Accounting for Free Stream Turbulence, *Proc 9th Int Heat Transfer Conf* Jerusalem, Israel, 303-308.

Maciejewski P and Moffat R (1992a). Heat Transfer with Very High Free-Stream Turbulence: Part I-Experimental Data, *ASME J Heat Transfer*, 114, 827-833.

Maciejewski P and Moffat R (1992b). Heat Transfer with Very High Free-Stream Turbulence: Part II-Analysis of Results, *ASME J Heat Transfer*, 114, 834-839.

Malkus W (1954), Discrete Transitions in Turbulent Convection, *Proc Roy Soc*, 225, 185-193.

Manno V and Azar K (1991). Impact of Heat Transfer Coefficient Models on the Simulation of Electronic Enclosure Thermal Response, *Proc NEPCON East*, Boston, MA, 349-360.

Mayinger F and Wang Z (1994). Experiments on Natural Convective Air Cooling of a PCBS Array in a Closed Casing with Inclination, in *"Cooling of Electronics,"* S Kakac et al (eds), Kluwer Academic Publishers, Dordrecht, the Netherlands, 165-178.

Meinders E, van der Meer T, Hanjalic K, and Lasance C (1996). Application of IR Thermography to the Evaluation of Local Convective Heat Transfer on Arrays of Cubical Protrusions, to be published in Int J of Heat and Fluid Flow.

Mergui S, Penot F, and Cadre M (1990). Rapport de Stage, Laboratoire d'Etudes Thermiques Poitiers, France.

Mikheyev M (1968). *Fundamentals of Heat Transfer*, MIR Publishers, Moscow, 73-81

Misale M (1993). Electronic Cabinet Cooling by Natural Convection: Influence of Vent Geometry, *3rd World Conf Exp Heat Transfer, Fluid Mechanics and Thermodynamics*, Honolulu, Hawaii.

Misale M (1994). Some Considerations on Practical Formulas Used for Thermal Design of Natural Ventilated Enclosures, in *"Thermal Management of Electronic Systems," Proc Eurotherm Seminar no 29*, C Hoogendoorn, R Henkes, and C. Lasance (eds), Kluwer Academic Publishers, Dordrecht, the Netherlands, 85-94.

Miyamoto M, Katoh Y, Kurima J, Kurihara S, and Yamashita K (1985), Free Convection Heat Transfer from Vertical and Horizontal Plates, *Int J Heat and Mass Transfer*, 28, 1733-1745.

Moffat R and Anderson A (1990). Applying Heat Transfer Coefficient Data to Electronics Cooling, *ASME J Heat Transfer*, 112, 882-890.

Moffat R and Ortega A (1986). Buoyancy Induced Forced Convection, in *"Heat Transfer in Electronic Equipment." ASME HTD-57*, 135-144.

Moffat R and Ortega A., (1988). Direct Air Cooling of Electronic Components, in *"Advances in Thermal Modeling of Electronic Components and Systems*, A Bar-Cohen A and A D Kraus (eds), 1, 129-282.

Moutsoglou A, Rhee J H, and Wong Y H (1992). Natural Convection-Radiation Cooling of a Vented Channel, *Int J of Heat and Mass Transfer*, 35(11), 2855-2863.

Mukutmoni D, Joshi Y, and Kelleher, M D (1995). Computations For a Three-by-Three Array of Packages Cooled by Liquid Immersion: Effect of Substrate Thermal Conductivity, *ASME J of Elec Pack*, 117, 294-300.

Mukutmoni D, Kelleher M D, and Joshi, Y K (1993). AMPHIB: A Users Manual, Tech Report NPS-ME-93-001, Naval Postgraduate School, Monterey, CA.

Nakayama W (1988). Thermal Management of Electronic Equipment: A Review of Technology and Research Topics, in *"Advances in Thermal Modeling of Electronic Components and Systems*, A Bar-Cohen and A D Kraus (eds), 1, 1-78.

Nelemans W, Ryckebusch M, de Pauw P, and Vandam L (1989). Thermal Design of Telecom Equipment towards the Limits of Natural Convection Cooling, *Proc 3rd Conf on Heat Transfer, Energy and Environment Applied to Communication and Computer Fields*, Perros-Guirec, France.

Nigen J and Amon C (1992). Concurrent Thermal Design: Balancing Accuracy with Time Constraints", *Proc 3rd ITHERM Conf*, Austin, TX, 5-14.

Noronha R (1965). Free Convective Cooling of Cabinets Containing Heat Dissipating Components, *Proc Inst Mech Engrs*, 179, 439-450.

Ntibarufata E, Hasnaoui M, Bilgen E, and Vasseur P (1993). Natural Convection in Partitioned Enclosures with Localized Heating, *Int J Num Methods for Heat and Fluid Flow*, 3, 133-143.

Oktay S, Hannemann R, and Bar-Cohen E (1986). Heat Transfer: High Heat Flux From a Small Package, *Mechanical Engineering*, 108, 36-43.

O'Meara T and Poulikakos D (1994). Natural Convection Cooling of an Array of Heated Plates Simulating Printed Circuit Boards, in *"Cooling of Electronics,"* S Kakac et al (eds), Kluwer Academic Publishers, Dordrecht, the Netherlands, 123-144.

Opstelten I (1994). Experimental Study on Transition Characteristics of Natural Convection Flow, *PhD Thesis*, Delft, Holland.

Ortega A and Kabir H (1991). Substrate Conduction Mechanisms in Convectively Cooled Simulated Electronic Packages, *Proc 7th IEEE SEMITHERM Conf*, Phoenix, AZ, 1-6.

Ortega A and Lall B. (1996), Natural Convection Air Cooling of Discrete Heat Sources on a Conducting Board in a Shallow Horizontal Enclosure, *Proc of 12th IEEE SEMITHERM Conf*, Austin, TX, 201-213.

Ortega A and Moffat R (1985). Heat Transfer from an Array of Simulated Electronic Components: Experimental Results for Free Convection With and Without a Shrouding Wall, in *Heat Transfer in Electronic Equipment," ASME HTD-48*, 5-15

Ortega A and Lall B (1992). A Clarification of the Adiabatic Heat Transfer Coefficient as Applied to Convective Cooling of Electronics, *Proc 8th IEEE SEMITHERM Conf*, Austin, TX, 1-3.

Ortega A, Lall B, Chicci J, Aghazadeh M, and Kiang B (1993). Heat Transfer in a Low Aspect Ratio Horizontal Enclosure for Laptop Computer Applications, *Proc of 9th IEEE SEMITHERM Conf*, Austin, TX, 42-49.

Ortega A and Moffat R (1986). Buoyancy Induced Convection in a Non-Uniformly Heated Array of Cubical Elements on a Vertical Channel Wall, in *Heat Transfer in Electronic Equipment," ASME HTD-57*, 123-134.

Ostrach S (1972). Natural Convection in Enclosures, in *"Adv in Heat Transfer,"* 8, Hartnett, J P and Irvine, T F (eds), Academic Press, Brookline, MA, 161-227.

Ostrach S (1982). Natural Convection Heat Transfer in Cavities and Cells, *Proc 7th Int Heat Transfer Conf*, Munich, FRG, 1, 365-379.

Ostrach S (1988). Natural Convection in Enclosures, *ASME J Heat Transfer*, 110, 1175-1190.

Ozoe H, Miyachi H, and Imai K (1988). The Development of a Computational Scheme for Natural Convection in an Enclosure with Multiple Isolated Heat Sources, in *"Cooling Technology for Electronic Equipment,"* W Aung (ed), Hemisphere Publishing Corp, New York, NY, 611-623.

Papanicolaou E and Gopalakrishna S (1995). Natural Convection in Shallow, Horizontal Air Layers Encountered in Electronic Cooling, *ASME J Elec Pack* 117, 307-316.

Park K A and Bergles A E (1987). Natural Convection Heat Transfer Characteristics of Simulated Microelectronic Chips, ASME J Heat Transfer), 109, 90-96.

Parry J and Tatchell D (1994). Thermal Analysis of a Telecommunications Rack System, in *"Thermal Management of Electronic Systems," Proc Eurotherm Seminar no 29*, C Hoogendoorn, R Henkes, and C Lasance (eds), Kluwer Academic Publishers, 55-64.

Parales, R (1993). Solving the Thermal Design Concerns of a Low Voltage Power Supply Using CFD Modeling, *Flowtherm Second International Users Conference.*, London, UK.

Patankar, S V (1980). *Numerical Heat Transfer and Fluid Flow*, Hemisphere Publishing Corp, New York, NY.

Pecht M (ed) (1991). *Handbook of Electronic Package Design*, Marcel Dekker Inc., New York, NY.

Peeters T and Henkes R (1992). The Reynolds-stress Model of Turbulence Applied to the Natural-Convection Boundary Layer along a Heated Vertical Plate, *Int J Heat and Mass Transfer*, 34, 377-388.

Pera L and Gebhart B (1975). Laminar Plume Interactions, *J Fluid Mech*, 68, 259-271.

Perna C, diEvangelisti A, Paroncini M, and Ricci R (1994). Natural Convection in a Rectangular Enclosure with an Array of Chips mounted on a Vertical Wall, in *"Thermal Management of Electronic Systems," Proc Eurotherm Seminar no 29*, C

Hoogendoorn, R Henkes and C Lasance (eds), Kluwer Academic Publishers, Dordrecht, the Netherlands, 97-106.

Peterson G and Ortega A (1990). Thermal Control of Electronic Equipment and Devices, in *"Advances in Heat Transfer,"* 20, Hartnett, J P and Irvine, T F (eds), Academic Press, Brookline, MA, 181-314.

Pnueli D (1982). Free Convection and its Applications, *7th Int Heat Transfer Conf*, Munich, FRG, 253-259.

Polentini M S, Ramadhyani S, and Incropera F P (1993). Single-Phase Thermosyphon Cooling of an Array of Discrete Heat Sources in a Rectangular Cavity, *Int J of Heat and Mass Transfer*, 36, 3983-3996.

Prakash C (1990). Application of Computational Fluid Dynamics for Analyzing Practical Electronics Cooling Problems, in *"Heat Transfer in Electronic and Microelectronic Equipment"*, A Bergles (ed), Hemisphere Publishing Corp, New York, NY, 923-938.

Prasad V, Keyhani M, and Shen, R (1990). Free Convection in a Discretely Heated Vertical Enclosure: Effects of Prandtl Number and Cavity Size, *ASME J Elec Pack*, 112, 63.

Przekwas A, Lai Y, and Agonafer D (1994). Multi-scale Thermal Analysis of Electronics Packaging, ASME EEP-9, CAE/CAD Application to Electronic Packaging New York, NY, 7-14.

Queipo N, Devarkonda R, and Humphrey J (1994). Genetic Algorithms for Thermosciences Research: Application to the Optimized Cooling of Electronic Components, *Int J Heat and Mass Transfer*, 37, 893-908.

Ramos R and Milanez L (1996). Analysis of the Influence of Baffles and Shields on the Cooling of a Digital Transmission Equipment, *Proc ITHERM*, Orlando, FL, 31-37.

Ravi M, Henkes R, and Hoogendoorn C (1994). On the high-Rayleigh-number Structure of Steady Laminar Natural Convection Flow in a Square Enclosure, *J Fluid Mech*, 262, 325-351.

Refai-Ahmed G and Yovanovich M (1992). Numerical Study of Natural Convection from Discrete Heat Sources in a Vertical Square Enclosure, *J Thermophysics*, 6, 121-127, 165-178.

Reynell M, Clark I, and Rosten H (1991). A Two-Level Micro/Macro Approach to the Simulation of Air Flow and Heat Transfer in Typical Convection Cooled Electronic Systems, ASME 91-WA-EEP-33, Atlanta, GA.

Rosten H and Lasance C (1994). The Development of Libraries of Thermal Models of Electronic Components for an Integrated Design Environment, in *Proc of the Conference of the Int Elec Pack Soc*, Atlanta, GA, 138-147.

Rosten H and Lasance C (1995). DELPHI:, The Development of Libraries of Thermal Models of Electronic Components for an Integrated Design Environment, in *"Model Generation in Electronic Design"*, J Berge et al (eds), Kluwer, Dordrecht, the Netherlands, 63-90.

Rosten H (1995). DELPHI - A Status Report on the ESPRIT funded Project for the Creation and Validation of Thermal Models of Electronic Parts, in *Eurotherm Seminar No. 45 on "Thermal Management of Electronic Systems,"* Leuven, Belgium, 7/1-10.

Rotem Z and Claassen L (1969). Natural Convection above Unconfined Horizontal Surfaces, *J Fluid Mechanics*,38, 173-192.

Sathe S and Joshi Y (1991). Natural Convection Arising From a Heat Generating Substrate-Mounted Protrusion in a Liquid-Filled Two-Dimensional Enclosure, *Int J Heat and Mass Transfer*, 34, 2149.

Semiconductor Industries Association Roadmap (1994), Semiconductor Industries Association, San Jose, CA.

Shaw H, Chen C, and Clever J (1988). The Effects of the Location of Microelectronic Equipment on Two-Dimensional Laminar Natural Convection in an Enclosure, in *"Cooling Technology for Electronic Equipment"*, W Aung (ed), Hemisphere Publishing Corp, New York, NY, 73-92.

Shen R, Prasad V, and Keyhani M (1989). Effect of Aspect Ratio and Size of Heat Source on Free Convection in a Discretely Heated Vertical Cavity, in "Numerical Simulation of Convection in Electronic Equipment Cooling, ASME HTD-121, 45-54.

Smith T, Beckermann C, and Weber S (1991). Combined Conduction, Natural Convection and Radiation Heat Transfer in an Electronic Chassis, *ASME J Elec Pack*, 113, 382-391.

Soylemez M and Unsal M (1994). Computation of Steady Laminar Natural Convective Heat Transfer from Localized Heat Sources in Enclosures, in ' *"Cooling of Electronics"*, S Kakac et al (eds), Kluwer Academic Publishers, Dordrecht, the Nether lands, 225-238.

Tang L and Joshi Y (1997). Integrated Thermal Analysis of an Indirect Air-Cooled Electronic Chassis, ASME J Elec Packaging, in press.

Temmerman W, Nelemans W, Goosens T, and Lauwers E (1995). Experimental Validation Methods for Thermal Models, *presented at Eurotherm Seminar No 45 on "Thermal Management of Electronic Systems"*, Leuven, Belgium, 9/1-12.

TNO Apeldoorn, The Netherlands, (1992). Personal Communication

Torok D (1984). Augmenting Experimental Methods for Flow Visualization and Thermal Performance Prediction in Electronic Packaging Using Finite Elements, in *"Heat Transfer in Electronic Equipment", ASME HTD-32*, 49-57

Townsend A (1959). Temperature Fluctuations over a Heated Surface, *J Fluid Mechanics*, 5, 209-241.

van Doormaal, J P and Raithby, G D (1984). Enhancements of the Simple Method for Predicting Incompressible Fluid Flows, *Numerical Heat Transfer*, 17, 147-163.

Vanlaer D and Lasance C (1994). An Experimental Method to Determine the Effective Thermal Conductivity of Printed Circuit Boards, in *"Thermal Management of Electronic Systems", Proc Eurotherm Seminar no 29*, C Hoogendoorn, R Henkes and C Lasance (eds), Kluwer Academic Publishers, Dordrecht, the Netherlands, 201-210.

Vikram S (1996). Correlation of Natural Convection Heat Transfer from a Horizontal Printed Circuit Board in an Enclosure, *MS Thesis*, University of Maryland, College Park, Md.

Vikram S and Joshi Y (1996). Natural Convection from a Horizontally Heated Printed Circuit Board in an Enclosure, *Ninth Int Symp on Transport Phenomena ISTP-9, in Thermal-Fluids Eng*, 1, 486-491.

Vinke H and Lasance C (1995). Thermal Characterization of Electronic Devices by means of Improved Boundary Condition Independent Compact Models, *presented at 2nd EUROTHERM Seminar on "Thermal Management of Electronic Systems*, Leuven, Belgium.

Vinke H and Lasance C (1997). Recent Achievements in the Thermal Characterization of Electronic Devices by means of Improved Boundary Condition Independent Compact Models, *IEEE SEMITHERM*, San Diego, CA, 32-39.

Wang Z (1997). Personal communication.

Wang Z and Mayinger F (1993). Natural Convective Air Cooling of the PCB's Array in a Closed Casing, *Proc 6th Int Symp Transport Phenomena*, Seoul, Korea.

Witzman S, Newport D, and Nicoletta T (1989). Free Convection Air Cooling of the Electronic Equipment: Still a Miracle Waiting to be Explored and Explained, *5th IEEE SEMITHERM Conf*, 98-103.

Witzman S, Nicoletta T, and Mack B (1992), Enhanced Heat Transfer for Electronic Devices Operating Inside Enclosures, 8th IEEE SEMITHERM Conf, Austin, TX, 95-96.

Worlton J (1996). Computational Physics Then and Now: A Cautionary View, *Computers in Physics*, vol 10, 309.

Wroblewski D E and Joshi Y (1992) Transient Natural Convection from a Leadless Chip Carrier in a Liquid Filled Enclosure: A Numerical Study, *ASME J Elec Pack*, 114, 271-279.

Wroblewski D E and Joshi Y (1993). Computations of Liquid Immersion Cooling For a Protruding Heat Source in a Cubical Enclosure, *Int J Heat and Mass Transfer*, 36, 1201-1218.

Wu W and Cengel Y (1994). Radiation Heat Exchange between Electronic Components on a Circuit Board and the Walls of its Enclosure, *Heat Transfer Engineering*, 15, 34-43.

Xia J and Zhou Z (1992). Natural Convection in a Partially-open Cavity with a Divider, *Int Comm Heat and Mass Transfer*, 19, 447-460.

Yang K T (1988). Transitions and Bifurcations in Laminar Buoyant Flows in Confined Enclosures, *ASME J Heat Transfer*, 110, 1191-1204.

Yang K T (1997). Natural Convection in Enclosures, *in "Handbook of Single Phase Convective Heat Transfer,"* S. Kakac (ed), John Wiley and Sons, New York, NY, 13.1-13.51.

Yang K T and Lloyd J R (1985). Turbulent Buoyant Flow in Vented Simple and Complex Enclosures, in *"Natural Convection: Fundamentals and Applications"* S Kakac et al (ed), Hemisphere Publishing Corp, New York, NY, 303-329.

Yang M and Tao W (1995). Three-Dimensional Natural Convection in an Enclosure with an Internal Isolated Vertical Plate, *ASME J Heat Transfer*, 117, 619-625.

Yang H Q, Yang K T, and Lloyd J R (1986). Flow Transition in Laminar Buoyant Flow in a Three-Dimensional Tilted Rectangular Enclosure, *Proc of 8th Int Heat Transfer Conf*, San Francisco, CA, 14.

Yousef W, Tarasuk J, and McKeen W (1982). Free Convection Heat Transfer From Upward-Facing Isothermal Horizontal Surfaces, *ASME J Heat Transfer*, 104, 493-500.

Yu E and Joshi Y (1995). A Numerical Study of Three Dimensional Laminar Natural Convection in a Vented Enclosure, *ASME HTD-303*, New York, NY, 113-123.

Yu, E and Joshi Y (1996). Air Cooling of a Vented Enclosure by Combined Conduction, Natural Convection and Radiation, *ASME HTD-333*, New York, NY, 323-330.

Yuki M, Parsons J, and Krane R (1990). An Investigation of the Passive Cooling of Table Model Television Receivers, *ASME J. Elec Pack*, 112, 279-286.

Zinnes A (1970). The Coupling of Conduction with Laminar Natural Convection From a Vertical Plate, *ASME J Heat Transfer*, 105, 528-535.

APPENDIX A: NATURAL CONVECTION HEAT TRANSFER IN GEOMETRICALLY SIMPLE ENCLOSURES

A.1 INTRODUCTION

The short literature review on geometrically simple enclosures, which is the subject of this section, clearly indicates some of the problems which confront the analyst when trying to solve the heat transfer behavior of more complex systems. However, the conclusions from simple enclosures are only indicative and it is not to be expected that the published correlations can be directly transferred to real systems. On the contrary, as Ostrach (1988) stated "natural convection is extremely sensitive to changes in the container configuration and the imposed boundary conditions so that the use of the results from 'similar problems' is dangerous." Ostrach, who pioneered research into natural convection in enclosures, presented, among other things, a historical overview. He sheds serious doubt on using the average Rayleigh number as a metric to experimentally validate numerical results and he argues that the concept of the Nusselt number in situations where conjugate heat transfer via the walls plays a role, may not be meaningful because the number also depends on the properties and dimensions of the wall.

Ostrach further points out that many numerical papers use the non dimensional, instead of the normalized velocity for scaling with the consequence that the Grashof or Rayleigh number appears as the coefficient of the buoyancy term. Because the buoyancy force often becomes the dominant term in the equations due to the high value of the Rayleigh and Grashof numbers, in some cases, the true physics are misrepresented because in some regions the buoyant forces are not dominant. He concludes by saying that he finds it difficult to avoid skepticism of purely numerical solutions. In a recent survey, Hanjalic (1994) refuted this criticism by remarking that Ostrach's statement did an injustice to many numericists who had been and still are very well aware of the many complexities. For the layman, it is unavoidable to observe that there is still no consensus between different research groups as to how to tackle the inherently complex physics in a confined system.

Two-dimensional cavities heated from the side, preferably with adiabatic top and bottom walls, characterized by the perpendicularity of the temperature gradient to the buoyancy force, have become very popular among numerical analysts for testing algorithms developed for the computation of incompressible viscous recirculating flows. As it turns out, the configuration is a real fluid mechanics lab in itself. Increases in computer power over the past 15 years or so have made it possible to study the transition from laminar to turbulent flow.

In a numerical "contest" held in 1980 and published by de Vahl Davis (1983), the largest Rayleigh number studied was 10^6. This is still in the laminar regime. Ten years later, in a similar benchmark study published by Henkes and Hoogendoorn (1993, 1995), the maximum Rayleigh number was raised by at least four orders of magnitude. The results of different research groups have been compared in the proceedings. Numerical solutions are presented for a square enclosure with differentially heated side walls and adiabatic top and bottom walls for $\mathrm{Ra} = 5 \times 10^{10}$ which is still somewhat too low for a fully developed turbulent flow.[1] Over this range of values, flows have lost one important property, steadiness, which also makes them very interesting from a practical point of view.

Apart from testing the numerical coding, another objective of the continuing research is to investigate which instability mechanisms are responsible for turning a laminar flow into a turbulent one. Or, what is the location of the transition? What are the effects of stratification? Which are the routes to chaos? And particularly, how are all these effects influenced by different geometries and boundary conditions? Taking air as the convective agent, given a maximum temperature difference of, say, 100 K, high Rayleigh numbers can only be achieved in practice by using length scales on the order of meters. With the possible exception of large telecommunication

[1]It should be noted that the use of the wording "fully developed" for these configurations is a bit tricky because, depending on the aspect ratio and the size, core regions can be distinguished with relatively steady flow. Moreover, laminar regions exist at the initial portion of both vertical walls (the so-called "relaminarization") even at considerably high Rayleigh numbers. This is quite different from the case of differentially heated horizontal plates, where the entire domain becomes turbulent beyond a certain critical Rayleigh number.

equipment, such a large length scale is very seldom to be found in the types of enclosures which are the main subject of this chapter on electronic systems.

Besides the side-heated enclosures with a medium-to-large aspect ratio, another type of configuration has received much attention. This is the low aspect ratio enclosure with differentially heated top and bottom walls. From a practical viewpoint, one should be aware of the large differences in these two cases. Unlike natural convection in side-heated cavities where the major fluid movement occurs in wall-jet type boundary layers independent of the Rayleigh number, the whole fluid is set into motion in low-aspect-ratio enclosures heated from below, but only at Rayleigh numbers higher than some critical value. At relatively low Rayleigh number, this transport is mainly via a row of two-dimensional cells filling the enclosure. At higher Rayleigh numbers, say above 10^5, the roll like motion gives way to irregular plume and knotlike large coherent structures. So, the air flow in enclosures dominated by horizontal heated walls becomes turbulent at much lower Rayleigh numbers.

For some time now, it has been realized that it is extremely difficult to validate benchmark type computational studies by experiments. Hence, an increasing number of studies is devoted to more "realistic" boundary conditions especially perfectly conducting horizontal walls which, obviously, introduce the effects mentioned above for heated horizontal plates. Furthermore, in real systems, significant influences of instabilities initiated by inlet grilles, plumes originating from relatively high powered electronic parts and wakes behind obstacles may be expected. Unfortunately, not much is known about these low Reynolds number turbulence effects on the local heat transfer in real systems.

It is not the purpose of this section to cover the literature on this subject extensively. For example, shallow cavities will not be discussed here because, for this class of problems, the emphasis is usually on the suppression of heat transfer instead of its augmentation. Interested readers are encouraged to consult the reviews published by Catton (1978), Ostrach (1982, 1988), de Vahl Davis (1986) and Yang (1988). The purpose of this section is mainly to review the highlights and conclusions of more recent research [Henkes (1990), Henkes and Hoogendoorn (1993), Opstelten (1994), Le Quere (1994), Fusegi and

Hyun (1994), and Hanjalic (1994)].

A.2 TURBULENCE MODELS - THE STATE OF THE ART

It is commonly believed that the physics governing the fluid flow and heat transfer in the laminar regime can be exactly described by solving the combined equations describing the conservation laws of mass, momentum and energy, while the physics of the turbulent regime can be approximated to any degree of accuracy, depending on the turbulence model (and the grid size) chosen.

Today, the numerical analyst has a choice of several alternatives: either via numerical field solutions of the differential transport equations - whether for the full turbulence spectrum (Direct Numerical Simulation or DNS), or only for the large structures - averaging the smallest eddies (Large Eddy Simulation or LES) or via statistical averaging of the equations and employing some model to close the equation set. DNS and LES will probably prove to be the future methods for the solution of such problems but formidable requirements for computing resources and many yet unresolved problems impede the applicability of these tools for other than the most simple geometries. For two dimensions and for Reynolds numbers $> 10^{10}$ which is still in the transitional regime, the computational efforts required for a DNS approach to solve the Navier-Stokes equations are still too demanding for any but the largest computers available today. A DNS analysis of real fully turbulent 3-D flows is thus still beyond reach. So, turbulence models are likely to remain for some time to come.

After Reynolds averaging of the equations, cross-term correlations defining the Reynolds stresses and turbulent heat fluxes are added and should be solved. This is referred to as the closure problem of turbulence. Many different models have been proposed over the years, ranging from very simple constant-turbulent-viscosity models, zero-equation (or algebraic) models, such as Prandtl's mixing length model, two-equation or eddy-diffusivity models [to which class the various $k - \epsilon$ models (KEM) belong], to algebraic or fully differential Reynolds stress models (RSM). Algebraic stress/flux models (ASFM), in which some of the resulting partial differential equations are simplified to algebraic equations, were once regarded as a convenient compromise, but turned out to bring little benefit at the expense of increased numerical difficulties. Hanjalic (1994) and Henkes (1990) identify several flaws in the standard $k - \epsilon$ model which is commonly incorporated in commercial codes, mainly because in many instances the model is a good compromise between computational efforts and physical accuracy. In particular, it does not seem very realistic that the turbulence extracts its energy from the mean velocity field but not from the mean temperature field.

Another limitation is that, because it treats the turbulent viscosity as a scalar, it cannot be expected to model flows with strong anisotropic turbulence accurately. This happens to be the case for natural-convection boundary layers. On the other hand, inaccuracies in the turbulence model do not necessarily lead to an erroneous prediction of the mean temperature and velocity field. Better turbulence characterization is acquired with the introduction of low Reynolds number modifications which account for the low turbulence levels in the flow, especially in the thin wall region. Second-order modifications, particularly those proposed by Hanjalic and Vasic (1993), show promising potential for a wide class of problems.

Although lower-order turbulence models have proven to be successful in predicting forced-convection-dominated turbulent flows, unfortunately this has not been the case for buoyancy-dominated flows, because such flows exhibit structures which cannot be described by statistical averaging. This is because the often large-scale and coherent structures play a dominant role in the heat transfer. In contrast to isothermal flows (such as flow past a vertical plate), in non-isothermal flows the simplest form of the two-equation standard eddy diffusivity model fails to account for two important effects. One is the failure to define the intensive turbulent heat flux in conditions where the mean temperature may be uniform or its gradient may have the same direction as the heat flux vector. The second, which is essentially a consequence of the first, is demonstrated in the case of heated vertical walls, where the major source of the buoyancy-induced turbulence production originates from the vertical heat flux. Because these two types of boundary conditions are frequently met in most complex configurations, the application of this simple concept does not give much promise of reproducing the physics to any degree of certainty. The application of the generalized gradient diffusion concept eliminates this shortcoming.

Full Re-stress models (RSM), as compared to low-Reynolds number $k-\epsilon$ models do not improve the mean field quantities like the wall heat transfer, but do improve the turbulent quantities like the anisotropy and the turbulent viscosity. Dol et al (1995, 1996) recently presented studies using a fully differential second-moment closure for the Reynolds stress and the turbulent heat fluxes. The results presented show that the profiles for the normal Reynolds stresses and the vertical turbulent heat flux are far better predicted and in line with boundary-layer experiments. These results are also in accrodance with the results of Peeters and Henkes (1992) using the same turbulence model for the analysis of a vertical plate. Large differences between KEM and RSM are found in the corners and the horizontal boundary layers which agrees with DNS results at smaller Rayleigh numbers. Noticeable differences (up to 20 percent) were also found in the high Rayleigh number range between the exponent in the Nu-Ra correlation between KEM and RSM models. These differences are largely caused by the modeling of the turbulent diffusion.

Despite the fact that second-moment closure models seem promising, it is argued that in more complex cases the level of uncertainty involved in modeling the full transport equations for all the variables at present exceeds the potential benefits. Such a complete second-order closure model, even in a simple form, may require not less than 22 empirical coefficients [Peeters and Henkes (1992)]. This currently leaves algebraic models for the turbulent heat fluxes as the most convenient level at which to test modeling ideas for the various terms in the transport equation. It is argued that natural convection in infinite vertical cavities can be well predicted by a low Reynolds number $k-\epsilon$ model while the analysis of square cavities requires second-moment closures.

A major problem for turbulent natural convection modeling arises from the boundary conditions. Scaling of the flow properties close to the wall in terms of wall functions has been widely employed in the computation of forced convection to bridge the near wall region with steep variation of all properties and to avoid integration up to the wall. These wall functions cannot be employed in buoyant flows because they use the wall friction velocity as the velocity scale, which, in buoyant flows is not directly relevant. In general, different and multiple scales govern the buoyant wall flows but the search for universal wall functions, even for a single class of buoyancy flows, has not resulted in a success comparable to forced convection. While designing universal scales for the mean temperatures seems straightforward, the velocity field poses more of a problem. It is difficult to find a universal scale which will satisfy, for example, both a buoyant wall-jetlike boundary layer along a heated vertical plate, where the velocity exhibits a sharp peak very close to the wall and a wall layer heated from below where the flow may be moving slowly, or not at all, as in a mixed layer.

Thermal boundary conditions also play an important role and different scales can be defined depending on whether the wall temperature or heat flux is specified. Another reason that "standard" wall functions are inapplicable is that in buoyant convection, irrespective of how large the bulk Rayleigh number may be, the molecular effects will remain important in a significant portion of the flow domain requiring the implementation of low-Reynolds number modifications. This requires integration straight up to the wall. Finally, the employment of any wall function in the computation of natural convection may result in a great deal of uncertainty and too few computational benefits, except perhaps for some crude computations of real complex situations, which, of course, may be a goal in itself.

A.3 ABOUT TRANSITION FROM LAMINAR TO TURBULENT FLOW

Although the boundary conditions are steady, the Navier-Stokes equations allow for time-dependent solutions, either periodic, quasi periodic or chaotic. The various solutions are characterized by different frequencies in the Fourier spectrum. Above some critical Rayleigh number, the originally steady-state solution becomes unstable, which means that disturbances no longer decay for increasing time. At a given Rayleigh number, a certain large-time solution branches to another large-time solution. This is called a bifurcation point. Various types of bifurcation points exist and two types that occur in enclosure flow are the pitchfork bifurcation, branching from one steady state to another, and the Hopf bifurcation, branching from a steady solution to an unsteady solution with a single-frequency oscillation.

For natural convection, the transitional regime spans a wide range of values of the Rayleigh number which are important in technical applications, more so than the fully turbulent regime. The transition to turbulence in natural convection occurs at relatively high Rayleigh numbers and in most cases of practical relevance the molecular effects and turbulent-molecular interactions remain significant in some regions of buoyancy driven flows. In other cases, such as in partially divided enclosures where the partitions damp the turbulence, the turbulent regime may be confined to only a few regions even at very high bulk Rayleigh numbers. For example, in typical building enclosures the values range between $10^{10} \leq \mathrm{Ra} \leq 10^{12}$. These values are just sufficient to promote a laminar-to-turbulent transition along the vertical walls.

Another example of the importance of molecular effects, irrespective of how high the Rayleigh number may be, is the turbulent natural convection along the vertical non-adiabatic walls with the largest buoyancy effects being encountered very close to the wall where the viscosity effects are always present. The computation of the flow and heat transfer requires an integration up to the wall even though the wall boundary layer may be fully turbulent. On the other hand, free convection over the horizontal heated surfaces becomes turbulent almost instantly, as soon as the Rayleigh number exceeds the critical value, and the heat transfer occurs through vertical penetration of the fluid irrespective of the temperature gradient in that direction.

In fully chaotic turbulent flow, a broadband spectrum of frequencies characterizes the power spectrum of the fluctuations of temperature and velocity. In the transitional regime distinct frequencies are present. These frequencies are introduced through bifurcations which are caused by instability mechanisms responsible for the onset to turbulence. The type of instability mechanism is greatly influenced by the geometry and boundary conditions. A variety of instability mechanisms are regarded as responsible for the transitional behaviour. Henkes (1990) presents an overview of the following types:

- Rayleigh-Benard instabilities which mainly occur when heated horizontal plates are present.

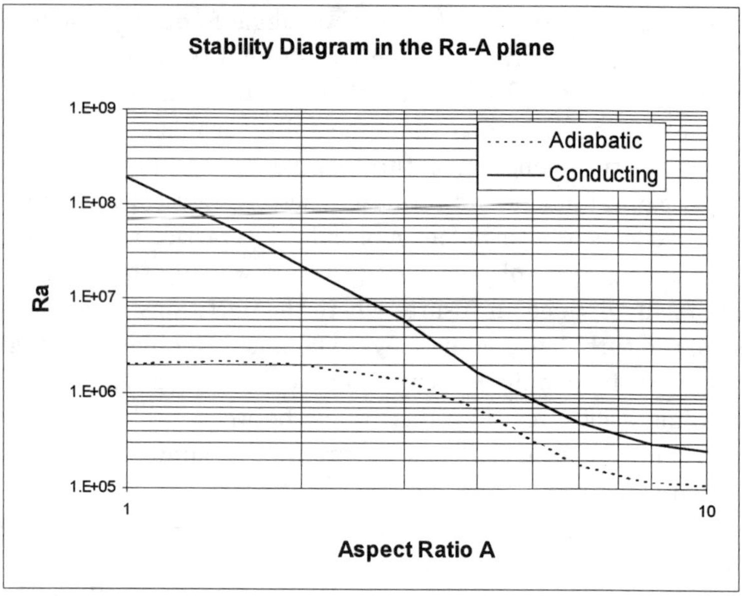

Figure A.1: Stability diagram of the Rayleigh number-aspect ratio plane [Le Quere (1994)].

- instabilities occurring mainly in the corner regions of the enclosure where the horizontal boundary layer has separated from the wall.
- Tollmien-Schlichting instabilities which generate travelling waves in the boundary layers.

Ravi et al (1994) showed that the separation of the boundary layer was not caused by a hydraulic jump, but by a thermal effect. Janssen (1994) and Janssen and Henkes (1995a) suggested that for the side-heated cavities with adiabatic horizontal walls, the observed low frequency of the instability could be caused by a shear-driven Kelvin-Helmholtz instability characterized by the appearance of vortices being convected with the flow. For side-heated cavities with conducting horizontal walls, Janssen and Henkes (1995b) showed that the transition is caused by a Rayleigh-Benard instability occurring in the unstably stratified horizontal boundary layers.

Figure A.1, presented by Le Quere (1994), shows an interesting phenomenon. It presents a stability diagram in the Rayleigh number -aspect ratio plane where the full line corresponds to adiabatic top and bottom walls and the dashed line to perfectly conducting top and bottom walls. It can be seen that perfectly conducting walls have a very large destabilizing effect as compared to the adiabatic case, especially for cubical enclosures. This indicates that the instability mechanisms are very different. The transition can be explained by three different mechanisms. For cubical enclosures with conducting walls the instability results from the horizontal flow driven by a horizontal pressure gradient becoming increasingly heated as it approaches the hot wall. On the other hand, in the case of adiabatic walls, it is the detached region close to the ceiling which becomes unstable first.

Le Quere also investigated the onset to turbulence in a cavity of aspect ratio 10 for both types of thermal boundary conditions of the horizontal walls for air and water. Very surprisingly, he discovered that the critical Rayleigh number for adiabatic walls was more than 30 times larger than for perfectly conducting walls. This is in sharp contrast to the case of air. One might ask whether the Rayleigh number is the proper scaling parameter and if it is not better to replace the Rayleigh number by the Grashof number, supporting the view of de Vahl Davis (1986) that for boundary-layer-type flow in enclosures the Grashof number should be the preferred choice, not the Rayleigh number.

A.4 ABOUT FULLY CHAOTIC BEHAVIOR

In his review, Le Quere also discussed several routes to chaos and demonstrated that the routes are heavily dependent on the aspect ratio. It became evident that the first onset to full chaos merely corresponds to temporal chaos and hardly affects the spatial structure of the flow. Fully chaotic motion can be obtained by increasing the Rayleigh number. To reach this state, statistically averaged steady-state solutions should be acquired by integrating for a sufficiently long time. One of the problems arising here, besides the enormous computer power necessary, is the large quantity of information generated, firstly on the dynamics of the solution, but also on all the time-averaged quantities that describe the turbulent flow and temperature fields. However, as argued before, full chaotic behaviour is not expected to occur in even the largest electronic systems.

A.5 SOME REMARKS ON THE HEAT TRANSFER

From a purely heat transfer point of view, it is interesting to note that the time- and space-averaged Nusselt numbers for Ra = 1010 are about 100 and 102 for aspect ratios of 1 and 4 respectively [Le Quere (1994)]. This indicates the minor influence of the geometry, at least in this study. All of the values closely follow the laminar Nu vs $Ra^{0.25}$ relationship which shows that the large fluctuations have only a minor influence on the mean heat transfer. This is also valid for local Rayleigh numbers and a graph is presented which indicates that the large eddies have a distinct but nevertheless small influence. It should be noted here, however, that this conclusion cannot be extrapolated to real systems, except perhaps for relatively large dissipating structures such as large transformers and picture tubes. The "turbulence" generated will undoubtedly have an impact on the heat transfer of smaller objects upstream. On the other hand, Henkes (1992) showed that the popular $k - \epsilon$ model overpredicts the heat transfer along a vertical plate by more than 50 percent as compared with experiments. More sophisticated turbulence models reduced this percentage considerably. For side-heated cavities, the simple KEM failed to predict the location at which transition occurs. This will obviously have an impact on the accurate prediction of the local heat transfer.

A.6 COMPARISON WITH EXPERIMENTS

The question has been raised as to how well do the idealized benchmark assumptions represent reality. Reviewing the literature, one thing became abundantly clear. There is no way of validating the numerical results by experiments when adiabatic boundary conditions have been applied in the numerical model, at least if air is used. Although such a benchmark is very suitable for testing numerical codes and comparing hardware performance, it could be argued that it is better practice to change the adiabatic boundary conditions to more realistic ones. One may wonder about the acquisition of a thorough understanding of the physical behavior of systems which do not exist in the universe and if it is possible to get the same insight using configurations with realistic boundary conditions which are much easier to verify. This point of view has also been advocated by Le Quere, who makes a plea for more realistic boundary conditions, other than the common adiabatic and perfectly conducting ones, and proposes to use Dirichlet-type boundary conditions which reproduce the measured boundary conditions.

In a thorough study, Opstelten (1994) investigates the characteristics of the flow in the transitional regime experimentally. The need for an extensive experimental study was felt because most of the numerical studies focused on differentially heated side walls and adiabatic boundary conditions for the other walls, which cannot be reproduced in an experiment. Another need for experimental validation lies in the fact that the two turbulent quantities (Reynolds stresses and turbulent heat fluxes) have to be approximated by some more or less complicated model in the numerical analysis, while the values can be obtained within experimental uncertainty. The main questions of the research addressed the following topics:

- How do the idealized 2-D numerical studies compare with realistic 3-D experimental configurations?
- What is the nature of the instability mechanisms?
- What is the experimental route to chaos?

LDA has been used for the velocity-field characterization, holographic interferometry for the temperature field and smoke visualization for the time-dependent behavior of the flow structures. Besides the problem of the adiabatic walls, other questions can be posed. Are the simplifications inherent to the numerical model, such as the Boussinesq approximation, 2-D flow, neglect of radiation, compressibility and viscous dissipation effects correct? In the following, an answer to some of these questions is given.

Begin with the consequences of the Boussinesq approximation. Le Quere states that the situation is different for gases or liquids, because for the latter compressibility is negligible and it is only necessary to take account of variations in physical properties with temperature while keeping a linear or quadratic relationship between density and temperature. Of course, care should be taken with water because of its anomalous behaviour at 4°C. In the case of gases, making an allowance for density variation requires the specification of an equation of state. The numerical treatment of this problem is not straightforward because of over-restrictive stability criteria and the possible occurrence of acoustic waves. By appropriate numerical measures, it has been shown that as the temperature difference expressed by $\Delta T/2T$ is increased from 0 to 0.6, the critical Rayleigh number is significantly affected. In practical terms, at $T = 350\,\mathrm{K}$ with $\Delta T = 70\,\mathrm{K}$, the ratio becomes 0.1 so no problems with the Boussinesq approximation are to be expected for electronic cooling conditions. In another study by Opstelten (1994), it is reported that for $\Delta T = 100°\mathrm{C}$ deviations up to a maximum of 15 percent were found in the most important variables. In case of doubt, the abandonment of the Boussinesq approximation is certainly an alternative considering the computer power currently available to almost everyone.

Comparison with experiments obviously raises the question of the adequacy of two-dimensional models of three-dimensional effects. Most of the work has been done for a cavity of aspect ratio of 4 with perfectly conducting top and bottom walls. There are two reasons for this. Unsteadiness sets in at relatively low Rayleigh numbers and top and bottom effects are appreciably less than for a cubical cavity. Some very well performed experiments that can serve as benchmarks are reported by Cheesewright and King (1986) for an aspect ratio of 5, Betts and Dafa'Alla (1986) for an aspect ratio of 30 and Le Quere and Perot (1987) for an aspect ratio of 4. By and large, it was found that the numerical and experimental results showed fairly good agreement. For an aspect ratio of 4, it was found that a major discrepancy, attributed to 3-D effects, was caused by the appearance of low frequency waves superposed on the high frequency travelling waves [Le Quere and Penot (1987)]. In the benchmark type enclosures the main flow characteristics are 2-D. In 3-D, when the added side walls are perfectly adiabatic, some 3-D effects appear when the flow is in the transitional regime. However, when these walls deviate largely from the adiabatic case, large 3-D effects may be expected. The occurrence of asymmetry in the velocity field has also been attributed to 3-D effects.

Recently, results of a study on three-dimensional transition effects were reported hy Henkes and Le Quere (1996). They concluded that three-dimensional perturbations turned out to be less stable than in 2-D in contrast to assumptions made earlier. Also, 3-D effects increased the maximum of the time- and depth-averaged wall-heat transfer by 15 percent.

As far as radiation effects are concerned, the following conclusions can be drawn. Kim and Viskanta (1984) showed that the coupled effect of wall conduction and radiation boundary conditions indicates that the Nusselt number is not a meaningful parameter for the correlation of the experiments. This is important, because in all practical cases such boundaries will be present. Although Opstelten succeeded in reducing the difference between the measured temperature profiles along the horizontal walls in his experimental set up and the profiles for the adiabatic case considerably, deviations are still apparent. This was mainly attributed to radiation effects despite the fact that the emissivity of the wall was kept at 0.08. Needless to say, it is virtually impossible to reduce this value further, and even more, to maintain such a low value for the duration of the experiments.

Another experimental problem has to do with the humidity of the gas under study. Water vapor absorbs and reradiates infrared radiation and the effect becomes significant in those areas where convection is negligible. A similar effect appears when using smoke visualization. Opstelten found out that cigar smoke significantly altered the effective boundary conditions, while a fog consisting of a mixture of triethylene glycol and propane-2-diol in water did not.

In discussing boundary conditions other than constant imposed temperatures and adiabatic walls, several observations can be made. Constant heat-flux heating is frequent in practical configurations and it is therefore of some interest to understand how this change of boundary conditions affects the unsteadiness. It is known that along a vertical plate heated with uniform flux in isothermal surroundings, the maximum velocity in the boundary layer increases as $z^{3/5}$ against $z^{1/2}$ in the isothermal case. Here, z, is the vertical height and substantial effects on the stability can be anticipated. Apart from this, the isoflux cavity is of interest. As was shown by Kimura and Bejan (1984), the equilibrium temperature of the walls increases

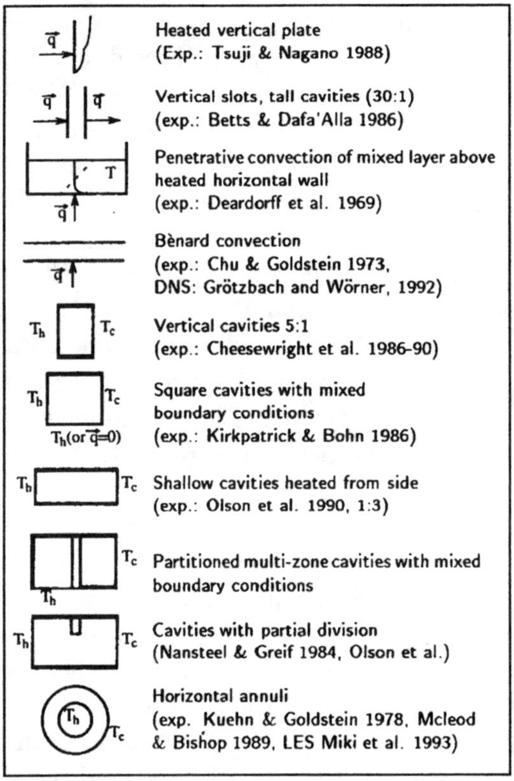

Figure A.2: Proposed test cases [Hanjalic (1994)].

linearly with height with the same gradient as the temperature in the stratified core region. This is exactly the condition required for the buoyancy layer to exist, and Kimura and Bejan showed that the flow structure consisted of boundary layers of constant thickness throughout the cavity height. Le Quere presents some arguments to support the view that the isoflux case is more difficult to solve because a finer mesh is needed, especially in the vertical direction.

To conclude this section, a set of test cases, suggested by Hanjalic (1994), is shown in Fig A.2, which can serve as a basis for the verification of modeling ideas and practice and which have been explored experimentally and in some cases numerically with DNS.

A.7 CONCLUSION

From the foregoing, it becomes evident that even cases with a very simple geometry are very difficult to solve by numerical means alone if the objective is to simulate all details of the flow. LES and DNS techniques are promising candidates to improve this situation. Experts predict the first application of LES to engineering applications within five years, but only for restricted types of flow. DNS is viewed as an even longer term development [Fisher and Rhodes (1994)]. Consequently, established models for turbulence will continue to be used for most real-life problems for some time to come. Progress has recently been reported by a few research groups active in this area [Kenjeres and Hanjalic (1995a, 1995b), Ince and Launder (1995)]. The ERCOFTAC/IAHR Workshop on Refined Flow Modeling (Paris, April 1996) compared numerical predictions of several research groups for two well-known cases: a natural-convection-boundary layer along a vertical plate and differentially-heated tall cavities, for Rayleigh numbers in the transitional regime. DNS data recently became available for the last case. For the first case only experimental data could be used for benchmarking. The conclusions of the workshop were that, on a term-by-term basis, the current state-of-the-art in turbulence modeling is not yet satisfactory but mean temperatures and heat transfer coefficients are well predicted by most advanced turbulence models. However, these models are not yet available in commercial codes.

At the last ERCOFTAC workshop (Delft, June 1997), some interesting trends could be noticed. For one, the domain of benchmark cases has been extended to more complex situations, such as fully developed flow in a matrix of protrusions on a flat plate, and the flow around a single surface-mounted cubical protrusion. The first case prescribed a channel-based Reynolds number of 13,000 (approach velocity 2 m/s), the second case an obstacle-height Reynolds number of 40,000 (about 20 m/s). The reported numerical results are worth considering. For the matrix case, only DNS could reasonably match the experimental data at a relatively moderate CPU time (about 100 hours on a one-processor CRAY), while sophisticated turbulence models failed. For the higher Reynolds number single-cube case, spectacular LES data were shown that resembled, to a great degree, the very complex flow field that showed up in flow visualization experiments. And again, even highly complicated turbulence models could not bring the same accuracy. Based on this evidence, some experts at the workshop concluded that the search for closing the Navier-Stokes equations by refining the turbulence models is a dead end. For relatively low Reynolds numbers, which cover, to a large extent, the entire domain of electronics cooling applications, DNS could be increasingly used, especially for geometries with a low degree of complexity. For higher Reynolds numbers, LES might offer a useful solution. When accurate prediction is at stake, the use of turbulence models of ever increasing complexity no longer offers an advantage, because of, on the one hand, the ever increasing computer power and, on the other hand, the vanishing difference in CPU time when compared to LES.

For the time being, the following strategy is recommended when accurate temperature and flow field analysis of a complex electronic system is the objective of the investigation. Due to the very fine meshes needed to model every boundary layer of interest, and the unsatisfactory capability of most turbulence models to match experimental data, the only way that remains to accurately predict the temperature of solid objects is to impose measured or other wise known heat transfer coefficients in the CFD-code.

The final conclusion to be drawn from this section is that even the numerical analysis of simple geometries gives rise to many problems and accurate data are not to be expected without the use of clever assumptions on how to reduce the complexity of the physics to be modeled while retaining the essential features.

Three additional remarks seem sufficient here in closing the discussion:

- Sophisticated turbulence models (any other than standard $k-\epsilon$) are not expected to be implemented in system-level analysis tools for quite some time.

- Trends are that these models will be taken over by LES and DNS analyses. For velocities up to say 2 m/s, DNS is expected to have an impact on a designer's environment in about ten years.

- On the other hand, there is no idea whatsoever about the errors originating from the use of laminar or zeroth-order turbulence models in a complex system level analysis, simply because there are no numerical data available.

It should be stressed, therefore, that despite the enormous progress in the capabilities of CFD codes, turbulence models and in computer power, fundamental research focused on simulation of the real physics associated with complex systems is still badly needed. This research must be benchmarked against very well-designed experiments, in order to be able to judge the predictability potential of the CFD codes that are currently in use at many design departments.

APPENDIX B: NATURAL CONVECTION HEAT TRANSFER ABOVE HORIZONTAL PLATES

When reviewing the literature, One of the most striking observations is the large discrepancy between the reported results, not only between different research groups, but also between experimental and numerical analyses. Lewandowski and Kubski (1983) compiled the results of about forty publications on the natural convection heat transfer behavior of both vertical and upward facing heated plates. For the vertical plates they found a scatter of approximately 25 percent. However, for horizontal plates, the situation is much worse. In the laminar region, differences of 45 percent were observed, and in the turbulent region about 100 percent. Besides this, different values of the exponent for the Nusselt-Grashof-Prandtl power law have been reported, varying from 0.2 to 0.3, and an additive constant has also been proposed such as the one proposed by Miyamoto et al (1985). On top of that, the critical Rayleigh or Grashof number which determines the onset to transition and to turbulence is a matter for dispute. Textbooks often cite the following Nusselt-Grashof correlations for heated or cooled horizontal plates:

upward facing	$\mathrm{Nu} = 0.54\mathrm{Ra}^{0.25}$
downward facing	$\mathrm{Nu} = 0.27\mathrm{Ra}^{0.25}$

Apparently, the conclusion seems to be that the heat transfer process for upward-heated plates is approximately twice as effective for downward-heated plates. However, this is not what computer experiments show. For example, Goldstein and Lau (1983) presented 2-D numerical results, studying the effects of horizontal extensions and vertical walls on the flow and temperature fields generated by upward-facing and downward-facing heated plates. Their studied geometries are shown in Fig B.1.

3-D experiments with square plates confirmed the analyses to a sufficient degree. However, the boundary conditions chosen were those for a relatively large enclosure and only one plate size was used in the numerical analysis. It was therefore decided to repeat the analysis with the CFD code FLOTHERM (Flomerics), based on

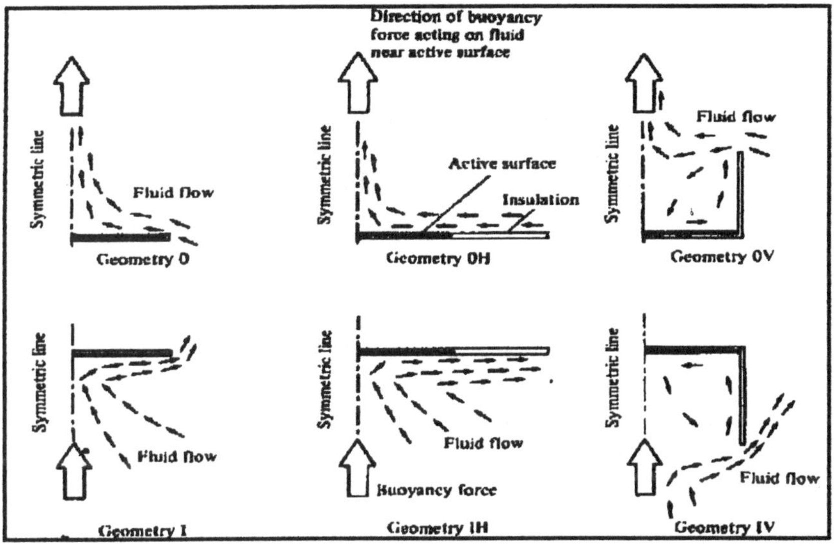

Figure B.1: Studied geometries of horizontal plates and plate-edge extensions [Goldstein and Lau (1983)].

essentially the same set of equations and solution techniques, but with a larger solution domain with free boundaries and a much finer grid. The FLOTHERM analyses were conducted, so that enlargement and refinement showed no further improvement. The plate size was varied between 12 and 160 mm, thereby extending the maximum Rayleigh number, based on the half width of the plate, from 8×10^3 to 8×10^3. In this way, the heat transfer behavior of a horizontal plate in a virtually infinite ambient could be studied.

Because only the power applied to the plate (with "infinite" thermal conductivity, which essentially imposes a constant temperature boundary condition, and the size were varied, the results are presented in the form shown in Fig B.2. For the purpose of comparison, Nusselt and Rayleigh numbers were also calculated, resulting in

$$\mathrm{Nu} = 0.64\mathrm{Ra}^{0.203}$$

for upward-facing heated plates and

$$\mathrm{Nu} = 0.56\mathrm{Ra}^{0.190}$$

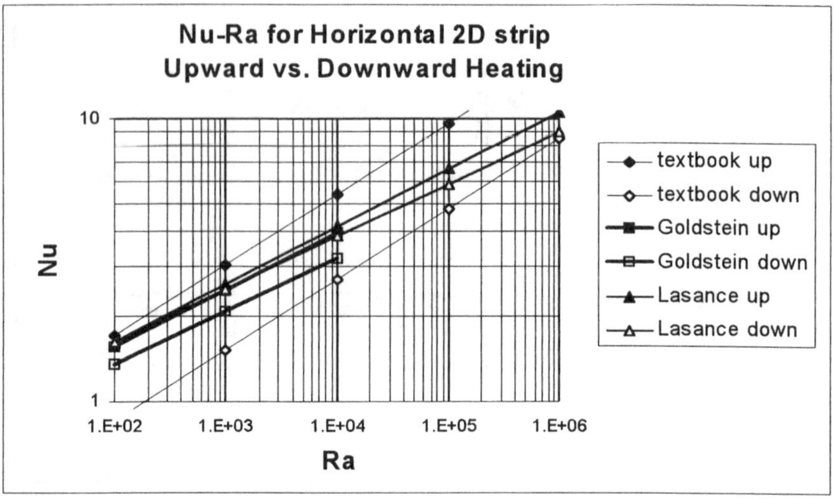

Figure B.2: Heat transfer results for upward and downward-facing plates.

for downward-facing heated plates using all data. The exponents in the correlations show very little disagreement with the results of Goldstein and Lau, ($Nu = 0.61Ra^{0.203}$ and $Nu = 0.56Ra^{0.190}$ respectively), thereby validating their numerical assumptions.

From these results, it is obvious that there is only a slight difference between the upward- and downward-facing plates. The reasons are not clear. Is the discrepancy caused by the unjustified extrapolation of large plates to smaller ones, or is it possible that the experiments do not obey the boundary conditions implicit in the numerical analysis? Is the 2-D approximation not allowed, or are the boundary layer simplifications of the usual analytical expressions not applicable? Further research should clarify these issues.

Many authors have tried to formulate answers in searching for the "truth" and many explanations have been suggested for the anomolies observed. Yousef et al (1982), in performing a very well designed experimental study on horizontal square plates, found random fluctuations in the laminar region that caused the average Nusselt number to vary by about 25 percent around the mean values and the

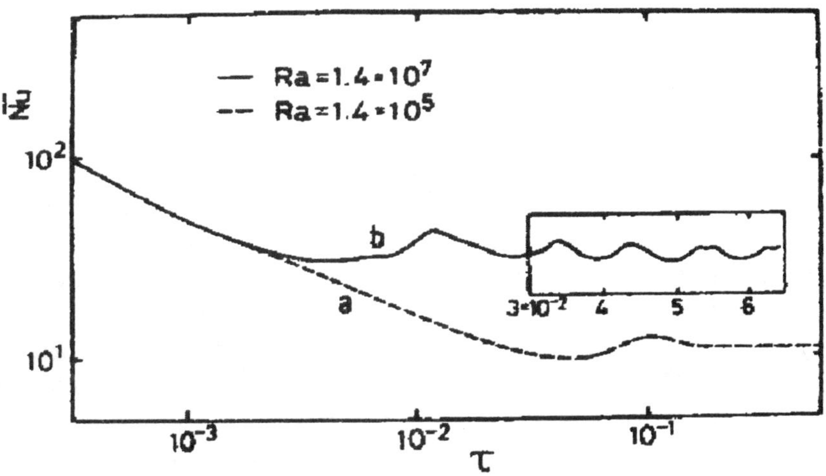

Figure B.3: Time dependence of the average Nusselt number for two different values of Ra.

local Nusselt number by as much as 45 percent. Furthermore, they concluded that theoretical solutions considerably underpredicted the average heat transfer coefficient. They attributed the fluctations to 3-D disturbances which were ignored in all of the theoretical analyses and concluded that a correlation which yields a single value for Nu or Nu_x is unrealistic. Gebhart et al (1988) pointed to a possible cause of the discrepancies between numerical and experimental analyses. Of course, the fluid motions should ideally be caused only by the effect of the heated surface. However, structural vibrations, flow disturbances in the ambient medium, fluid circulations and stratification caused by a finite ambient medium may influence the data.

Another major question is how the thermal boundary conditions at the surface are precisely attained experimentally. Ishiguro et al (1978) detected periodic, not random, oscillations in a time-dependent 2-D numerical study for the Rayleigh number well below the transition limit. Fig B.3 displays the time dependence of the average Nusselt number for two different values of the Rayleigh number. After step-wise heating of a narrow rectangular strip on a large isothermal horizontal plate, for a small value of the Rayleigh number, the Nusselt number monotonically decreased toward a stationary value. However, if the Rayleigh number exceeded some critical value, a periodic value of the Nusselt number results. These observations were justified by experiments with a heated strip in water. The 3-D fluid lumps originating at the higher Rayleigh numbers could be the reason behind the dicrepancies between the reported 2-D analytical and experimental results.

Rotem and Claasen (1969), in studying semi-infinite plates, found that the flow is considerably less stable for a horizontal configuration than for flow over slightly inclined plates. In relation to this, Gebhart et al (1988) remarked that horizontal flows are not a special case of inclined flows because the tangential buoyancy force vanishes. Clausing and Berton (1989) studied mixed convection and reported that very small drafts can influence the natural convection results.

Husar and Sparrow (1968) were the first to publish flow visualization studies studying square, rectangular, triangular and circular plates. In all but the case of the circular plate, partitioning of the flow field along the diagonal became evident as was the formation of a central plume. From the flow fields it could be concluded that generalization to other geometries may involve more than simply using the short side as the characteristic dimension. Computer simulations of the square plate experiment showed an interesting effect. The left part of Fig B.4 clearly reproduces the experimental flow patterns shown in the center. Notice the diagonal patterns, the main central plume and several small plumes. However, if the direction of the inflow of air is changed from horizontal to vertical, the patterns change completely as indicated in the right side of Fig B.4. From a practical point of view, it is to be envisioned that such effects might have an influence on the optimal placement of components on a circuit board.

Another study that could be related to optimal placement was published by Pera and Gebhart (1975). They investigated interactions between laminar natural convection plumes generated by line and concentrated heat sources. As shown in Fig B.5. flows with free boundaries were found to be affected by nearby surfaces. It was also concluded that reducing the space between sources enhanced the plume interaction.

Afzal and Hussain (1984) studied mixed convection over isothermal plates, and, for a certain combination of parameters, dual so-

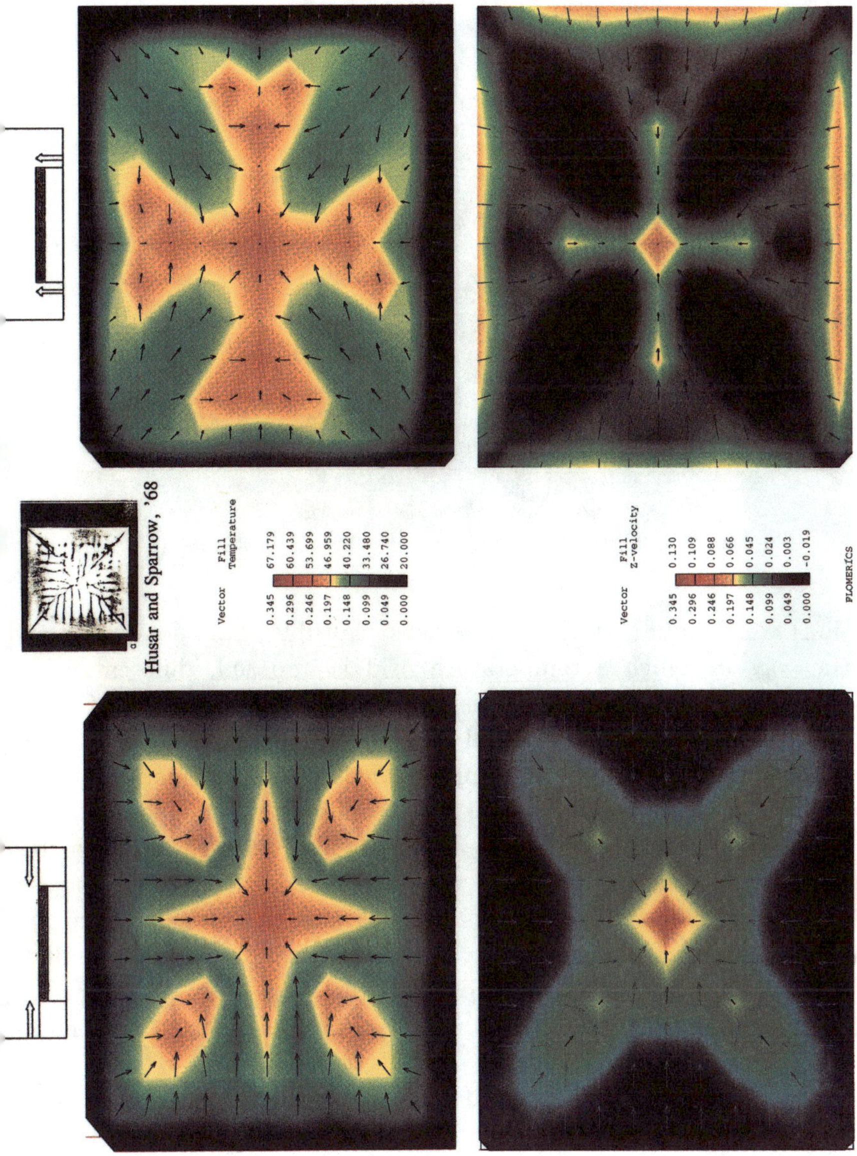

Figure B.4: Natural Convection Heat transfer; The temperature and z-velocity just above a horizontal square plate.

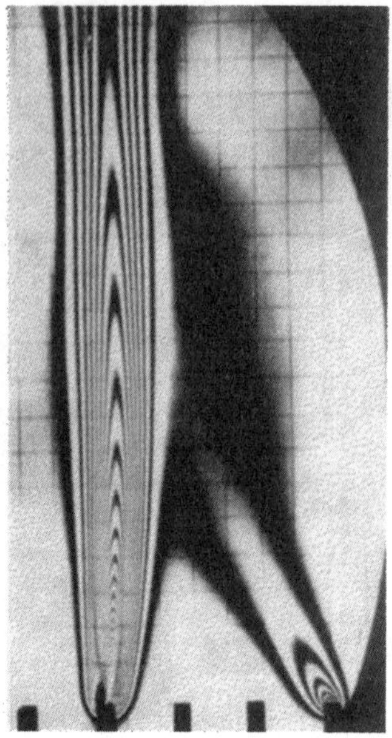

Figure B.5: Plume interactions [Pera and Gebhart (1975)].

tions were found. It therefore seems plausible that, depending on the way in which a temperature field is imposed, different solutions could be obtained. It is of interest to observe that Cohen and Reshotko (1956) also found dual solutions in a study concerning laminar boundary layers in compressible flow with non-zero pressure gradients, and, according to them, the solution that emerges in an experiment is determined by the initial conditions which establish the pressure field acting on the developing boundary layer. Mikheyev (1968) concluded that the heat transfer from upward-facing horizontal plates was 30 percent higher than for vertical plates under similar conditions, while for downward-facing plates the heat transfer was 30 percent less. However, Gebhart et al (1988) reported that horizontal flows are always weaker than vertical flows.

Accurate measurements and flow visualizations over an extremely wide range of Rayleigh numbers (10^6 to 10^{15}) were performed by Kitamura and Kimura (1995). Temperature patterns were observed on the constant heat flux surface, depending on the distance from the leading edge as follows: two-dimensional, steady and streaky, unsteady and irregular, small and dense. Flow visualization experiments revealed various flow patterns: laminar boundary layers, transitional flows, fully turbulent flows and colliding flows near the centerline. Local heat transfer coefficients showed significant variations. Other recent experimental work in the range $2 \times 10^8 \leq \mathrm{Ra} \leq 7 \times 10^{10}$ was reported by Goldstein and Volino (1995). One of the interesting conclusions was that the experimentally determined value of the critical Rayleigh number depends strongly on the measurement technique employed.

Another interesting field from the viewpoint of the cooling of electronics is the closed or partly open enclosure with its bottom at the higher temperature. Chu and Goldstein (1973), pursuing the pioneering work of Malkus (1954) in discussing turbulent convection in a horizontal layer of water, also discovered discrete transitions, showing themselves as abrupt changes in the Nusselt-Rayleigh number slope. Temperature reversals were observed for relatively low Rayleigh numbers. The development of thermals (rising 'mushroom'-like fluid elements) played a decisive role in the heat transfer. It was also noted that the transition to turbulence occurred at Rayleigh numbers about four times lower for air than for water. Hetsroni et al (1996) recently proposed a mechanistic model for turbulent heat transfer from a horizontal wall to a fluid. The simple model is based on the assumption that the dominant mechanism is the bursting of coherent structures from the wall region into the mainstream. The burst leads to the emergence of zones with very small thermal resistance in the near wall flow.

Turbulent convection experiments in a cavity open at the top were conducted by Townsend (1959). He observed comparatively infrequent "active" periods, during which the temperature oscillates wildly between wide limits with extended periods of quiescence, during which the temperature is low and varies very little. A satisfactory explanation could not be given, but he suggested that regions of hot rising air, though very different from the plumes rising above localized heat sources, were the cause of the chaotic behaviour measured. Pnueli (1982) addressed the mathematical aspects in a review paper dealing with natural and mixed convection. He argued that, in many cases, where forced convection in channel flow is thought to be dominant, in reality, mixed convection occurs. The magnitude of the mixed convection depends on the location of the temperature difference. Although this seldom has much effect on the average flow field, the temperature field is affected and so is the heat transfer. Pnueli's approach proceeds by making a clear distinction between boundary value and eigenvalue problems. External flows come close to the first category, while the second category encompasses most internal flows and also large horizontal plates.

Boundary-layer approximations are doubtful in the treatment of internal flows. An eigenvalue problem permits the solution at zero velocity, thereby giving way to a critical Rayleigh number. An implication is that even in slow flows the various possible patterns may vary in time, never quite repeating themselves. Unsteady flow cases have been reported in laminar situations quite removed from turbulence, and the prediction of heat transfer is still an open question. Pnueli (1982) then makes a classification based on boundary values, eigenvalues, and mixed problems and concludes with some engineering advice: only boundary-value problems are reliable. Once a number of experimental results are known, they can be extrapolated, because the results are reproducible. This does not apply to eigenvalue problems. Engineers cannot apply them with confidence to systems even minutely different from the original work.

Only recently, it has become feasible to study numerically turbulent thermal convection in slender enclosures heated from below. Kenjeres and Hanjalic (1995b). They found appreciable differences between numerical and experimental results, especially in the predicted organized roll patterns at relatively high Rayleigh numbers which were not found in experiments. It is argued that long-time averaged structures and wavelengths may still exist in reality, though difficult to detect by experimental techniques at such high Rayleigh numbers.

Stratification effects could possibly play a role in those applications in which, for some reason, the top of the enclosure is hotter than the bottom. Such a situation could, for example, occur in car radios or CD players with an amplifier on top, or sun-irradiated systems or via wall conduction. In such systems, the thermal flow can become "stratification-trapped" [Gebhart (1979)]. If the stratification-induced density variation is of the same order as the density variation imposed by some relatively large object such as transformers, heat sinks, or electrolytic capacitors, heat transfer during the transient condition can be seriously hampered, resulting in much higher temperatures than expected. Eventually everything will be mixed, due to thermal diffusion, and the heat transfer coefficient will regain its "old" value.

In closure, in reviewing the literature it can safely be stated that the heat transfer above horizontal plates is not very well understood from a physical or a numerical viewpoint. Even more so than in the case treated in Appendix A, it can be argued that the possible effects on the temperature of electronic parts are not completely known. And again, there is a need for very-well-designed experiments to address the issue of using CFD codes for the analysis of complex geometries encompassing heat-dissipating horizontal plates.

CHAPTER 2
THERMAL MODELING OF AIR COOLED COMPONENTS MOUNTED ON PRINTED CIRCUIT BOARDS

Sorin Witzman
Northern Telecom Ltd (NORTEL)
Ottawa, Ontario, Canada

2.1 INTRODUCTION

A frequently encountered electronic packaging configuration is one in which an array of electronic components is mounted on one or both side(s) of a printed circuit board. Many such boards are then stacked in a parallel configuration within an enclosure. The cooling of heat dissipating components is accomplished by forcing ambient air through the channels formed between adjacent boards. An effective air-cooling system for such applications is one which keeps all of the electronic components at or below the allowable operating temperature level.

The thermal performance of electronic components is commonly expressed as the sum of two independent thermal resistances, named "internal" and "external", according to their general region of influence. The former is defined as the temperature difference between the semiconductor junction, or the active layer in the silicon, and a point on the package surface, per unit power dissipation. The latter is similarly defined in terms of that same surface location and the average local air temperature. Internal thermal resistance, for all practical purposes, is an inherent property of the package geometry, materials and the fabrication process. Even though the value of the internal thermal resistance might vary in time due to changes in the thermal contact resistances between different layers of the package structure, in most practical applications the internal thermal resistance is assumed constant. External thermal resistance is a function of the air flow characteristics, the substrate thermal properties and, in most cases, of the board lay-out.

In general, the determination of the temperature rise of a component operating under field conditions, requires more than just calculating the component rise above its local air temperature. The reason is that the local ambient temperature inside the electronic enclosure can vary significantly from one location to another. The temperature of the air inside the enclosure is a function of the total power dissipated by all the heat sources inside the enclosure, the mass flow rate of the air, the geometry of the cabinet and the ambient temperature. As the air passing through cabinet must absorb the total power dissipated inside the enclosure, a significant temperature difference can be created between the inlet and the outlet. In hermetic enclosures, in which no "flow through" is present, buoyant flows resulting from temperature variations inside the cabinet still exist. Buoyancy-induced flow must be relied on to transfer the heat from the components to the external surfaces of the enclosure.

Using the principle of superposition, a general equation describing the temperature of an electronic component (at the chip level) operating inside an electronic enclosure is

$$T_d - T_o = \Delta T_a + \Delta T_{\text{int}} + \Delta T_{\text{ext}} \tag{2.1}$$

where

T_d is the temperature of the active surface of the silicon chip (the die temperature),

T_o is the temperature of the air outside the cabinet,

ΔT_a is the temperature rise of the air surrounding the component, (from the external ambient to the local temperature),

ΔT_{int} is the internal temperature rise (die to package surface) and

ΔT_{ext} is the external temperature rise (package surface to air).

It is the purpose of this chapter to provide the designer and/or the applications specialist with some rules that should be followed when the superposition method for temperature prediction is implemented. In addition, some simple formulations that circumvent the mathematical difficulties associated with this type of prediction, are presented. Even though analytical by nature, these expressions involve some experimental correction factors to ensure acceptable accuracy of the results. The analytical formulation of these equations allows for a better understanding of the physical phenomena associated with the convective cooling of electronic components. Moreover, they allow for a rapid estimation of the impact that the change of different factors (size, materials, and board lay-out) has on the final result.

The methodology and the solutions proposed in this chapter provide a rapid and accurate prediction of the die temperature when the following conditions are met:

- Convection from the component surface and the board adjacent to the component is the dominant cooling phenomena. This is especially valid in natural convection and very often for all other air cooling techniques. The benefit of using this methodology when conduction cooling is the dominant cooling mode is marginal.

- The terms in the sum in eq (2.1) can be estimated or bounded using previous experience or the existing requirements or specifications associated with thc final product. First estimates of the different terms of this equation will allow the designer to focus attention on the factors that have the most impact on the final result. Other factors, that might be very difficult to compute accurately, but which are relatively small (smaller by an order of magnitude), can be estimated with simple expressions and relatively low accuracy. Due to the fact that the contribution of these terms is small, the error transferred to the final result will be negligible. Imposed limits from the design specifications can also be used to quantify particular terms in eq (2.1)

- The variations in geometry and technology, from the mechanical point of view, are relatively small. This is usually the case when upgrades of existing equipment are made in order to introduce new electronic functionality. In these cases, most of the time, the configuration of the enclosure and other parts of the hardware are kept the same, while the electronics is modified. The presented methodology allows for the use of similarity relationships in dealing with the inherect complexity or when physical (material) parameters are not available. For example, this methodology can provide very fast and accurate estimates of the impact on chip temperature of changing the board characteristics, the device size or the local flow conditions, if previous experimental results are available.

Some examples showing the problems that can be solved using the superposition algorithms are presented at the end of this chapter.

2.2 THERMAL BACKGROUND

Because the chip package is typically connected to the printed circuit board (PCB), a significant amount of heat first diffuses inside of the solids surrounding the device, before it is absorbed by the passing air by convection. Considering the phenomena that generate the temperature field under discussion, the terms of eq (2.1) cannot be independent from each other. On the other hand, considering the complexity of the equations involved, computation of the die temperature, achieved by integration of the physical equations describing the fully-conjugate problem, is practically impossible. Analytically "decoupling" the temperature field at the package/fluid interface by applying boundary conditions of the third kind at the this interface, makes it possible to avoid the conjugate problem. Thus, at the package surface,

$$q = hA(T_c - T_o) \tag{2.2}$$

By splitting the conjugate problem in this manner, distinct and more detailed models for the convection and the conduction part of the heat transfer phenomena can be created. One can also use experimental data for solid blocks, which match the external geometry

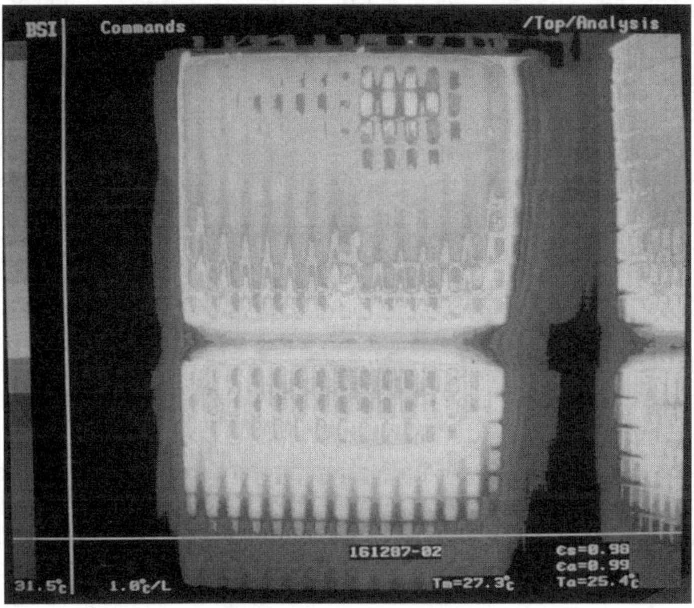

Figure 2.1: Temperature field in boards inside a two shelf open cabinet (20 W per 265 × 385 mm boards. The board spacing is 20 mm and there are 135 uniformly powered components).

of the components in contact with the flow field, as boundary conditions to a conduction model. The main problem associated with this methodology is the selection of the convection heat transfer coefficient, h, and the extent of thermal diffusion in the solids.

The first silicon chips had relatively low levels of integration and low power dissipation. In order to perform complex electronic tasks, large numbers of these packages were mounted in standard configuration arrays on relatively large PCB's. The low level of power dissipation per component produced a thermal field similar to a uniformly heated surface. The temperature level achieved was primarily related to the total power dissipation, the configuration of the channel and the air flow characteristics. For this reason, the first thermal work in this subject focused on determining the convection heat transfer coefficients for flow through channels with rugged, isothermal, congruent walls [Aung (1972), Miyatake and Fujii (1974), Bar-Cohen and Rohsenow (1980), Wirtz and Stutzman (1982), Landis (1986) and Moffat and Ortega (1988)]. The best known semianalytical solution for this problem was achieved by Elenbaas in 1942. The thermal profile of a printed circuit board carrying uniformly heated electronic components, is shown in Fig 2.1.

In considering free convection cooling of electronic components, Aung (1972) successfully modified the Grashof (Gr) number to allow for non-equal heating on the two walls. Bar-Cohen and Rohsenow (1980) have shown how the limiting relation for heat transfer in parallel plate channels could be combined to analytically predict the average Nusselt (Nu) numbers for the full range of PCB spacing and boundary conditions. The studies of Miyatake and Fujii (1974) are notable in that local plate Nusselt numbers were computed. R. Wirtz and Stutzman (1982) measured local plate Nusselt numbers inside symmetrically heated channels.

As the component power increases, the error associated with computing the temperature of the component, assuming that the thermal field is produced by a uniformly heated plate, becomes significant. More and more research is dedicated to developing models that can estimate the temperature of individual electronic components mounted on printed circuit boards [Buller and Kilburn (1981), Furkay (1984), Ashiwake et al (1983) and Aung (1988)]. Buller and Kilburn (1981), through experimental research work, found that the geometric characteristics (package size and configuration) had more impact on the local heat transfer coefficient than the characteristics of the flow surrounding the component. In addition, their experimental results suggest that the characteristic dimension for the determination of the convection heat transfer coefficient is less dependent on the geometry of the channel than it is on the component geometry.

A number of publications [Luikov (1973), Hollworth and Fuller (1987), Davalath and Bayazatoglu (1987), Zebib and Wo (1989) and Godfrey et al (1990)] contain studies of conjugate heat transfer in free or forced convection. These studies are mostly based on the results of numerical integration of the fully elliptic form of the Navier-Stokes equations. Due to computer memory and CPU time limitations, most of these studies are limited to idealized problems. Godfrey et al (1990) presented an interactive procedure to decouple the conjugate problem and determine the temperature distribution on a printed circuit board. An important step forward was achieved by a team of researchers working at Stanford University [Arvizu (1981), Arvizu

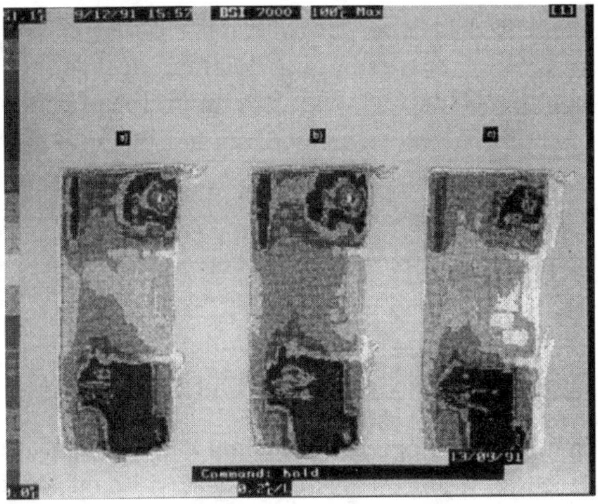

Figure 2.2: Temperature field for digital multiplex switch line card, a) Open Space, b) Inside Enclosure and c) Forced Convection (Vertical Flow).

and Moffat (1982) Ortega and Moffat (1985), Moffat (1990), Moffat and Anderson (1990) and Anderson (1994)]. The *adiabatic heat transfer coefficient* was initially introduced to facilitate separation of the sensible temperature rise in the cooling fluid surrounding the component from the local temperature rise across the thermal boundary layer. However, its use has also made it possible to decouple the contribution of convection and conduction and to correctly apply superposition of these thermal fields for most situations encountered in the electronics industry. This notion will be explained later in this chapter.

In the 1980's, a group of researchers from Northern Telecom [Witzman et al (1989, 1990) and Graham and Witzman (1988)] started a large scale experimental program intended to produce experimental correlations for use in the thermal design of electronic equipment for the telecommunication industry. By this time, the degree of integration of the circuitry and the resulting power dissipation on the silicon chip was high enough to make previous analysis methods obsolete. Figure 2.2 shows the thermal field produced by a typical, populated printed circuit board in three different enclosures: open space, inside an hermetic enclosure and inside a "flow through" cabinet. One can see that for these LSI components, the power is concentrated in a relatively small number of components, surrounded by parts that dissipate little or no heat at all. There is very little resemblance between the thermal profile of this board and a uniformly heated plate or an array of similar discrete heat sources, shown in Fig 2.1.

2.3 EXPERIMENTAL APPARATUS AND PROCEDURE

2.3.1 THERMAL FIELD VISUALIZATION

The main difficulty in mapping the temperature field produced by a powered electronic component inside an electronics enclosure is associated with the need to mount the board inside the actual enclosure. The exact flow conditions that the components will see in real applications must be reproduced. The local flow around a component is a strong function of a large number of variables, such as cabinet inlet/outlet configuration, fan characteristics, electromagnetic interference screens, mechanical hardware and cable arrangement. The problem is more severe when buoyancy induced flow is the cooling agent because any modification made to the actual enclosure (to facilitate temperature measurement) could modify the flow conditions and, in turn, affect the component temperature.

Temperature measurements are typically performed using thermocouples as temperature sensors. However, due to strong variation in the local temperature field, a misplaced thermocouple will generate misleading data. In order to compensate for and/or dilute the errors created by relatively small errors in thermocouple positioning, a very large number of measurement points is necessary. Unfortunately, the many thermocouple connections will themselves modify the temperature field. As an alternative, a good indication of the temperature field on printed circuit boards can be achieved by taking infrared pictures of the board immediately after its removal from the cabinet. The board thermal inertia is large enough to maintain the temperature field for the time required to remove the board and expose it to the camera. This methodology can help locate expected "hot spots" on the board, but often it is not accurate enough for scientific research purposes.

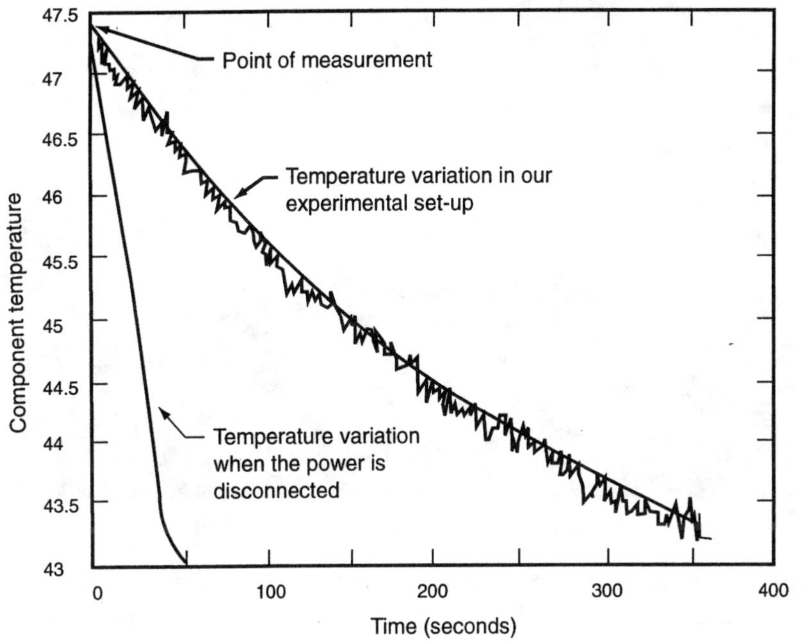

Figure 2.3: Component temperature variation with cabinet open.

2.3.2 APPARATUS

In order to reduce the time required to expose the board to the external ambient and avoid the cooling effect generated by the quick movement of the board, a special shelf stand was developed at Northern Telecom (NORTEL) The shelf is split into two sections and the stand rotates part of the shelf, exposing the test board to the infrared camera. In this way, the boards remain powered during the experiment and, because the boards are kept stationary during the measurements, rapid forced convection cooling of the boards is avoided.

All of the boards contained in the cabinet are powered and remain powered during the procedure. Due to the thermal inertia of the board itself, the changes in the temperature field, created by opening the shelf, are negligible during the first 5 seconds after the opening. Because of the limitations imposed by the need to remove the heavy mechanical parts that obstruct the view of the infrared camera, the actual time required to capture the thermal image is about one second. This reduces the measurement error to less than 0.5°C, which is the allowable limit for such measurements (see Fig

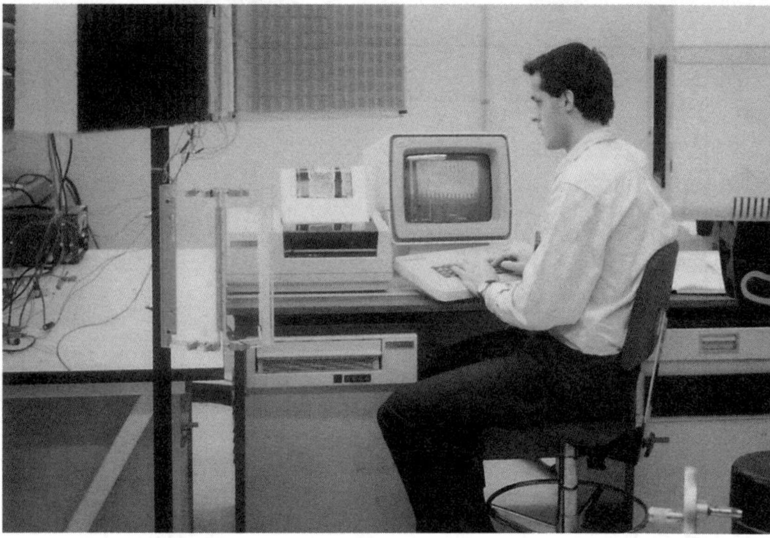

Figure 2.4: The experimental set-up.

2.3). A picture of the set-up with the shelf open and the board exposed is shown in Fig 2.4. The PCB temperature field, shown in Figs 2.1 and 2.2, was taken using this experimental set-up.

In all of the NORTEL experiments, arrays of components were mounted on printed circuit boards which were installed inside different cabinet configurations. The studied enclosures were single and multiple-shelf flow-through cabinets, baffled and hermetic enclosures. The enclosures were air cooled by free or forced convection. The impact of sealing the air flow channels formed between adjacent boards was also studied. The studied enclosures were selected to cover the entire line of telecommunication products, from the smallest hand held terminals to the largest switching cabinets.

There is very little work that provides experimental data related to the formation and the characteristics of the thermal plume created around the small heated bodies inside of the surrounding cooling fluid. Some tentative work of this kind employed interferometry to visualize the thermal boundary layer around simple geometries [Moffat (1990)] or used the thermal "footprint" of heated fluids on solids covered with liquid crystals [Moffat and Anderson (1990)].

The use of interferometric techniques has been found to be too expensive and the use of liquid crystal techniques is laborious, very slow and lacks the accuracy necessary for producing useful empirical equations. Consequently, it was decided to use infrared measure-

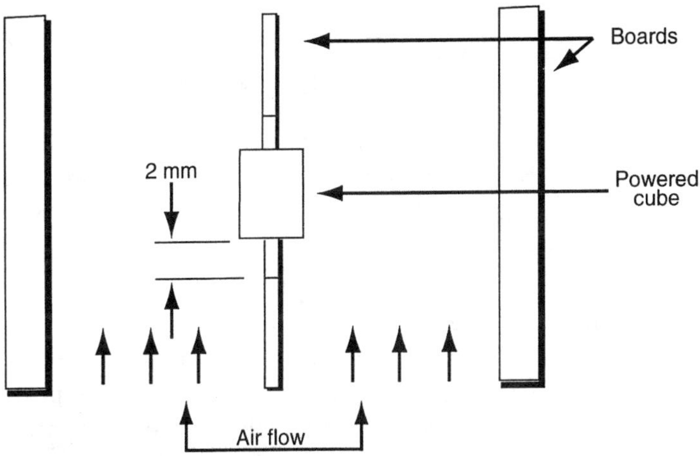

Figure 2.5: Experimental set-up for the determination of the Nusselt-Reynolds correlations.

ment techniques which can produce fast and accurate thermal field measurements. The results can be stored and analyzed numerically. Obviously, infrared thermometry cannot be used to directly monitor the temperature field inside the cooling fluid, especially in the case of air cooled objects. What can be monitored is the thermal "footprint" that hot gases leave on thin, non conductive membranes immersed inside the hot fluid.

In order to study the thermal field created in the proximity of small heated three dimensional bodies (smaller than one inch cubes), an apparatus similar to the one shown in Fig 2.5 was fabricated. The experimental apparatus consisted of a very thin foil of thermally non conductive material (polyethylene) with a window in which a cubical body of copper was suspended. The foil did not touch the body which was heated by a resistor mounted on the inside. Due to the high thermal conductivity of the copper, the surface could be considered isothermal. No direct conduction was considered between the copper and the foil. Due to the thickness and low thermal conductivity of the foil, the heat spreading inside the foil was minimal. The copper and the foil were painted black to provide an emissivity of 0.92. Under such conditions it was considered that the thermal "footprint" on the foil, monitored by an infrared camera, represented the cross-section of the thermal boundary layer created around the heated body.

2.3.3 EXPERIMENTAL RESULTS (FLOW THROUGH CABINETS)

In all of the cases studied, the temperature varied along the width as well as along the height of the board, creating a high temperature area in the center of the shelf. At the same time, the temperature did not increase linearly with board height, which would have been expected for a constant heat transfer convection coefficient.

The component temperature variation per uniform component power dissipation is shown for the central column of populated boards in Fig 2.6, for several power dissipations and component positions. If the first and last components in the column are eliminated, the component temperature rise, for 20 mm board spacings and 10 to 30 Watt/board power dissipation, can be estimated from the empirical correlation

$$\Delta T_n = T_c - T_o = 76.1 q_d^{0.8}[1 + 0.23(n-2)^{0.62}] \tag{2.3}$$

where all of the components dissipate equally and where

n is the position of the component ($n = 1$ is the bottom row),

ΔT_n is the average temperature rise of the nth component above the ambient temperature, (°C), and

q_d is the component power dissipation, W.

The numerical values of the experimental constants are valid only for the cabinet configuration studied.

Different constants, but very similar correlations were achieved when, for example, board spacing, was varied. The approximate square root correlation with the position of the component on the board and the 0.8 power of the heat dissipation is to be noted. This variation can not be explained by the inflow of cold air from the lateral sides of the channel because eq (2.3) was determined for laterally sealed channels.

The most important feature of eq (2.3) is that in all of the cases studied, the component temperature (the maximum value of the temperature of the external surface of the package) could be expressed as a sum of two functions, one of which is independent of the position of the component on the board. Besides being a function of the

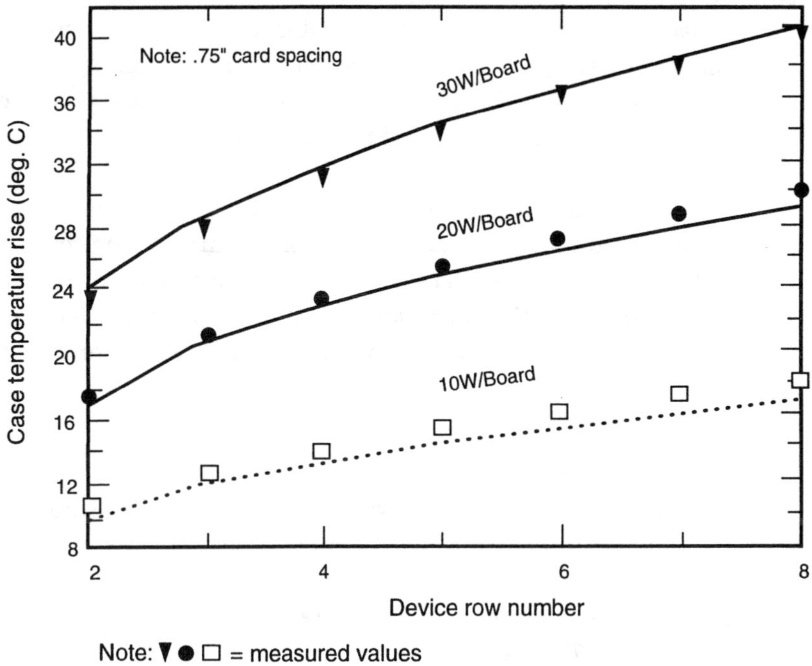

Figure 2.6: Component temperature rise.

component power, the first term of this expression was later found to be a function of the component/board size and technology as well as the local flow conditions. A very weak variation of this term with the board layout was found. A larger variation was found for the case in which the power per component was unequal and the studied component was mounted upstream from a much higher power dissipating component.

When eq (2.3) is compared with eq (2.1), there is a strong motivation to associate the two terms of eq (2.3) with ΔT_{ext} and ΔT_{int} respectively from eq (2.1). This can also explain the temperature field patterns shown in Figs 2.1 and 2.2. For situations in which the power dissipated per component is small compared to the total power dissipated inside the channel (say, one hundredth of the total), ΔT_{ext} is small when compared with ΔT_a. In this case, the thermal pattern seen on the board is almost the same as that predicted numerically for the temperature field of the air inside the channel (Fig 2.1). When the power is dissipated by a small number of widely separated elements, the temperature field is similar to the one associated with the thermal boundary layer forming along the surface of the component (Fig 2.2). The variation of the air temperature, for the different cabinet situations, will modify the absolute value of the temperature without modifying the local appearance of the isotherms.

The experimental program showed that, for the situations studied, the influence of the upstream (or downstream) channel conditions on the component temperature can be superimposed on the temperature rise in the boundary layer of the component. Defining T_c as the temperature of the external surface of the component or "case" and using eq (2.1)

$$T_c = (T_o + \Delta T_a) + \Delta T_{\text{ext}} \tag{2.4}$$

This expression is similar to the expression used by the Stanford University team, as discussed by Witzman et al (1990), in which the external surface temperature of the heated elements is expressed as a summation of the *adiabatic temperature*, T_{ad}, and the adiabatic temperature rise, ΔT_{ad}, i.e.

$$T_c = T_{\text{ad}} + \Delta T_{\text{ad}} \tag{2.5}$$

The *component adiabatic temperature* is the temperature achieved when no power is applied to the element and the rest of the system is activated [Arvizu (1981) and Arvizu and Moffat (1982)]. By comparing eqs (2.4) and (2.5), one can conclude that the component adiabatic temperature is equal to the sum of the ambient temperature, T_o and the air temperature rise, ΔT_a. In other words, this temperature represents the temperature of the environment in which the component is located. This is not to be confused with the component adiabatic temperature rise, which is the increase in the temperature of the powered component on top of the component adiabatic temperature. The component adiabatic temperature rise represents the temperature rise across the component thermal boundary layer or ΔT_{ext}.

When there are relatively large distances between the heat sources, (the IC packages), very little interaction takes place between the thermal boundary layers of different components mounted on the same board. This interaction is mainly in the form of a thermal wake produced by the high power components mounted upstream from the subject device. The results are similar for forced and free convection cooled electronic cabinets, and, to a certain extent they are also valid for large hermetic (re-circulation) enclosures when the distance between a package and the enclosure wall is larger than 15 mm.

2.3.4 COMPONENT TEMPERATURE RISE INSIDE SEALED ENCLOSURES

As previously noted, experimental measurements of the temperature rise on single, populated boards have revealed that the thermal fields are similar to those for powered boards located inside enclosures and in free space (see Fig 2.2). Thus, the temperature rise above the local ambient at any point on the board is identical for an enclosed board and a freely suspended board of identical configuration. The ambient temperature in small enclosures (smaller than $0.25\,\mathrm{m}^2$ external surface area) was found to be a function of a single variable - the external area of the enclosure - for all cases where the components were at least 15 mm from the enclosure wall.

Considering the average enclosure temperature as equivalent to the bulk air temperature for a component mounted inside a ventilated enclosure, the component temperature rise can be calculated from the equation

$$T_c = T_{\mathrm{cf}} + \Delta T_{\mathrm{ae}} \tag{2.6}$$

where T_c is the package "case" temperature, T_{cf} is the component temperature measured when the board is tested in free space (outside of the enclosure) and ΔT_{ae} is the average temperature rise of the air inside of the enclosure.

The average temperature rise of the air inside the enclosure can be computed from

$$\Delta T_{\mathrm{ae}} = \frac{q_T}{U_{\mathrm{enc}} A} \tag{2.7}$$

where A represents the external surface area of the enclosure and U_{enc} is the heat transfer coefficient based on the enclosure surface area. This is an overall heat transfer coefficient, as it involves convection and radiation from the inside fluid to the enclosure wall, conduction through the enclosure material and conduction plus radiation from the enclosure surface to the environment. In most situations, the heat transfer limits are established by the capability of the enclosure

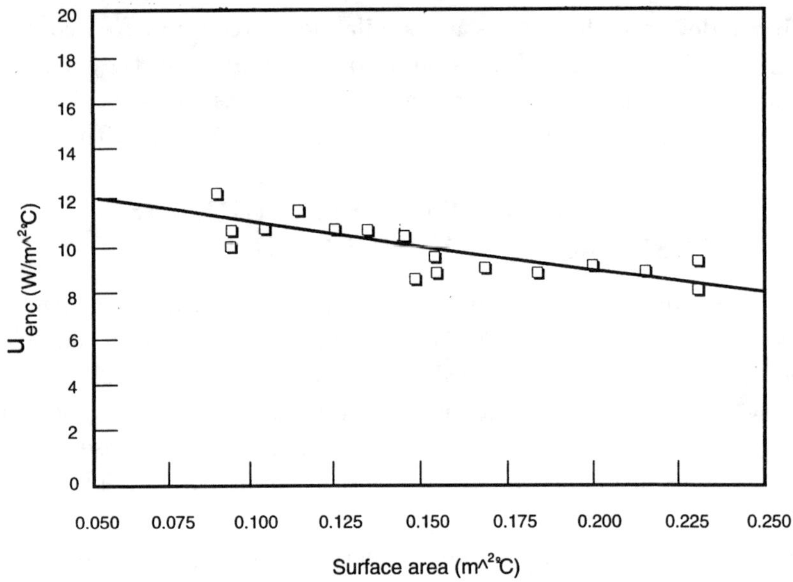

Figure 2.7: Variation of the convection heat transfer coefficient for enclosures as a function of the enclosure external area.

to dissipate the heat from the external surface to the ambient air. This resistance is usually much higher than the internal and "through the material" thermal resistances.

The experimental study produced sufficient data to create an empirical correlation for the value of the overall enclosure coefficient, U_{enc}, through small sealed enclosures

$$U_{\text{enc}} = 12.696 - 16.774A \tag{2.8}$$

This correlation is valid for enclosures with a surface area between 0.05 to $0.250\,\text{m}^2$ and it cannot be extrapolated safely outside this interval. A comparison between the correlation and the experimental results is shown in Fig 2.7. For cabinets with external surface areas between 0.25 to $1\,\text{m}^2$, U_{inc} can be considered a constant ($U_{\text{enc}} = 8\,\text{W/m}^2\text{-K}$). It should be noted that the contribution of thermal radiation was included in this value.

Another important parameter associated with cooling of electronic components inside enclosures is the local (approaching) air velocity for buoyancy-induced flow. Neutrally buoyant helium bubbles were used to visualize and monitor the flow inside enclosures.

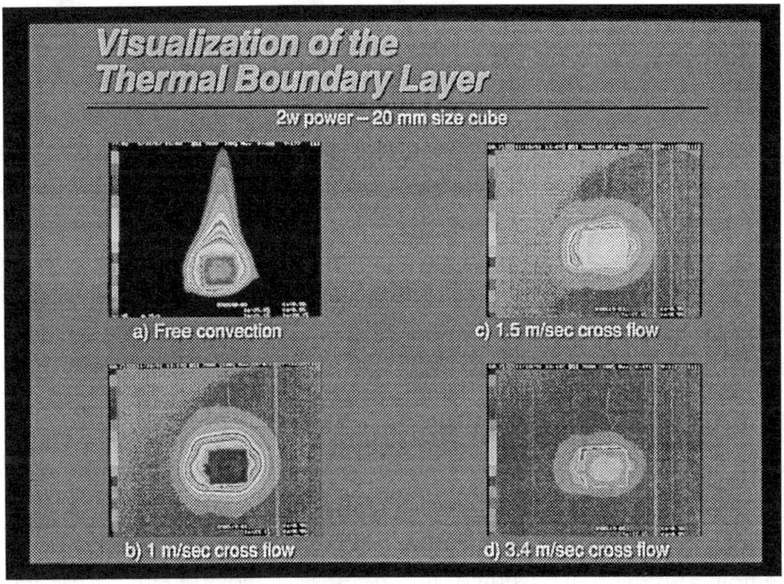

Figure 2.8: Visualization of the thermal boundary layer, for a 20 mm Cube dissipating 2 W, (a) free convection, (b) 1 m/s in cross flow, (c) 1.5 m/s in cross flow and (d) 3.4 m/s in cross flow.

The bubbles were produced by trapping helium inside soap bubbles. Using centrifugal separation, bubbles having a density similar to that of the air inside the cabinet were selected. These bubbles were introduced inside the cabinets and their movement was recorded and analyzed using a special high speed video camera. For simple cabinets, carrying vertical parallel boards, the approaching air velocity was found to be contained within a relatively small domain. The local air velocity associated with this situation is 0.4 to 0.7 m/s; the higher values (0.7 m/s) are associated with cabinets taller than 0.5 m.

For complex cabinets containing horizontal boards and/or baffles, the flow could become more complex and special experimental or numerical computational methods might be necessary in order to determine the local air velocity.

2.3.5 VISUALIZATION OF THE THERMAL FIELD AROUND A SMALL COMPONENT PACKAGE

The preliminary test results presented here justified focusing attention on the thermal-fluid phenomena in close proximity to the package, rather than detailed consideration of heat transfer and fluid flow throughout the entire channel.

The results of the visualization studies of the thermal boundary layer of a single heated body in a large enclosure (Fig 2.8) were similar to the results achieved in the past for the plume produced by single small bodies in free convection [Moffat (1990)]. The visualization of the plume formed in the presence of flow across heated bodies, at ever increasing velocities parallel to the flow direction, was even more impressive. For very low velocity (lower than 0.6 m/s), the plume vector rotated in the flow direction but decreased very little in size and temperature. For higher velocities. the plume was aligned with the flow and decreased slowly in size until the local velocity exceeded 3m/s. A further increase of the local air velocity above this value produced no visible change in the size or shape of the thermal plume.

The temperature and velocity data for the foegoing experiment have been non-dimensionalized and are presented in Fig. 9. The results were corrected to account for the radiation between the body and the environment. Following Yovanovich and Vanoverbeke (1988), the local convection coefficient can be computed from the adiabatic Nusselt number which is expressed as the sum of the effects of thermal diffusion and convection from the component surface

$$\mathrm{Nu}_{\mathrm{ad}} = \mathrm{Nu}_{\mathrm{D}} + \mathrm{Nu}_{\mathrm{fc}} + \mathrm{Nu}_{\mathrm{nc}} + \mathrm{CR} \tag{2.9}$$

where

$$\mathrm{Nu}_{\mathrm{D}} = \frac{h_{\mathrm{ad}}\sqrt{A_{\mathrm{ext}}}}{k} \tag{2.10}$$

in which h_{ad} is the adiabatic convection heat transfer coefficient associated with the temperature rise through the thermal boundary layer or plume, A_{ext} is the external surface area of the heated element, and k is the thermal conductivity of the cooling fluid.

Moreover,

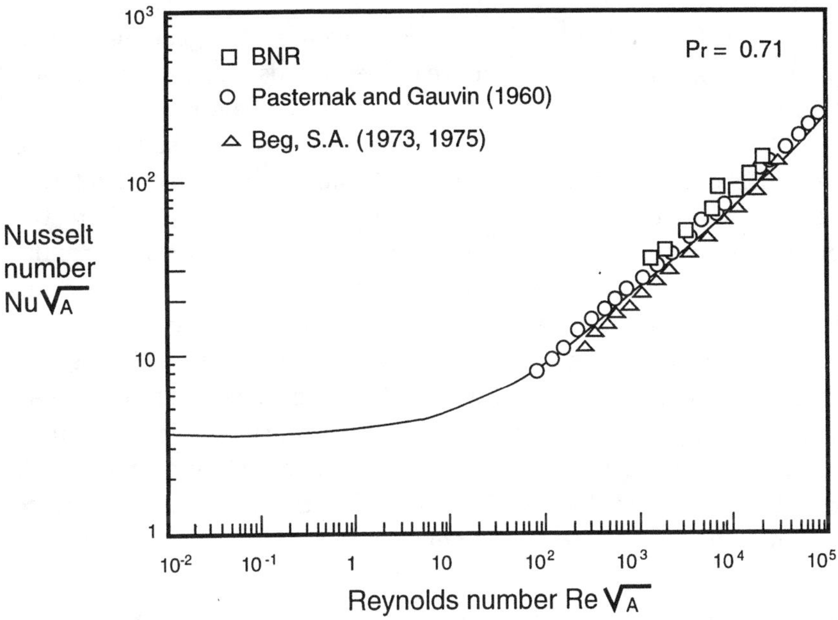

Figure 2.9: Variation of the convection heat transfer coefficient for a suspended small heated body, as a function of the approaching air velocity.

Nu_D is the diffusive limit

$$Nu_D = 2\sqrt{\pi} \tag{2.11}$$

Nu_{fc} is the pure laminar forced convection limit

$$Nu_{fc} = 0.825 Re^{0.5} Pr^{0.33} \tag{2.12}$$

Nu_{nc} is the pure laminar natural convection limit

$$Nu_{nc} = 0.694 Ra^{0.25} \tag{2.13}$$

and CR is the mixed convection, cross flow correction

$$CR = 0.86 - 3.54 \left(\frac{Ra}{Re^2} \right)^{0.25} \tag{2.14}$$

The value of the correction parameter, CR has been determined as the "best fit" curve for the experimental results, measured for the interval $0.10 \leq Gr/Re^2 \leq 10$. As the value of CR is generally small, approaching zero when the air velocity exceeds 1 m/sec, extrapolation outside of the interval will produce negligible error.

The characteristic length in these expressions is the square root of the surface area of the heated body. [Yovanovich, (1988)]. Equation (2.9) covers the entire interval between completely static air to the beginning of the turbulent flow regime. The experimental results and practically all of the results encountered in the literature are bounded within ± 10 percent from the predictions by eq (2.9).

The correlation only applies for solid bodies without holes and with an aspect ratio close to unity. It is suggested that the results can be extrapolated to packages attached to printed circuit boards, because electronic components are relatively small heat sources which create independent, local heat plumes. The contribution of substrate conduction is examined in the next section.

2.4 CONJUGATE HEAT TRANSFER

2.4.1 PHYSICAL BACKGROUND

As noted in the previous section, when chip packages are attached to a circuit board, in addition to convective heat transfer from the surface, they exchange heat with a substrate of a finite conductivity. Figure 2.10 shows an electronic component mounted on a printed circuit board. There are several paths for heat flow from the chip to the surrounding fluid (air). Heat is conducted through the component to the external surface, and is convected away by the fluid. Heat is also transferred to the substrate, by conduction, convection and radiation, conducted through the component substrate and convected away from a larger area of the board.

For low conductivity printed circuit boards, as in FR-4 technology, only a small fraction of the heat is transferred from the component to the printed circuit board. Consequently, the plume of heated air is maintained and there is very little interference among the thermal boundary layers of the individual components. The adiabatic convection heat transfer coefficient (between the component and its adiabatic surface temperature) can be considered similar to the convection coefficient around a small heated body of unity aspect ratio,

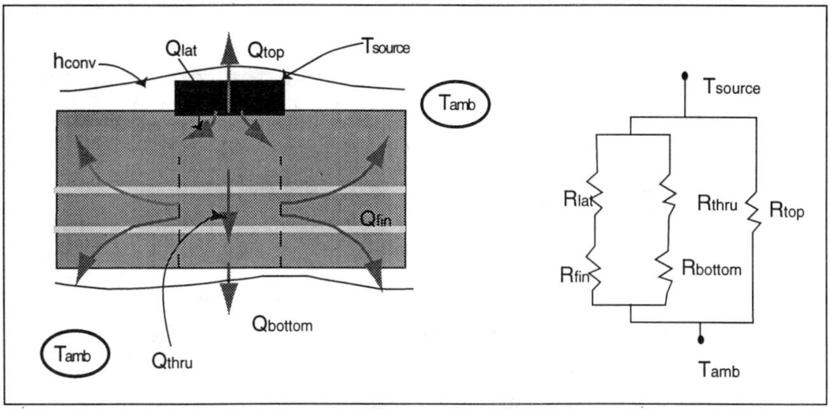

Figure 2.10: Combined heat flow paths for a component and circuit board.

and can be computed using eq (2.9). The relatively low Reynolds number flow (under 2000 when the characteristic dimension is the square root of the external area of the heated body) has very little impact on the component thermal boundary layer in this domain. The plume or wake from the upstream components will affect the adiabatic temperature of the component without modifying the adiabatic heat transfer coefficient. For substrates with relatively high thermal conductivity (ceramic technology), the spreading of the heat in the board is significant and corrections that take this into account should be considered. For very high conductivity substrates such as beryllia or diamond, the thermal footprint of a single component can spread over the entire available surface, creating a nearly isothermal plate boundary condition.

Even though the methodology of solving the conjugate (convection-conduction) heat transfer problem described in this chapter is valid for all types of substrates, attention will be focused on the low conductivity board technologies (FR-4). Not only is this situation the most frequently encountered in the electronic industry, but it is the most difficult to approach using conventional heat transfer methodologies. In this case, the problem can be solved by treating the board as a cooling fin associated with an internally heated small body which represents the component.

A heat balance written for any specific component will become

$$q_d = q_a + q_b \tag{2.15}$$

where q_a is the heat dissipated directly from the component package into the cooling fluid (air), q_b is the heat dissipated through the board, and q_d is the total power dissipated by the component (silicon die). Considering that the package heat dissipation occurs inside a relatively small volume occupied by the silicon die, a thermal resistance model describing the main thermal paths for the heat dissipation, shown in Fig 2.10, can be used. Recognizing that each thermal path has an associated finite thermal resistance, the equivalent external thermal resistance, commonly known as the "case-to-ambient" thermal resistance, R_{ext} can be written as

$$R_{\text{ext}} = \frac{R_a R_b}{R_a + R_b} \tag{2.16}$$

where R_a is the thermal resistance from the surface of the package to the air and R_b is the thermal resistance between the printed circuit board and the air. The thermal path associated with q_b, (R_b), can be represented by a sum of three distinct thermal resistances

$R_{\text{b,g}}$, the resistance of the gap between the component and the board,

$R_{\text{b,sp}}$, the resistance associated with the heat spreading inside the board and

$R_{\text{b,a}}$, the convection thermal resistance from the board to the air flow.

The equivalent overall package thermal resistance, commonly known as the "junction-to-ambient" thermal resistance, becomes

$$R_{\text{ja}} = \frac{R_a R_b}{R_a + R_b} + R_{\text{int}} \tag{2.17}$$

where R_{ja} is the equivalent thermal resistance between the silicon chip and the local ambient (or the adiabatic thermal resistance). Here, R_{int} represents the internal thermal resistance of the component, also known as the "junction-to-case" thermal resistance. In most cases, in the absence of internal heat spreaders or other structures that create preferential thermal paths inside the package, the heat can be considered as spreading radially from the die to the external surface of the package.

In practice R_{int} is experimentally measured following the procedure described in MIL-STD-883C, Method 1012 and SEMI Std DOC #1321. According to this experimental method, the internal thermal resistance of the package is measured by mounting the component package on a 5×5 cm test coupon which is immersed in a highly agitated fluorocarbon fluid bath. The mounting of the component package directly on the test coupon simulates all the internal conduction paths that will be experienced in a real situation and the highly agitated liquid minimizes any external temperature gradient inside the fluid and the temperature difference between the fluid and the package. The temperature change is determined by monitoring the voltage drop of a calibrated diode on the test chip. For most components, this parameter is supplied by the component vendor and it can be found in the component catalog.

A correct computation of all of these thermal resistances might be required in some cases in which large, complex packages or heat spreaders are involved. In most cases (simple package configuration, qualified technologies, repeat use of similar configuration) acceptably accurate and faster results can be achieved by employing a simplified technique for estimating the terms in eq (2.17). This methodology is based on the experimental finding shown in Fig 2.11 that, for most of the actual air cooling technologies, the thermal footprint on the package side of the printed circuit board is nearly identical to that on the bottom of the board. This can be explained by the dominance of the convective resistance over the conductive resistance for these packages, mounting techniques and PCB's. This dominance is brought about by the relatively high thermal conductivity of ceramic packages and the use of thermally conducting pads inside plastic packages, as well as the role played by the pins, leads and plated-through-holes (vias) in conducting the heat through the printed circuit board.

For the configuration shown in Fig 2.11, one can consider that the heat is dissipated inside a barrel composed of the package itself and the volume contained in a circular disc of the board, underneath

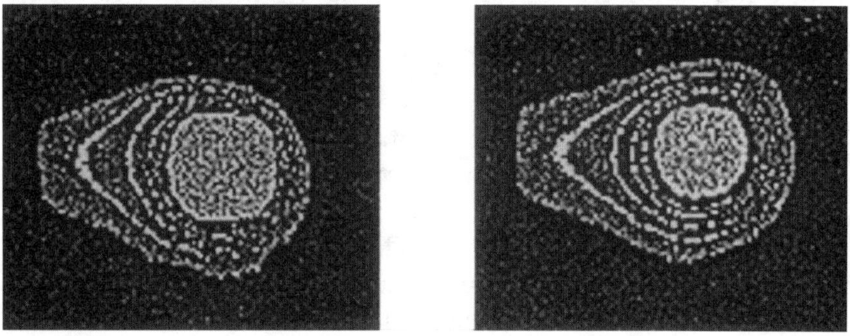

Figure 2.11: The thermal print, frontside and backside of a 28 Pin, Leadless Ceramic Carrier, mounted on a FR-4 printed circuit board.

the package. The external area of this heated volume, $A_{\rm ext}$, is

$$A_{\rm ext} = \frac{\pi a^2}{4} + A_{\rm a} \tag{2.18}$$

where $A_{\rm a}$ is the wetted surface area of the package, a is the square root of the package projected area and $\pi a^2/4$ represents the heat transfer area underneath the component. It is important to note that this surface was selected as the reference surface area for the definition of the adiabatic convection heat transfer coefficient. As the value of the convection (adiabatic) heat transfer coefficient is dependent on the surface of reference, any change in the area selected requires a modification of the convection heat transfer corefficient. Accordingly, the characteristic geometric dimension in eq (2.9) should be $\sqrt{A_{\rm ext}}$. However, only relatively small errors are introduced if this value is replaced by the square root of the component external surface.

As the thermal resistance through the board (in the thickness direction) is always much smaller than the equivalent lateral (radial direction) printed circuit board thermal resistance, the conjugate heat transfer problem is reduced to solving the problem of heat transfer from a small heated body connected to a large cooling fin. The thermal field images also support the idea that the use of cylindrical coordinates is appropriate for this geometry.

Under these assumptions, R_a, representing the thermal resistance for the component surface to the air flow, can be computed from a relation of the form

$$R_a = \left[h_{\text{ad}} \eta \left(\frac{\pi a^2}{4} + A_a \right) \right]^{-1} \tag{2.19}$$

Here h_{ad} is the local adiabatic heat transfer coefficient and η represents the thermal efficiency of the package surface defined as the the ratio of the actual power dissipated by the package to the power dissipated by the package if operating at the maximum surface temperature. For quick estimations, η can be taken as varying between 0.85 to 0.95 for ceramic packages and 0.70 to 0.90 for plastic components.

A good approximation for h_{ad} can be obtained using eq (2.9). It is possible to correct for errors introduced by the simplifications of the physical model by introducing an experimental factor, C_i. This experimental factor compensates for changes in the boundary layer caused by the non-cuboid geometry for which eq (2.9) was determined. With this correction, eq (2.9) becomes

$$\text{Nu}_{\text{ad}} = \text{Nu}_D + C_i(\text{Nu}_{\text{fc}} + \text{Nu}_{\text{nc}}) + \text{CR} \tag{2.20}$$

Experimental results obtained at Northern Telecom show that C_i is practically constant for a particular packaging technology (DIP's, CLC's, QUAD's and PLCC's), varies very little with the size of the component within the same family, and is practically independent of the board characteristics. For most of the studied technologies, the value of C_i ranges between $0.85 \leq C_i \leq 1.15$ indicating that relatively small errors will result from using unity as the default value. Even though this was the case for most of the technologies used in the past, in order to avoid possible errors, the determination of the Ci value at the introduction of any new packaging technology is strongly recommended.

Assuming heat dissipation inside the board from an isothermal heated barrel, and that the board thickness is small enough to neglect the variation of the temperature in the z direction (the direction perpendicular to the board), R_b, the thermal resistance between the board and the air, can be expressed in the form of the modified Bessel functions

$$R_b^{-1} = \pi mabk \left[\frac{I_1(\alpha)K_1(\beta) - K_1(\alpha)I_1(\beta)}{I_1(\alpha)K_0(\beta) + K_1(\alpha)I_0(\beta)} \right] \tag{2.21}$$

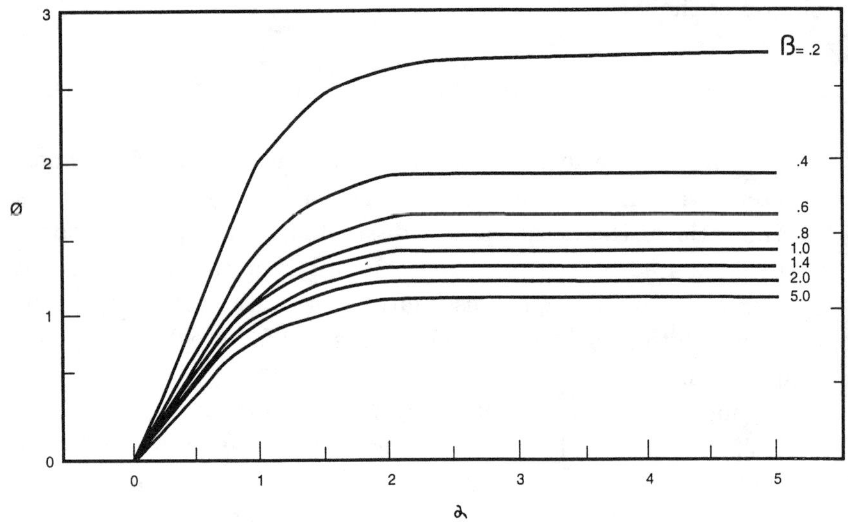

Figure 2.12: The parameter Φ as a function of α and β.

where

$$\alpha = Lm \quad \text{and} \quad \beta = \frac{am}{2}$$

and where

$$m = \left(\frac{2h_{\text{ad}}}{kb}\right)^{1/2} \tag{2.22}$$

In eqs (2.21) and (2.22), a, as before, is the square root of the package projected surface area, b is the board thickness and L is the radius of the unit board cell, to be considered as a circular cooling fin surrounding each component. For homogenous materials, k is the board thermal conductivity and for composite materials (FR-4 technology), k is an effective board conductivity. The effective board conductivity is set equal to the conductivity of an homogeneous board of the same thickness that produces the same temperature rise on the package surface as the real composite board. The values of k in this equation vary between 0.4 W/m-K for single layer boards to 2 W/m-K for multilayer (up to 8 layers) boards with at least two solid copper ground planes. It is important to mention that the effective board thermal conductivity for this computation is different than the physical value of the lateral board thermal conductivity, which is usually a much higher value. The explanation is associated with the fact that, while effective heat spreading occurs across the board through the copper

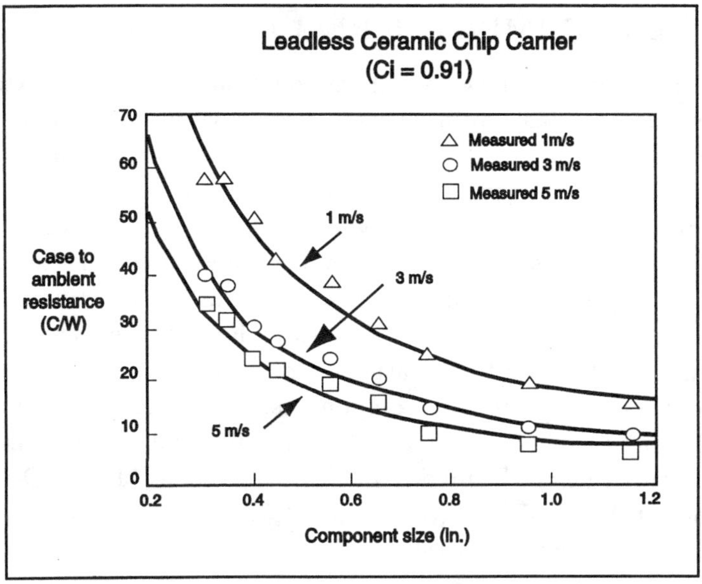

Figure 2.13: Comparison between predicted and measured values of the case to ambient thermal resistance (adiabatic temperature rise) for Leadless Ceramic Carriers.

layers, the heat must then return to the board surface in order to be absorbed by the air flow. The epoxy layers separating and covering the copper layers act as a significant barrier in the thickness direction, reducing the effective board conductivity. Additional aspects of this issue are discussed in subsequent sections.

The Modified Bessel functions in eq (2.21) may be combined into a parameter, Φ

$$\Phi = \frac{I_1(\alpha)K_1(\beta) - K_1(\alpha)I_1(\beta)}{I_1(\alpha)K_0(\beta) + K_1(\alpha)I_0(\beta)} \tag{2.23}$$

Observe that Φ is a function of α and β which include the heat source dimension, a, the substrate size, L, and the conduction-convection parameter, m. The variation of Φ with $\alpha = mL$ and $\beta = am/2$ is shown in Fig 2.12. For most practical applications, the value of the parameter, Φ lies in the range, $0.80 \leq \Phi \leq 1.20$.

The values of R_a and R_b, as computed by eqs (2.19) and (2.21) are combined, as shown in eq (2.17), in order to find the equivalent adiabatic thermal resistance (junction to ambient), R_{ja} for a specific component mounted on the printed circuit board.

2.4.2 EXPERIMENTAL RESULTS

In order to estimate how well these equations predict the external temperature rise (case to ambient), a comparison between the predicted value of the case to ambient thermal resistance using the external adiabatic temperature rise, and measured values for two different package technologies is shown in Figs. 13 and 14. Because the case temperature is strongly non-uniform, and direct measurement of the package surface temperature is rather difficult, calculated case temperatures are used in these figures. These values were derived from measured silicon temperatures and known internal thermal resistances for the considered components. The silicon temperature was measured using a thermal die, and the internal temperature rise was measured using the fluid immersion technique (MIL-STD-883C, Method 1012 and SEMI Std. DOC#1321.). These techniques were discussed in a previous section.

As can be observed from Figs 2.13 and 2.14, the correlations accurately predict the temperature of single components mounted on multiwire single-sided FR-4 boards. It is interesting to note that in obtaining the Nusselt correlations for these packages types, the C_i coefficient was found to be approximately 12 percent higher for plastic components than for ceramic packages of similar geometries. This anomaly is thought to reflect the way in which the case temperature is defined and measured and the conservative estimate of the surface thermal effectiveness. In this case, the C_i value is an indication of the way in which heat flows from the silicon chip to the case rather than an indication of the overall thermal efficiency of the package .

2.5 PREDICTIONS FOR MULTIPLE COMPONENT ARRAYS

2.5.1 WAKE EFFECTS

Few electronics enclosures contain just a single electronic component. However, for situations where the component spacing is substantially larger than the component size and the equivalent board thermal conductivity is lower than 2 W/m-K (FR-4 technology), the components can be considered as locally independent entities, linked only through their cumulative contribution to the temperature of the air.

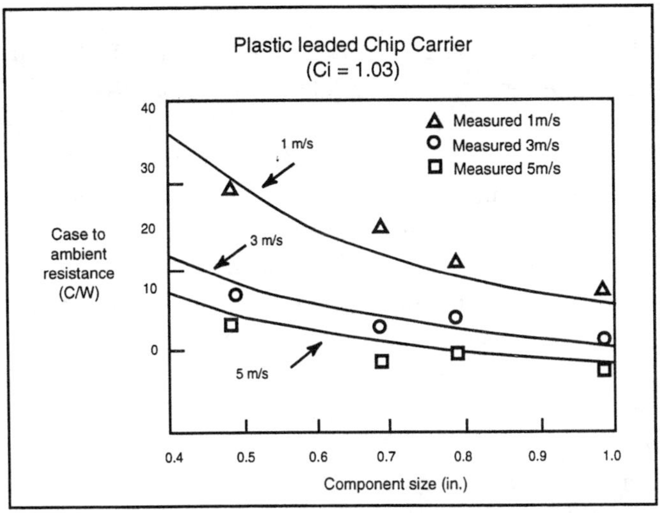

Figure 2.14: Comparison between predicted and measured values of the case to ambient thermal resistance (adiabatic temperature rise) for Leaded Plastic Carriers.

Each one, when powered, produces an independent thermal plume that will determine the temperature rise above the component adiabatic temperature. For these cases, eq (2.17) can be used to predict the temperature rise of the silicon die. When a populated board is mounted inside an enclosure, it is necessary to account for the contribution of the other components in the system to the local air temperature rise of the considered component.

By definition, the component adiabatic temperature refers to the temperature the component achieves when its own power is shut "off," but the remainder of the system is "on." A non-powered component, mounted on the inside of a cabinet will reach the temperature of the location in which it is mounted. As no heat exchange can take place between the component and the cooling media when this equilibrium is reached, one can say that the adiabatic temperature of the component is equivalent to the local ambient temperature (the air temperature at the location of the component). Considering that the power dissipated by one single component is usually much smaller than the total power dissipated inside the enclosure, a variety of analytical and numerical techniques, as well as commercial software, can be used to determine the fluid temperature at the component location and, hence, its adiabatic temperature.

For best results, the thermofluid simulation should be performed with the component under consideration unpowered and the value of the adiabatic temperature should be selected as the temperature of the board at the location of the component. The value of the adiabatic temperature computed in this way should be added to the adiabatic temperature rise computed according to the method described in a the previous section. In some cases, such as enclosures with baffles and separating walls with complex flow patterns, this might be the only way of making a satisfactory estimate of the component adiabatic temperature. In many cases, however, not enough information is available to make efficient use of a numerical fluid flow program.

For simple flow through cabinets, a heat balance may be sufficient to calculate the component adiabatic temperature. Based on the assumption that all the heat is absorbed by the air stream, the mean (bulk) air temperature inside the cabinet, at the location at which the component is mounted, is given by

$$T_{\mathrm{a,m}} = T_o + \frac{\sum q_{\mathrm{up}}}{\dot{m}_a c_{p,a}} \tag{2.24}$$

where T_o is the temperature of the air outside the enclosure, $\dot{m}_a$ is the air mass flow rate, $c_{p,a}$ is the air specific heat and $\sum q_{\mathrm{up}}$ is the total power dissipation inside the enclosure, upstream from the component location.

Considering that the heat is dissipated non uniformly inside the cabinet, an element situated immediately downstream from a heated element will be exposed to air that is hotter than the mean air temperature because the thermal energy will not have diffused across the channel. These components are thus immersed in the thermal wake produced upstream and a wake contribution should be added to the mean air temperature. Thus

$$T_{\mathrm{ad}} = T_{a,m} + \Delta T_{\mathrm{TW}} \tag{2.25}$$

where ΔT_{TW} is due to thermal wake of the upstream package, and $T_{\mathrm{a,m}}$ is defined by eq (2.24). It is to be noted that with this correction, ΔT_a, the local air temperature rise in eq (2.1), is thus equal to $T_{\mathrm{a,m}} + \Delta T_{\mathrm{TW}} - T_o$.

There are no known analytical expressions for the value of ΔT_{TW}. Arvizu and Moffat (1982) presented data describing a thermal wake function behind a single heated element in a regular array. In their original studies and presented results, the temperature wake function data were obtained with only one of 100 elements heated. Therefore, there was no discernible rise in the mean temperature, and there were only three temperatures to observe: the heated element, the observed element and the air temperature. Hence, the original definition for the thermal wake was the temperature increment above the air inlet temperature, and not above the mean air temperaure. A discussion of the Arvizu and Moffat results and the way in which these results can be used in predicting the adiabatic temperature of an electronic component was presented by Moffat and Anderson (1990). The original definition was modified so that ΔT_{TW} was now defined as the temperature increment above the local mean air temperature. In this work, the thermal wake is expressed as the difference between the component adiabatic temperature and the mean temperature of the cabinet cross-section at which the component is located.

The thermal wake factor decreases rapidly with the distance from the heated element. Moreover, the value of this correction factor is usually small when compared with the mean air temperature rise above the external ambient. For components up to 25 mm in size, placed on the board in in-line arrays, with array spacing approximately twice the component dimension, the proposed simple empirical correlation for ΔT_{TW} was provided by Witzman et al (1990)

$$\Delta T_{\mathrm{TW}} = \frac{(q_d - 0.5)C_{\mathrm{TW}}}{D} \tag{2.26}$$

where D is the number of "rows" the component of interest lies downstream of the dissipating component, q_d is the power dissipation in watts of the upstream component which influences the downstream components and C_{TW} is an experimental constant expressed in units of °C/W. The value of C_{TW} has been found to decrease with increasing air velocity (typical values are 3 for $v = 1\,\mathrm{m/s}$, 2 for $v = 2\,\mathrm{m/s}$ and 1 for velocities higher than $v = 3\,\mathrm{m/s}$).

2.5.2 CONDUCTION EFFECTS

The thermal wake correction factor accounts for the interaction between components that takes place via the air flow. At the same time, with more than one component mounted on the same board, a significant interaction between components can result from conduction in the board. There are two preferred ways in which the interaction due to conduction with other components on the board can be taken into account.

The first of these is computing, at the location of the considered component, the adiabatic temperature using the superposition of temperature fields produced by multiple sources on the same substrate. According to this methodology, the adiabatic temperature of the component is represented by the temperature distribution $T(x, y)$ at any point along the component border. In turn, this temperature is equal to the sum of two temperature distributions, $T_1(x, y)$ and $T_2(x, y)$. In this formulation, $T_1(x, y)$ represents the temperature field produced on the board by the component under consideration and $T_2(x, y)$ represents the temperature field produced by the surrounding components. Each component has its local convection heat transfer coefficient computed from eq (2.20). The temperature field created by each component can be expressed in terms of the modified Bessel functions. The values of the temperatures obtained along the component border are superimposed to produce the case temperature of the package. Due the rapid increase in computational complexity with the increase in the number of heat sources, the use of this methodology is viable only when a small number of components are interacting. The second method for addressing the conduction effect is based on estimating the "cooling area" of the component as a portion of the total board area. In this approach, a circular fin of radius, L, is associated with every package, and it is used to compute the value of Φ for each component using eq (2.23).

Consider a component mounted on a PCB carrying a geometrically uniform array of n components dissipating identical power and cooled by convection. Because of the symmetry, for the determination of the temperature of a single component, it is sufficient to solve for the temperature field for one single component having a "cooling fin" the size of the total board area divided by n. If the local ambient temperature is selected as the reference temperature in this computation, the result is the adiabatic temperature rise of the component, ΔT_{ad} or ΔT_{ja}. Because the air temperature varies along the surface of the board, the adiabatic component surface temperature is found by summing the adiabatic temperature rise and the air temperature at the component location.

The same idea can be extrapolated to arrays that are non uniform with respect to geometry or power [Witzman et al (1990)]. Each component can be treated as an independent entity, a heat dissipater mounted on a board "coupon". The size of the "coupon" should be proportional to the ratio between the component power dissipation and the total power dissipated on the board. Each coupon has an associated heat transfer coefficient that is determined for the component mounted on the coupon. Considering that the temperature decreases very rapidly with the distance from the component (Fig 2.11), relatively large errors in predicting the coupon size (L) will have only a minor impact on the overall accuracy of the end result.

When a number of components dissipating significant power are concentrated in a small area of the board, a more correct calculation of the coupon size is needed. A better estimate of the coupon size can be made by considering only the components situated in the immediate proximity of the associated package. The length of the coupon [L in eq (2.23)] will equal the distance between the two components, multiplied by the ratio between the powers dissipated by each component.

$$\frac{L_1}{L_2} = L_T \frac{q_{d1}}{q_{d2}} \tag{2.27}$$

where L_T in this equation represents the distance between the two components (center to center) and q_{d1} and q_{d2} are the power dissipations and L is the coupon size for components-1 and -2 respectively.

Decreasing the distance between components will have a relatively small impact on the C_i values. But for accurate predictions these values should be corrected for different board population densities. It is important to note that a unique empirical value for C_i should be determined for a component mounted in the first row of the system (at the inlet), and for non-square components. The value of this correction factor is also sensitive to the component orientation.

2.5.3 AN EMPRICIAL CORRELATION

A pragmatic, but very efficient approach, especially for designers who deal with a relatively small number of similar products, is to approximate the adiabatic temperature rise by a relation of the form

$$\Delta T_{ad} = \frac{q_d B \left(v^n a^x k^y b^z L^w\right)}{C_i} \tag{2.28}$$

and experimentally determine the values of the parameters B, n, x, y, z and w, for this equation. It is to be noted that C_i is the same parameter which appears in eq (2.20). The determined values can be extrapolated to a very large number of applications. The accuracy of this relation is dependent on the degree of similarity between the test conditions and the field conditions.

For vertical, single sided, boards carrying sparse component arrays, where the distance between packages is larger than the package size, and where the components are mounted inside electronic cabinets, the values

$$B = 0.078,\ n = -0.334,\ x = -0.815,\ y = -0.581,$$

$$z = -0.5 \quad \text{and} \quad w = 0$$

for the experimental constants are recommended. All of the physical properties and dimensions used in eq (2.28) are expressed in SI units. For densely populated component arrays, where the distance between packages is smaller than the package size, the comparable values are

$$B = 2.127,\ n = -0.574,\ x = -0.815,\ y = -0.253,$$

$$z = -0.5 \quad \text{and} \quad w = -0.58$$

The proposed correlations are valid for air velocities between 0.5 and 4 m/s. The expected error, for most of the existing technologies, are very small (within a ±3°C range), unless major difference exist between the component or substrate technologies used in the test conditions and the field conditions. It is recommended to test for accuracy, and eventually calibrate, these correlations, for any new component and/or substrate technology introduced into the manufacturing line. A similar procedure could be used to correlate the results of several numerical thermofluid simulations for the configuration of interest.

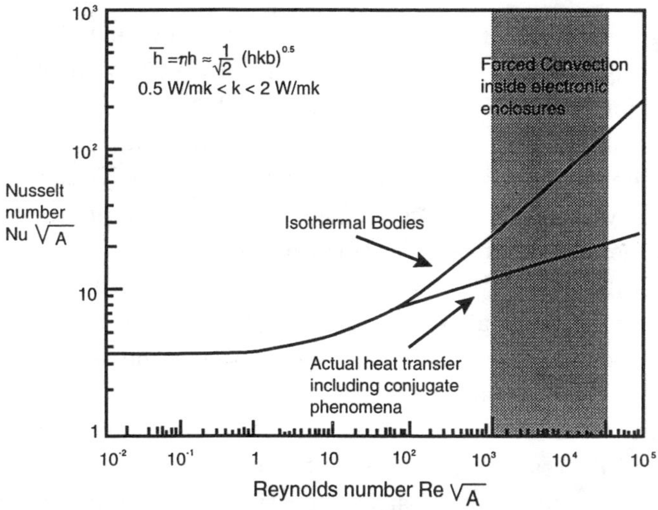

Figure 2.15: Actual (effective) heat transfer coefficient, including conjugate phenomena, as a function of air flow.

2.6 ENHANCEMENT METHODOLOGIES

If the idea that the component heat dissipation is controlled by the thermal plume around the device is accepted, it is reasonable to consider that dispersion, spreading or cooling of this plume, will reduce the component thermal resistance. Simple ways of achieving this result include the use of air movement in the form of forced convection and air impingement jets, the use of conductive substrates and extraction of heat from the plume by a non-contact heat sink. These techniques are easy and inexpensive to implement. In most cases they can produce a reduction of 5 to 10°C in chip temperature. This is often the difference between success or failure for a specific product design.

2.6.1 INCREASING THE LOCAL AIR VELOCITY

It has been shown that the heat transfer coefficient from a component to the air stream is a local phenomenon and is practically independent of the channel flow conditions. The characteristic dimension which best correlates the local heat transfer conditions is the square root of the external area of the component. For suspended heated bodies, the sensitivity of Nusselt number to increasing air velocity is quite high; it is proportional to the square root of Reynolds number, when the local Reynolds number varies between 0 to 3000. Unfortunately, for components mounted on a substrate with an effective thermal conductivity between 0.7 to 2 W/m-K, as is the case for printed circuit boards, the sensitivity of the local Nusselt number to the air velocity is reduced. Figure Fig 2.15 indicates that for air velocities from 1 to 4 m/s, this sensitivity is reduced to the 0.3 power [Graham and Witzman (1988)]. The decrease in sensitivity of the package thermal resistance with the increase in local air velocity should be associated with the decrease of the surface thermal efficiency of the assembly consisting of the board and component surface. As the convection coefficient increases, the thermal resistance of the fluid decreases and more heat is dissipated directly from the package. For a constant substrate effective conductivity, less heat is available for spreading into the substrate and therefore the heated area of the board in the vicinity of the component is reduced. For velocities above 5 m/s, any increase in the convection coefficient due to an increase in air velocity is practically matched by a decrease in the heated substrate area that participates in heat transfer. Thus, the overall component adiabatic thermal resistance remains practically the same, even though the local air velocity is increased.

In components showing large internal thermal resistance, silicon chip temperatures can be expected to show an even lower sensitivity to the increase in local air velocity. An example of the variation of silicon temperature with the increase of approaching air velocity is shown in Fig 2.16 where A, B, C and D represent different size packages of different technologies and dissipation, mounted on the same board that is tested inside of a wind tunnel. As the air velocity increases, the silicon temperature decreases significantly for the B and D packages that have a relatively low internal thermal resistance (under 9 °C/W). The temperature remains practically unchanged for the A and C packages that have a high internal thermal resistance (higher than 40 °C/W). It is important to mention that the measurements presented here refer to the component adiabatic temperature rise (junction to local ambient temperature rise) only. In most situations, the local component ambient temperature inside of the cabinet (the component adiabatic temperature) does not remain constant, but varies from one location to another, as a function

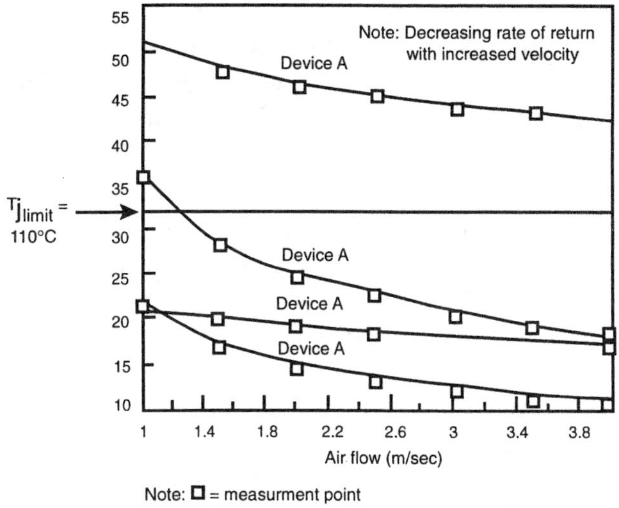

Figure 2.16: Variation of the chip temperature with approaching air velocity.

of the total power dissipated inside the enclosure (upstream from the studied component).

Even though increasing the air velocity is not an efficient way of improving the local heat transfer, in some cases, such as when power density on the board exceeds 90 W/m^2, large quantities of air must be circulated through the system in order to reduce the cabinet air temperature and the adiabatic temperature of the component. Impingement cooling, a method of blowing air at high velocities over a component, has the advantage of using cool air from outside the cabinet instead of air that has been heated upstream. Laboratory results with air jets directed at the top of the components show an improvement of at least 15 percent in the local heat transfer coefficient when compared with channel flow of the same velocity. Unfortunately, the mechanical complexity of jet impingement cooling and the additional volume occupied by the manifold restrict the use of this option to very limited applications.

2.6.2 CONDUCTIVE SUBSTRATE

Except for a few applications (notably in avionics) where conduction is the primary cooling mode, the conductivity of the substrate is rarely high enough to remove the heat from the component to the ultimate heat sink solely by conduction. For sparse arrays, the most common benefit of a conductive substrate is to spread the heat or thermal footprint created by the component to a larger area from which it can be more readily transferred to the passing air. In this way, the substrate can be considered as a cooling fin to which the rules for conjugate convection-conduction can be applied [Witzman et al (1990)]. Experimental, as well as analytical studies, indicate that for sparse arrays, the component external thermal resistance decreases with approximately the square root of the increase in board thermal conductivity [Graham and Witzman (1988)]. This rule of thumb is applicable to arrays in which the distance between components dissipating heat is larger than the size of the package and when the board effective conductivity varies between 0.7 to 2 W/m-°C. For densely populated boards and highly conductive substrates, the impact of increasing board conductivity is less obvious and sometimes controversial.

A significant problem associated with the thermal contribution of substrate conduction, is the lack of data for the effective thermal conductivity of the board. The existence of continuous copper tracks or even copper layers inside of FR-4 boards allows for relatively easy spreading of the heat laterally, inside of the substrate. Lateral conductivities of 30 or even 40 W/m-°C are easily achieved. On the other hand, the heat has to overcome relatively high thermal resistances (conductivities of 0.3 W/m-°C or less) in order to penetrate the board or return to the board surface from which it is convected to the air stream. An additional degree of complexity is added if more than one heat dissipater is mounted on the board. In this case, the copper layers can become almost isothermal and, in some cases, approach the temperature of the package under consideration. For these reasons, the equivalent conductivity of the board, as defined here, is much lower than the lateral thermal conductivity of the board.

Existing numerical software can be used in order to predict the board participation in the heat dissipation process. An example of such computation is presented by Nakayama and Park (1996). In this example, the authors use homogeneous substrate materials, but similar computational techniques can be used for multilayer or composite materials. A possible technique for predicting the effective thermal conductivity would be to predict the component temperature using a similar numerical simulation in which the real board structure is accurately modeled and to then determine the equivalent thermal conductivity of an homogeneous substrate that would produce the same temperature rise under the same flow conditions.

A simplified method for the estimation of the effective board conductivity is to assume that only the copper planes, usually the ground planes, are contributing to the heat dissipation from the component. In this computation, the contribution of the epoxy layers to the lateral spreading of the heat is ignored with these layers considered as conductive in the z direction only. Each layer of copper is individually connected to the heat source (the component), through a chain of thermal resistances. These resistances represent the resistance of the air gap and the resistance(s) of the low thermally conductive layers that are covering the conductive layer under consideration. The conductive layers that are closest to the surface will receive a larger quantity of heat than the conductive layers imbedded inside the substrate which are closer to the surface opposed to the side to which the component is mounted. This means that the heat is spreading laterally inside of the conductive layer while being transferred by convection to the air surrounding the component.

If samples of the proposed substrate are available, the effective thermal conductivity of the board can be estimated using a simple experimental model. A coupon carrying a component in which a thermal die is packaged should be tested inside of a wind tunnel and the temperature of the thermal die monitored. Using the methodology presented in this chapter, the effective thermal conductivity of the board under consideration can be estimated. Some recommended values of the equivalent (actual) board thermal conductivities that should be used in conjunction with this computation method are shown in the tabulation where the values have been computed and have been experimentally corrected for boards having embedded copper ground planes. Significant improvements (up to 30 percent) in substrate effective thermal conductivity (FR-4 technology) can be achieved by bringing the ground plane closer to the board surface.

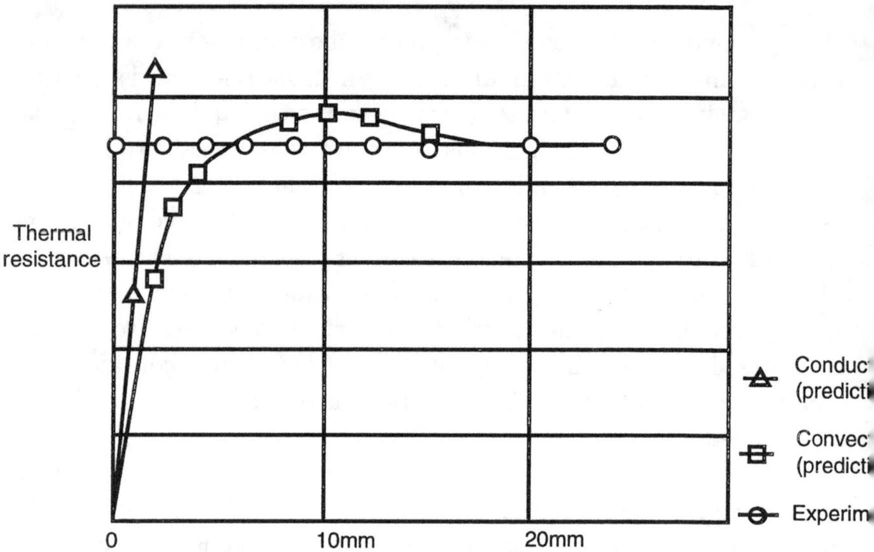

Figure 2.17: Heated plate to ambient thermal resistance, as a function of the distance between the plate and isothermal surface.

Number of Layers	Number and Type of Ground Planes and Copper Weight	Board Conductivity W/m-°C
2	none	0.4
4	1, Solid (2 oz)	1.4
8	1, Solid (1 oz)	1.25
8	1, Lattice (2 oz)	0.8
10	2, Solid (1 oz each)	1.9
10	2, Lattice (1 oz each)	1.0

When copper plating or a lattice is left on the surface a printed circuit board to enhance the heat spreading capabilities, the effective conductivity will be different than the experimental values presented. The effective conductivity for FR-4 based boards having significant unetched copper islands in the area in which the component is placed can be estimated using the correlation

$$k_{\text{eff}} = \frac{(\Lambda k b)_{\text{copper}} + (kb)_{\text{FR-4}}}{b_{\text{total}}} \tag{2.29}$$

where

> Λ is the percent copper surface coverage (for lattices), dimensionless
>
> k is the thermal conductivity and
>
> b is the material thickness.

The values of the effective thermal conductivity for FR-4 technologies presented in the foregoing table are informative only. Slight variations in the amount of copper or the position of the ground planes inside of the structure might produce different values for the effective board thermal conductivity. Even though it is not expected that these variations will produce significant errors in the preliminary estimations of the component temperature rise, it is recommended that a check be made to see if these values are applicable for the board technology intended to be used in the specific application.

2.6.3 THE THERMAL PLUME

Most of the package external (case to ambient) thermal resistance is concentrated inside of the thermal plume surrounding the package. It is logical to assume that that the introduction of a conducting body inside of the plume will reduce the overall resistance by conducting the heat away from the component. As previously noted, experimental results have indicated that this plume is not readily affected by the low velocity flows encountered inside electronic enclosures. Conceptually, it is expected that conducting the heat from the boundary layer with a non-contact thermal spreader might be a more efficient thermal management approach.

As shown in Fig 2.17, the thermal resistance between two isothermal plates (one hot and one cold) immersed in a fluid is a function of the distance of separation, up to a certain critical value. This behavior can be understood by considering the combination of conduction and the convection that takes place inside the fluid gap. It is expected that electronic components mounted on the board and brought close to a cold enclosure wall will show similar behavior. When the distance between the component and the wall is decreased below the 15 mm limit, the component temperature increases slightly

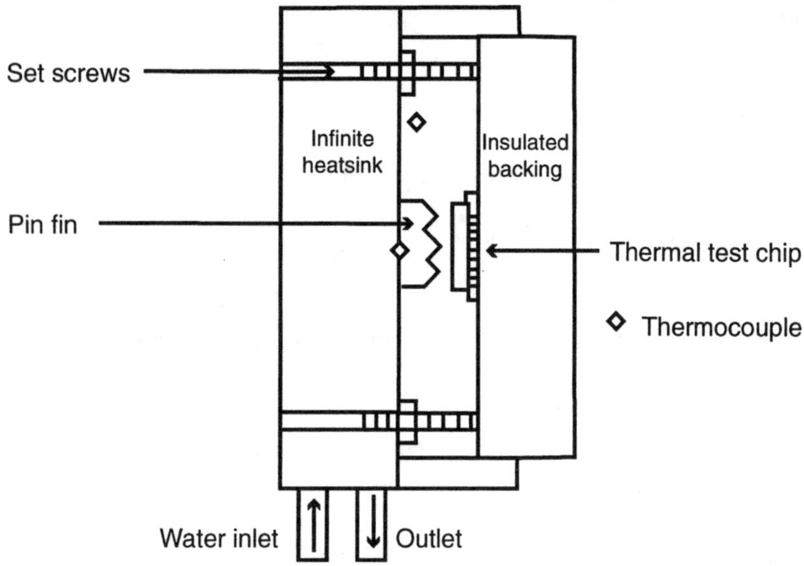

Figure 2.18: Non contact heat spreader testing apparatus.

before decreasing sharply when the separation drops below 5 mm. This cooling phenomenon is associated with a similar drop in the adiabatic surface temperature rise when conductive bodies are inserted inside the package thermal plume. Alternatively, the increase in plate temperature when the distance between the heated wall and the isothermal surface is 5 to 15 mm, can be explained by the increase in air temperature between the plates due to reduced convection.

An experimental program in which different isothermal finned surfaces were introduced inside the thermal plume of chip packages was carried out in order to determine the optimum configuration in terms of spacing, size and orientation. The non-contact heat sinks were maintained at a constant temperature of 20°C by circulating water through internal channels machined into their bodies. The materials used for the heat sinks were aluminum and copper. A sketch of the experimental set-up is shown in Fig 2.18. The experimental results show that the flat plate having the same size as the package top surface outperformed all other geometries for the condition of identical minimum distance between the component and the heat sink. This indicates that the resistance of the air gap increases very rapidly with the distance from the heated surface as the increase in the total surface area of the heat sink is unable to compensate for the decrease in the gap thermal conductivity. This result also proves that diffusion is the dominant phenomena controlling the component thermal resistance.

2.7 EXAMPLES

The methodology described in this chapter can be used to provide rapid and relatively accurate estimates of the silicon temperature rise and other key thermal parameters in a multitude of situations for which, otherwise, complicated programming and computational tools might be required. Moreover, to obtain meaningful results, these programs must be supported by a large library of experimental values and physical constants. These are usually not readily available to the designer at the time the predictions are required. The situations of interest involve packaging selection, design optimization, modifications of existing systems and concurrent design problem solving.

2.7.1 EXAMPLE 1

Intent: The determination of whether a selected chip package will meet the thermal specification.

Thermal Input Data (From the Product Specifications)

- Maximum air inlet temperature: 40°C.
- Minimum air flow through the channel between the boards: $0.5 \times 10^{-3}\,\mathrm{m^3/s}$.
- Minimum air velocity through the enclosure: $\hat{V} = 1\,\mathrm{m/s}$.
- Board size & location: $100 \times 100\,\mathrm{mm}$, 20 mm from other surfaces.
- Estimated board power dissipation: 10 W, almost uniformly distributed.
- Board characteristics: two layers, no ground plane, FR-4, vertical, 1.4 mm thick.

- Die power dissipation (for the component under consideration): 0.8 W.
- Maximum power dissipation by the nearest upstream component: 1 W.
- Maximum allowable silicon temperature: 100°C.

Proposed Solution: The distance between the studied board and other parallel surfaces inside the cabinet, such as external walls or other boards, is larger than 15 mm. Therefore the impact of adjacent surfaces on the component thermal boundary layer is very small and it can be ignored. Considering the die size, the electrical requirements and the board real estate, a 14 mm, 64 lead plastic-quad-flat-package (PFQP) might be a possible candidate. For a die size of 250 mils (6.35 mm) mounted face-up on a 300 mil (7.62 mm) pad copper lead frame, the internal temperature rise, measured using the fluid immersion technique (with the device mounted on a 2-layer FR-4 50 × 50 mm coupon) shows an internal thermal resistance of 19 ± 1.5°C/W. In most cases, the die to case thermal resistance is provided by the vendor as part of the the product packaging specification. Here, the vendor's provided documentation shows an expected junction-to-case thermal resistance of 20°C/W.

Using eq (2.1), the device temperature can be expressed as

$$T_d - T_o = \Delta T_a + \Delta T_{\text{int}} + \Delta T_{\text{ext}} \tag{2.1}$$

As the maximum acceptable temperature of the die is 100°C when the external ambient temperature is 40°C, an extimate of the maximum acceptable temperature rise through the component boundary layer is

$$\Delta T_{\text{ext}} = (100 - 40) - \Delta T_a - \Delta T_{\text{int}}$$

The reader is reminded that ΔT_a, the temperature rise of the local air surrounding the component under consideration, is the sum of the mean air temperature rise and the thermal wake factor given by eq (2.24). As a first approximation, the mean air temperature, ΔT_a, designated as $\Delta T_{\text{a,m}}$, can be estimated using a simple heat balance equation that assumes that the heat dissipated inside the channel formed between the boards is absorbed by the volumetric flow rate, $\hat{V}$, of the air with density, $\rho = 0.998\,\text{kg/m}^3$ and specific heat, $c_p = 1008\,\text{J/kg-K}$.

$$\begin{aligned}\Delta T_{\text{a,m}} &= \frac{q_b}{\hat{V} c_p \rho} \\ &= \frac{10}{(0.5 \times 10^{-3})(1008)(0.998)} \\ &= 20^{\circ}\text{C}\end{aligned}$$

This value represents the maximum temperature rise inside of the channel. Devices mounted toward the entrance of the channel on the bottom side of the board will see a lower mean air temperature. It is recommended that, at this stage of the design, the maximum air temperature rise inside of the cabinet is taken as $\Delta T_{\text{a,m}}$. This choice is based on the fact that the layout of the board is not known and that the component may be placed in the hottest region of the cabinet. For situations in which the geometry of the cabinet is known and the power dissipation inside of the cabinet is given, a more accurate estimation of this term can be obtained by using a subroutine that predicts the temperature of the region in which the component willl be placed. Returning now to eq (2.25)

As the thermal wake is expected to decrease significantly after the second row of components, in this first estimate, it can be assumed that only the thermal wake coming from the nearest component upstream has a significant impact on the subject component. Consequently, for a power dissipation of 1 W per component and $C_{TW} = 3$ for an estimated air velocity of 1 m/s, using eq (2.26)

$$\begin{aligned}\Delta T_{TW} &= \frac{(q_d - 0.5)C_{TW}}{D} \\ &= \frac{(1 - 0.50)3}{1} \\ &= 1.50^{\circ}\text{C}\end{aligned}$$

The internal temperature rise of the package is found, in similar fashion, to equal

$$\Delta T_{\text{int}} = q_d R_{\text{jc}} = (0.80\,\text{W})(20\,^{\circ}\text{C/W}) = 16^{\circ}\text{C}$$

By inserting these values into eq (2.1) and rearranging terms, the allowable external temperature rise is found to

$$\Delta T_{\text{ext}} = 100 - 40 - 20 - 1.50 - 16 = 22.5°\text{C}$$

Consequently, ΔT_{ext} must be smaller than 22.5°C to meet the targeted chip temperature of 100°C for an external ambient temperature of 40°C.

Using the methodology described in the foregoing and represented by eqs (2.20) and (2.11) through (2.14), one can estimate the value of the external temperature rise of the component. As the exact value of C_i might not be available, the default value of unity is used. The non-dimensional adiabatic heat transfer coefficient or Nusselt number for this configuration is given by eq (2.20)

$$\text{Nu}_{\text{ad}} = \text{Nu}_{\text{D}} + C_i(\text{Nu}_{\text{fc}} + \text{Nu}_{\text{nc}}) + \text{CR} \tag{2.20}$$

Begin with eq (2.11)

$$\text{Nu}_{\text{D}} = 2\sqrt{\pi} = 3.545$$

To calculate the value of Nu_{nc}, it is necessary to find the value of the Prandtl number of air, as well as the values of β and ν which are part of the Grashof number. For air at 47.5°C = 320.5 K

$$\begin{aligned} \text{Pr} &= 0.70 \\ \beta &= 3.12 \times 10^{-3}\,\text{K}^{-1} \\ k &= 0.0272\,\text{W/m-K} \end{aligned}$$

and

$$\nu = 17.95 \times 10^{-6}\,\text{m}^2/\text{s}$$

The Grashof number is

$$\text{Gr} = \frac{g\beta(T_c - T_{\text{a,m}})x^3}{\nu^2}$$

where x is the characteristic length of the device. As already considered, x is the square root of the device total area (including the bottom) which is 22.5 mm for a 14 mm square package with a 2 mm height. With an expected $T_c - T_{\mathrm{a,m}}$ of 30°C,

$$\mathrm{Gr} = \frac{9.8(3.12 \times 10^{-3})(30)(22.5 \times 10^{-3})}{(17.95 \times 10^{-6})^2} = 32,440$$

and the Rayleigh number is

$$\mathrm{Ra} = \mathrm{GrPr} = (32,440)(0.70) = 22,700$$

Hence, by eq (2.13)

$$\begin{aligned} \mathrm{Nu_{nc}} &= 0.694\mathrm{Ra}^{0.25} \\ &= 0.694(22,700)^{0.25} \\ &= 8.52 \end{aligned}$$

Turning to forced convection, for the stated conditions, the Reynolds number is

$$\mathrm{Re} = \frac{\hat{V}x}{\nu} = \frac{1(22.5 \times 10^{-3})}{19.95 \times 10^{-6}} = 1253$$

and by eq (2.13)

$$\begin{aligned} \mathrm{Nu_{fc}} &= 0.825\mathrm{Re}^{0.50}\mathrm{Pr}^{0.33} \\ &= 0.825(1253)^{0.50}(0.70)^{0.33} \\ &= 25.93 \end{aligned}$$

The overall adiabatic Nusselt number is found from eq (2.20) with $C_i = 1.00$ and CR $\rightarrow 0$

$$\begin{aligned} \mathrm{Nu_{ad}} &= \mathrm{Nu_D} + \mathrm{Nu_{fc}} + \mathrm{Nu_{nc}} \\ &= 3.55 + 25.93 + 8.52 \\ &= 38.00 \end{aligned}$$

and the associated heat transfer coefficient is

$$h_c = \frac{k}{x}\mathrm{Nu_{ad}} = \frac{0.00272}{0.0225}38.00 = 45.94\,\mathrm{W/m^2\text{-}K}$$

As the surface of the component is estimated to be approximately 30°C above the mean temperature of the enclosure, a significant amount of heat is transferred between the component and the enclosure through radiation via a radiation heat transfer coefficient

$$h_r = \frac{5.67 \times 10^8 \epsilon F \left[\left(\frac{T_c + 273}{1000} \right) - \left(\frac{T_{\text{a,m}} + 273}{1000} \right) \right]}{T_c - T_{\text{a,m}}}$$

where ϵ is surface emissivity and F is the surface view factor.

For $\epsilon = 0.80$ and $F = 1.00$, the value of the radiation heat transfer coefficient is evaluated as

$$h_r = 7.64\,\text{W/m}^2\text{-K}$$

and when this value of h_v is combined with h_c, the total heat transfer coefficient becomes

$$h_{\text{ad}} = h_c + h_r = 45.94 + 7.64 == 53.58\,\text{W/m}^2\text{-K}$$

and the thermal resistance associated with the heat transfer directly from the component surface to the air stream can be determined from eq (2.19) with a surface efficiency of $\eta = 0.85$, a component surface area of $A_a = 308 \times 10^{-6}\text{m}^2$ and a component size of 14 mm. Thus,

$$R_a = \left[h_{\text{ad}} \eta \left(\frac{\pi a^2}{4} + A_a \right) \right]^{-1} \tag{2.19}$$

can be evaluated to show that

$$R_a = 47.6°\text{C/W}$$

At this design stage, the lay-out of the board has not yet been determined. Therefore, there is not enough data to correctly estimate the thermal influence area of the device. A recommended practice in cases such as these is to consider that the packages are as close together as the board technology might allow. In this situation, the manufacturable best array will be with the components spaced one component length apart. Considering that the power dissipation is almost the same for all of the components on the board, the distance of separation, L, is equal to

$$L = a = 14\,\text{mm}$$

As mentioned previously, unless the components are very tightly packed or the substrate has a high thermal conductivity (values above 3 W/m-K), relatively large errors in estimating the value of the parameter, L, will have a relatively low impact in the overall accuracy of the results.

The board thickness is 1.4 mm and the effective board conductivity for this technology, as shown in Section 2.6.2 is estimated at 0.4 W/m-K. The PCB "fin" performance parameter is

$$m = \left(\frac{2h_{\mathrm{ad}}}{kb}\right)^{1/2} = \left(\frac{2(53.58)}{0.40(0.014)}\right)^{1/2} = 437.7\,\mathrm{m}^{-1}$$

so that

$$\alpha = mL = 6.12 \quad \text{and} \quad \beta = \frac{am}{2} = 3.06$$

Using these values and Fig 2.12 (or Tables of Bessel Functions) allows Φ to be estimated as $\Phi = 1.07$. Then, the package thermal resistance for heat flow to the PCB is in accordance with eq (2.21)

$$\begin{aligned} R_b &= (\pi mabk\Phi)^{-1} \\ &= [\pi(437.7)(0.014)(0.0014)(0.40)(1.07)]^{-1} \\ &= (0.011535)^{-1} \\ &= 86.7\,\mathrm{K/W} \end{aligned}$$

By combining R_a and R_b as two parallel paths from the die to the air, the total local thermal resistance of the component can now be estimated according to eq (2.16) as

$$\begin{aligned} R_{\mathrm{ext}} &= \frac{R_a R_b}{R_a + R_b} \\ &= \frac{47.6(86.7)}{47.6 + 86.7} \\ &= 30.73\,\mathrm{K/W} \end{aligned}$$

and for a component dissipating 0.80 W, the sought after external temperature rise is now seen to equal

$$\Delta T_{\mathrm{ext}} = q_d R_{\mathrm{ext}} = 0.80(30.7) = 24.6^{\circ}\mathrm{C}$$

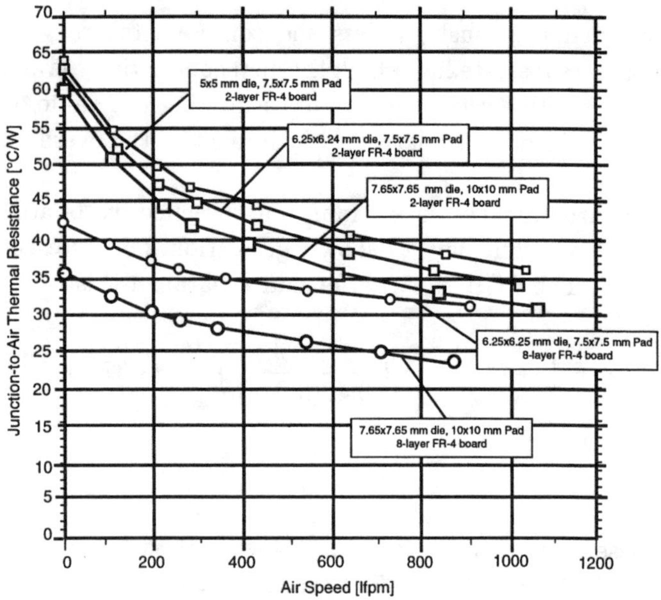

Figure 2.19: 14 mm PQFP thermal characteristics.

As established at the beginning of this example, in order to satisfy all of the imposed requirements, this value should be smaller than 22.5°C. Therefore, this package does not meet the requirements of a maximum die temperature of 100°C under ther specified conditions. The computation also points out that, if the cabinet conditions cannot be modified, a larger package (18 mm) or a package with a lower internal thermal resistance (10°C/W) should be selected.

It is important to mention that without using the C_i experimental correction factor, all of the packages with the same dimensions will yield the same value for R_{ext}. This is inconsistent with experimental results. In reality, due to the differences in how heat spreads on the component surface, small differences ususally exist between different packaging technologies The measured value for the junction to ambient thermal resistance for a similar package working in similar conditions (see Fig 2.19) was 49°C/W. Considering an internal thermal resistance of 20°C/W, the measured case to to ambient thermal resistance was 29°C/W where the C_i value for this technology package was estimated as 1.11. For 0.8 W power dissipation, the measured value for ΔT_{ext} was 23°C.

The value of C_i is close to 1.00 for ceramic and thin plastic packages containing large die or internal heat spreaders. Even though the accuracy achieved is acceptable for a first estimate, and it is generally small when dealing with components dissipating less than 1 W, significant errors (±15 percent) might occur if technologies involving small packages with highly non-uniform surface temperature and high power dissipation are considered. If experimental correction factors are available, the accuracy of the computation can increase to ±0.5°C for any existing packaging technology. Another source of error is also associated with the estimate of the effective thermal conductivity. Test boards tend to have more copper than real PCB's and, therefore, a higher effective thermal conductivity than was assumed for this computation.

A very common way of avoiding the lengthy computation for the external temperature rise is to produce diagrams showing the variation of the component thermal resistance with the variation of air velocity. An example of such a diagram is provided in Fig 2.19. The diagrams show the variation of the junction to air thermal resistance as a function of the approaching air velocity and are usually available in the vendor catalogs.

A note of warning with regard to the use of such diagrams produced by another party is associated with the lack of knowledge of the experimental conditions under which they were produced. For example, most of the measurements are made at higher power and in open cabinets or inside wind tunnels. Under such conditions, the contribution of the radiation factor may be quite different from the conditions inside of the real cabinet. Indeed, in most cases the type of substrate used when the measurements were performed is not specified. This might be the source of significant errors unless computational methods to account for such differences are available.

2.7.2 EXAMPLE 2

Intent Investigation of the minimum air velocity requirement.

In most applications, especially in those involving cabinets with forced air cooling, it is simpler to modify the air flow inside the cabinet than to change package size or technology. The beneficial effect of increasing the amount of air passing through the cabinet is related to two aspects:

1. Reduction of the mean air temperature inside the cabinet. This phenomena is associated with the increase in the heat capacity of the air flowing through the cabinet.

2. Increase of the adiabatic heat transfer coefficient, followed by a reduction of the case-to-ambient thermal resistance. This is usually associated with an increase of local air velocity in the area in which the component is located. A further benefit associated with the increase in air velocity is the reduction of the size and temperature of the thermal wake due to the more thorough mixing of the of the air adjacent to the component.

A consideration of the results from the previous example can be used to reduce the temperature of the component by increasing the air velocity around the device under consideration. It is thus necessary to satisfy the condition that the device temperature not exceed 100°C when the selected package is a 14 mm PQFP and the power dissipated by the component is 0.8 W.

For simplicity, it is assumed that no increase in the volumetric flow of the air is possible. This can be a real constraint that is associated with limitations placed upon the air handling system. Under such limitations, some improvements can be achieved by diverting the air towards the component under consideration with the help of baffles placed inside the cabinet. This will result in a higher air velocity in the vicinity of the component. Assume, for example that this velocity doubles from 1 to 2 m/s. If there is no mass flow change, the maximum mean air temperature rise will remain unchanged at $\Delta T_{\mathrm{a,m}} = 20°\mathrm{C}$.

The thermal wake will decrease with the increase in local air velocity. Under the assumption that the local air velocity is doubled to 2 m/s, the new estimated value for ΔT_{TW} is

$$\Delta T_{\mathrm{TW}} = \frac{2(1 - 0.50)}{1} = 1°\mathrm{C}$$

A computation that is similar to the one in Example 1 yields a case to ambient thermal resistance of $R_{\mathrm{ext}} = 26°\mathrm{C/W}$ instead of the

$R_{\text{ext}} = 30.7°\text{C/W}$ for a velocity of 1 m/s. The resulting temperature rise is now 21°C and, with this value, the temperature rise of the die dissipating 0.8 W is expected to be

$$T_d - T_o = \Delta T_a + \Delta T_{\text{CW}} + \Delta T_{\text{int}} + \Delta T_{\text{ext}} = 20 + 1 + 16 + 21 = 58°\text{C}$$

For an ambient temperature of 40°C, the silicon die can now be expected to reach a temperature of about 98°C which is acceptable.

If a diagram similar to the one presented as Fig 2.19 is available, one can easily estimate the necessary local air velocity for meeting certain imposed conditions. In Fig 2.19, the improvement in R_{ja} from 47.5 K/W at 1 m/s to 42 K/W at 2 m/s is clearly visible. It can also be seen that continued increase of the local air velocity leads to a resistance asymptote. An increase above a certain air velocity (say 3 m/s) will produce an insignificant reduction of the external temperature rise of the component.

2.7.3 EXAMPLE 3

Intent Estimation of the impact of design modifications in existing systems.

Recognizing the common desire to upgrade existing equipment, on numerous occasions the designer must decide if certain types of components or electronic packs can be installed in existing enclosures. The modifications might involve the replacement of a certain component with a new one, usually of higher dissipation, replacement of the printed circuit board with another substrate, increasing the power dissipation inside the channel or changes in the fan system. Most of the time, more than one change is performed.

A very common question asked under these circumstances is how these modifications will impact the component temperature. Obviously, the methodology presented here can be used to produce a new prediction for the entire cabinet or for a single component. This is not a very efficient way of solving this problem, especially when experimental data related to the thermal performance of the existing cabinet/component version already exists. Under these circumstances, the fastest way of achieving accurate results is the use of a relationship in the form of eq (2.28).

$$\Delta T_{\text{ad}} = \frac{q_d B\left(v^n a^x k^y b^z L^w\right)}{C_i} \tag{2.28}$$

In this equation, B is a system parameter that is the most difficult to estimate and it is a function of numerous factors which are flow and cabinet geometry dependent. However, if the case temperature of the existing component was or can be measured, then the case temperature of the replacement device can be estimated as

$$\Delta T_{\text{ext},2} = \Delta T_{\text{ext},1}\left[\frac{v_1^n a_1^x k_1^y b_1^z L_1^w}{v_2^n a_2^x k_2^y b_2^z L_2^w}\right]\left(\frac{q_{d,1}}{q_{d,2}}\right)$$

In the following example, an upgrade of an existing design requires a design change related to the board discussed in Example 1. The modifications required are:

- A new component, packaged in a 18 mm PQFP package.
- The power dissipation for the new component is 1.5 W.
- The internal (junction to case) thermal resistance for the new device is 17°C/W.
- The board technology is changed from two layer with no ground to six layers with two solid ground planes with of one ounce copper.
- The minimum air velocity is 2 m/s.

With very few exceptions, the spacing between packaged components is equal to or larger than the size of the package. If the components mounted on the board have similar power dissipations, then the length of the thermal coupon area is equal to the device spacing. In these situations, the relationship for sparsely populated boards is recommended. Using eq (2.28) for sparsely populated boards leads to the selection of the following exponents

$$n = 0.334$$
$$x = 0.815$$
$$y = 0.581$$
$$z = 0.500$$
$$w = 0.000$$

The data for the problem is based somewhat on Problems 1 and 2 and is summarized in the tabulation where it may be noted that the estimated board thermal conductivity for the new application is 1.25 W/m-K.

	Original Component	**Replacement Component**
Package size, a, mm	14.0	18.0
Air velocity, $\hat{V}$, m/s	1.0	2.0
Board conductivity, k, W/m-K	0.40	1.25
Board thickness, b, mm	1.4	1.4
Device spacing, L, mm	14.0	18.0
Power dissipation, q_d, W	0.8	1.5
External Temp. Rise, ΔT_{ext}, °C	25	?

The new value of ΔT_{ext} is obtained in a straightforward manner

$$\frac{\Delta T_{\text{ext},2}}{\Delta T_{\text{ext},1}} = \left(\frac{1}{2}\right)^{0.334} \left(\frac{14}{18}\right)^{0.815} \left(\frac{0.40}{1.25}\right)^{0.581} \left(\frac{1.4}{1.4}\right)^{0.50} \left(\frac{14}{18}\right)^{0.0} \left(\frac{1.50}{0.80}\right)^{1.0}$$

or

$$\Delta T_{\text{ext},2} = 25(0.80)(0.81)(0.51)(1.00)(1.00)(1.875) = 15.5°\text{C}$$

Because the configuration of the system is practically the same and the power dissipated inside of the channel is practically unchanged, it can be assumed that the maximum mean air temperature rise is still 20°C. The thermal wake factor computed for a 1 W upstream component power dissipation and a 2 m/s air velocity is 1°C. The internal temperature rise for the new device is

$$\Delta T_{\text{int}} = 17(1.50) = 22.5°\text{C}$$

and for an external ambient temperature of 40°C, the maximum silicon temperature will be

$$T_d = 40 + 20 + 1 + 25.5 + 15.5 = 102°\text{C}$$

Even though the accuracy of the computation is somewhat less than in the previous examples, the calculation technique is highly recommended when some experimental values from previous applications are available.

A note of caution: Although the equations used in Examples 1 to 3, with the default value of unity for the experimental factor, C_i, are expected to produce acceptably small errors for most of the packaging technologies used today, it is recommended that these equations be calibrated to specific conditions if accuracy better than ± 15 percent is required. The errors are expected to be significant for very large packages containing relatively small dies with strongly non uniform package surface temperatures.

2.7.4 EXAMPLE 4

Intent Other Applications of the Proposed Methodology.

If the board described in Example 1 is to be placed in a sealed module, the adiabatic temperature rise $\Delta T_{\text{ad}} = \Delta T_{\text{int}} + \Delta T_{\text{ext}}$ for the component under consideration must be computed for buoyancy induced flow. The local air velocity associated with this situation is 0.4 to 0.7 m/s. Assuming a 0.5 m/s air velocity and following the same methodology used in Examples 1 and 2, the adiabatic temperature rise for a 20 mm CerDIP package, mounted on a single sided, no ground plane board, dissipating 1 W, is estimated to be 29°C. Here, the thermal wake is found to contribute 2°C.

For an external surface area of $A = 0.5\,\text{m}^2$ and total power dissipation inside the enclosure of $q_T = 100\,\text{W}$, the expected mean temperature rise of the air inside the hermetic enclosure is estimated at

$$\Delta T_{\text{ae}} = \frac{qT}{U_{\text{enc}}A} = \frac{100}{8(0.5)} = 25°\text{C}$$

For a maximum external ambient temperature of 40°C (outside the enclosure), the estimated silicon temperature for the component under consideration is:

$$\Delta T_d = 40 + 25 + 2 + 29 = 96°\text{C}$$

Considering that the component maximum acceptable temperature for the worst conditions was set at 100 C, the proposed solution is considered acceptable. This result is expected to be bounded by a ±5°C error.

If complex geometry enclosures are involved, and highly accurate temperature predictions are required, a hybrid computation can be used. The temperature field of the air inside the enclosure (T_{air}) can be determined using appropriate numerical CFD software. To simplify the computation, the heat can be considered to be dissipated uniformly by the boards that hold the electronic components. As the numerical CFD software is used to produce a map of the air temperature inside the enclosure, in order to achieve the component temperature, the adiabatic temperature rise (ΔT_{ad}) computed using the method presented in this chapter should be added to the air temperature at the component location. This method will considerably increase both the speed (when compared with computing the temperature field inside enclosures carrying discrete heat sources) and the accuracy of the computation unless a very fine mesh is used.

2.7.5 EXAMPLE 5

Intent Consideration of Concurrent Thermal and Electrical Optimization of High-Speed Packages and Systems.

A suitable problem formulation for multidisciplinary optimization is the physical partitioning and placement of packages on a common substrate, that is, MCM's or PCB's. The aim is to find a placement configuration which satisfies interconnect delay and noise specifications, as well as package thermal constraints.

While a variety of techniques exist for both electrical and thermal analysis, optimization of designs for thermal performance is commonly done without regard to electrical performance and vice versa.

The consequence is a design optimized for the constraints of one discipline, while suffering from inferior performance in the other domain.

In order to formulate a suitable methodology, two fundamental analysis capabilities are required. From the electrical discipline, a technique developed for interconnect delay estimation is used. Similarly, from the thermal discipline, a methodology is required to obtain package temperature for a given cooling technology. The goal of this optimization is to minimize the thermal/electrical cost function by repeatedly perturbing the placement configuration. Following each perturbation, the cost function is evaluated to determine the feasibility of the solution.

The most important factor in selecting a computational method for the electrical or thermal performance is the required computation time. Taking into account that the computer must search for an optimum (or acceptable) solution from an extraordinary large number of possible combinations, the time allocated to a single performance computation cannot exceed a small fraction of a second. This is far above the computation capability of most of the tools and programs existing today.

The superposition equation, eq (2.1) is an ideal candidate for the efficient computation of the die temperature in an electronic component

$$T_d - T_o = \Delta T_a + \Delta T_{\text{int}} + \Delta T_{\text{ext}} \tag{2.1}$$

The terms ΔT_o and ΔT_{int} are independent of the package position on the board, while ΔT_a, ΔT_{TW} and ΔT_{ext} are dependent upon the placement of the package. The position of adjacent packages causes the dependency of ΔT_{ext} by affecting the available surface area of the board, close to the package, which acts as a cooling fin for the package.

The gain in computation time is associated with the fact that this methodology involves a single computation for the air temperature field ($\Delta T_{\text{a,m}}$). This computation can be done using software that allows for prediction of the temperature field of the air inside the cabinet when the real boards are replaced by uniformly heated plates. Under the assumption that the total power dissipated by the entire cabinet is much larger than the power dissipated by one single component, ($\Delta T_{\text{a.m}}$), at a fixed location, stays practically unchanged when the components are permutated on the board. All other terms in this equation that are dependent on the component location are computed independently, for any new component location, and superimposed on the mean air temperature at the new component location. This reduces the computation time associated with the estimation of the silicon temperature for a new proposed board lay-out to a fraction of a second.

An example of an optimization algorithm for thermal and electrical performance in a high speed electronics package is given by Mihan et al (1995). Although this paper primarily focuses on two disciplines, electrical signal delay and thermal performance, it should be noted that the method is valid for any number of disciplines. Furthermore, the application of this method is geared towards the PCB domain. In general, the method can be applied to any level of the physical hierarchy.

In the example considered by Mihan et al (1995), an area of a printed circuit board containing eight packages was selected as an example. These packages were already placed according to some initial configuration. All of the packages were selected for placement optimization and thermal analysis. In addition, eighteen receiver pins were selected for delay analysis. The required thermal and delay limits were set according to the design specifications. Considering that meeting both the electrical and thermal performance were mandatory, the "weighting factors" were set to unity prior to optimization.

The electrical parameter under consideration was the signal delay. The delay values of interest are referenced to the diver pin of the signal net and are seen at the input of the receiver pins. The delay is defined as the time taken for the voltage at the input of a given receiver pin to reach ninety percent of it's steady state value. This delay is a function of the electrical characteristic impedance of the transmission line which is primarily a function of the geometrical parameters of the line and electrical properties of the materials. As the location of the component on the board is changed, at least one of the geometrical parameters influencing the value of the characteristic impedance, the length of the electrical path, is modified.

Equation (2.1) was used for the estimation the thermal performance of the components. Each term in this equation was computed for all of the components mounted on the board using the methodology presented in this chapter. Some of the terms of this equation (the internal and the external component temperature rise) are independent of the component location, while the air temperature rise is dependent on the board layout and the component location on the board.

An adaptive heuristic optimization algorithm such is the "annealing algorithm" was used to optimize the component placement. The simulated annealing algorithm requires a perturbation function to generate a new random solution starting from an initial solution. In this problem, the perturbation function randomly selects positions for the packages on the board. It is during the implementation of this function that the manufacturing constrains such is component spacing, height restrictions and package alignment must be addressed. In addition, functional and logical grouping of packages must be handled by the perturbation function. This would insure that the optimum solution is indeed realizable by the manufacturing process.

2.7.6 EXAMPLE 6

Intent To consider thermally induced stresses in electronic components.

In the successful design of plastic integrated circuit packages for high performance, and relatively high power, VLSI chips are a crucial element in the development of a cost effective packaging technology. Unfortunately however, most common plastic encapsulating materials provide relatively low thermal conductivities, as well as low mechanical strengths and a limited operating temperature range. As a consequence, plastic IC packages are susceptible to a large number of thermally induced package failure modes. The package degradation that can occur due to the thermal strain and resulting stress might not produce a package failure directly. But, a small amount of degradation is still undesirable. For example, partial delamination at an interface might be sufficient to trigger other failure mechanisms such as wire bond failure, corrosion, adhesive chemical bond breakdown or even secondary voltage breakdown (latchup).

Most of the existing stress prediction tools predict the stress associated with the mismatch in thermal expansion coefficient between different materials inside the package. Due to the complexity of the problem, the temperature field inside the package is considered uniform, identical to the temperature of the ambient media. In real applications, the temperature field is strongly non uniform and the stresses are generated by both mismatch in expansion and temperature gradients. Even though the existing computation tools do not have the computational power to produce accurate predictions of the stress/strain fields inside electronic packages, the present method for computing the component temperature, can be used to produce better predictions than using a constant package temperature field.

To determine the steady-state temperature on the active surface of the chip, the temperature difference across the chip, the die-bond, the "heat spreading" element (half thickness) and the case temperature differences are summed. For one dimensional heat conduction, this summation is

$$T_{s,i} = q_d \left[\left(\frac{\Delta x}{k_i A_i} \right)_s + \left(\frac{\Delta x}{k_i A_i} \right)_{d/a} + \left(\frac{\Delta x}{2k_i A_i} \right)_{h/s} + R_{\text{ext}} \right] + T_{\text{air}}$$

where A_i is the cross-sectional area normal to the heat flow path of Δx_i and k_i is the thermal conductivity of the ith component. The subscript s refers to the silicon layer, d/a refers to the die attach layer and h/s refers to the heat spreader layer. Moreover, R_{ext} is computed using the algorithm given by eq (2.16)

$$R_{\text{ext}} = \frac{R_a R_b}{R_a + R_b} \tag{2.16}$$

A first order approximation of the average temperature of the encapsulant is the half-thickness temperature of the encapsulant. For a package that is uniformly cooled by an equivalent effective heat transfer coefficient, where it might be assumed that the heat flow off the upper and lower surfaces of the lead frame is equal and uniform, it is possible to express the half-thickness temperature of the encapsulant as:

$$T_{h/e} = q_d \left[\left(\frac{\Delta x}{2k_i A_i} \right)_{\text{encaps}} + R_{\text{ext}} \right] + T_{\text{air}}$$

Finally, the die-bond through-thickness temperature gradient is also of interest. Because the steady-state analytical model, for the active surface of the silicon, considers the case were all the heat dissipated in the chip flows through the thickness of the chip and then the die-bond before spreading across the entire plane of the package, it is possible to express the die-bond temperature gradient as

$$T_{d/a} = q_d \left(\frac{\Delta x}{k_i A_i} \right)_{d/a}$$

It is interesting to note that in convectively cooled packages where a high thermal conductivity lead-frame, substrate, or heat spreader is present, the thermal resistances in the silicon and die-bond materials are approximately two orders of magnitude smaller than the effective convective resistance, while the resistance of lead-frames or heat spreaders is found to be about four orders of magnitude smaller.

An example of an analytical model for estimating the thermally induced stress in convectively cooled packages, during the steady-state and transient conditions is presented by Witzman et al (1994).

One can determine the mid-point temperature of each material layer of the die packaging through the use of the thermal algorithms presented in this chapter. If the thermal field and the material themo-mechanical properties are known, a mechanical stress model can be employed in order to evaluate the stress/strain levels that are developing inside the package structure.

Very useful information concerning the stress level inside of a package can be obtained by using the "tri-material plate structure" configuration and assuming that all the materials involved are in the elastic regime. This configuration consisted of a chip adhesively bonded to a substrate composed of a lead frame and underlying encapsulant. This configuration is representative of the situation where a void (delamination) is present at the top of the chip surface, between the chip and the encapsulant. This configuration constitutes a worst case shear condition for the adhesive layer at steady-state. Since the thermal expansion coefficients of the leadframe and encapsulant are approximately equal, and since it is anticipated that the in-plane strain will be more prominent than bending, the stiffnesses of the leadframe and encapsulant can be combined to form a single substrate with an effective Young's modulus. In comparing the stiffness of the "plates" constituting the tri-material structure, it is observed that the intermediate layer, the adhesive, is far less stiff than the chip or substrate, thus the thermal expansion of the adhesive layer itself can be neglected.

With these assumptions, and using the tri-material configuration, one can estimate the maximum stress/strain levels (both in shear and normal direction) on the chip surface or die-attach layer. The comparison with the 2D numerical computation results show that the results obtained using this simplified model were in agreement with the FE results within a factor of 2 (the analytical model results are conservative, predicting the worst possible situation). For proper estimation of the stress/strain level inside plastic encapsulated components, one cannot ignore the residual stresses created inside the package during the encapsulant curing stage, and the impact of the humidity and temperature in releasing these stresses.

2.8 CONCLUSIONS

Many of the experimental results collected over the last few years have lent support to the idea that the heat transfer from a component to its environment is a local phenomenon controlled by the thermal plume surrounding the package. It is practically independent of the upstream/downstream conditions and it is a very weak function of the flow pattern developed inside the channel or enclosure. For all the cases studied, involving free or forced convection inside open or sealed enclosures, the chip or die temperature can be expressed as a summation of the component adiabatic temperature and the adiabatic temperature rise, [combining eqs (2.1) and (2.25)]

$$T_d = \Delta T_{a,m} + \Delta T_{\mathrm{TW}} + \Delta T_{\mathrm{int}} + \Delta T_{\mathrm{ext}} \tag{2.1}$$

To a very good approximation, the adiabatic temperature rise is independent of the component position inside the array. The local adiabatic heat transfer coefficient (for a specific component mounted on the board) can be estimated using eq (2.9)

$$\mathrm{Nu}_{\mathrm{ad}} = \mathrm{Nu}_{\mathrm{D}} + C_i(\mathrm{Nu}_{\mathrm{fc}} + \mathrm{Nu}_{\mathrm{nc}}) + \mathrm{CR} \tag{2.9}$$

The characteristic dimension which best describes the thermal phenomena at the component level is the square root of the component external surface area.

This work also provides a correlation that describes the external temperature rise of an electronic component package, mounted on a printed circuit board.

$$R_{\text{ext}} = \frac{R_a R_b}{R_a + R_b} \tag{2.16}$$

where

$$R_a = \left[h_{\text{ad}} \eta \left(\frac{\pi a^2}{4} + A_a \right) \right]^{-1} \tag{2.19}$$

$$R_b^{-1} = \pi mabk\Phi \tag{2.21}$$

and

$$\Phi = \left[\frac{I_1(\alpha)K_1(\beta) - K_1(\alpha)I_1(\beta)}{I_1(\alpha)K_0(\beta) + K_1(\alpha)I_0(\beta)} \right] \tag{2.23}$$

This correlation takes into account the conjugate effect of the convection/conduction and, at the same time, eliminates the time consuming effort of defining the value of the convection heat transfer coefficient for each and every specific application or component.

The proposed correlation expresses the temperature rise of a component in an integral form, showing explicitly the contribution of different factors on the overall result, as well as the sensitivity to changes in the design parameters. By rational use of this equation, the designer can make intelligent decisions about the factors that can/should be modified in order to achieve the design goals.

Thermally conductive substrates help spread the heat concentrated inside the thermal plume by acting as cooling fins for components. The film thermal resistance (the adiabatic thermal resistance), for most practical applications, is reduced proportionally with the square root of the increase of the effective substrate thermal conductivity.

Spreading the thermal plume by low Reynolds number flow, similar to the flows existing inside electronic enclosures, is inefficient and costly to achieve. The sensitivity of the component adiabatic thermal resistance to the local air velocity is proportional approximately the cube root of the velocity. This makes forced convection cooling a very inefficient solution, especially where air velocities higher than 2 m/sec are required.

Introduction of conductive bodies (non-contact heat sinks) inside the component thermal plume is a viable cost effective solution which can reduce the component adiabatic free convection thermal resistance by about 50 percent. This solution is particularly advantageous when the product specification requires sealed enclosures or unpackaged chips on multichip modules. Although this solution is less effective than the conventional heat spreaders mounted on the package surface which can reduce the adiabatic thermal resistance by a factor higher than five, non contact heat spreaders do not produce any mechanical stress on the component leads and can be used for cooling large surface mounted components.

If the packaging configuration has been pre-selected, the effects of thermofluid parameter variations on the chip temperature can be determined from eq (2.28).

$$\Delta T_{ad} = \frac{q_d B \left(v^n a^x k^y b^z L^w\right)}{C_i} \tag{2.28}$$

The coefficients and exponents for sparsely and densely populated vertical printed circuit boards, respectively, are reported and the proposed computational algorithms allow for early predictions of the impact of design decisions on the temperature level of electronic components. Thus, eqs (2.9) and (2.28) can be used at a time in the design cycle when insufficient data is available to use more precise thermal prediction models. Moreover, this methodology allows for fast and easy computations to determine the consequences of various perturbations and, thereby, creates the opportunity to shorten the design cycle.

2.9 NOMENCLATURE

Roman Letter Symbols

a	characteristic length of package, m
A	surface area, m^2
A_a	wetted surface area of the package, m^2
A_{ext}	external surface area of chip package m^2
b	thickness of printed circuit board, m
CR	mixed convection cross flow correction factor [see eq (2.14)], dimensionless

Roman Letter Symbols (Cont'd)

C_i	correction factor for non-cuboid geometry [see eq (2.20)], dimensionless
C_{TW}	emprical constant for wake function, dimensionless
$c_{p,a}$	specific heat of air at constant pressure, J/kg-K
D	number of rows downstream from component of interest, dimensionless
F	radiation view factor, dimensionless
Gr	Grashof number, dimensionless
h	heat transfer coefficient, $\mathrm{W/m^2\text{-}K}$
h_{ad}	adiabatic heat transfer coefficient, $\mathrm{W/m^2\text{-}K}$
I_0, I_1	modified Bessel Functions, dimensionless
k	thermal conductivity, W/m-K
K_0, K_1	Bessel functions dimensionless
L	radius of circular fin, m or radius of unit cell on PCB, m
L_T	center-to-center spacing between components on PCB, m
n	position of component relative to inlet ($n = 1$ at bottom or exponent on velocity [see eq (2.28)], both dimensionless
Nu	Nusselt number, dimensionless
$\mathrm{Nu_D}$	Nusselt number at the diffusive limit, dimensionless
$\mathrm{Nu_{ad}}$	Nusselt number based on the adiabatic heat transfer coefficient, dimensionless
$\mathrm{Nu_{fc}}$	Nusselt number for pure laminar forced convection, dimensionless
$\mathrm{Nu_{nc}}$	Nusselt number for pure laminar natural convection, dimensionless
q	heat flow, W
q_a	component heat dissipated directly to air, W
q_b	component heat dissipated through the PCB, W
q_d	total heat dissipated by a packaged device, W
q_T	total heat generated in an enclosure, W
$\sum q_{\mathrm{up}}$	sum of heat dissipated upstream of a component, W
Ra	Rayleigh number, dimensionless
Re	Reynolds number, dimensionless

Roman Letter Symbols (Cont'd)

R	thermal resistance, K/W or °C/W
R_a	thermal resistance between the package surface and air, K/W or °C/W
R_b	thermal resistance between the PCB and the air, K/W or °C/W
$R_{b,a}$	convection thermal resistance between the PCB and the air, K/W or °C/W
$R_{b,g}$	gap thermal resistance between the component and the PCB, K/W or °C/W
$R_{b,sp}$	thermal resistance associated with spreading in the PCB, K/W or °C/W
R_{ext}	external or "case-to-ambient" package thermal resistance, K/W or °C/W
R_{int}	internal or "junction-to-case" thermal resistance, K/W or °C/W
R_{ja}	total or "junction-to-ambient,, thermal resistance, K/W or °C/W
T_a	air temperature, °C
T_{ad}	adiabatic temperature of package surface, °C
$T_{\mathrm{a,m}}$	average air temperature at specified location, °C
T_{cf}	temperature of component case in free space, °C
T_c	temperature of component surface, °C
T_d	die temperature, °C
T_o	external ambient air temperature of outside cabinet, °C
ΔT_a	temperature rise (above external ambient) of local air, K or °C
ΔT_{ad}	temperature rise of the case above the adiabatic temperature, K or °C
ΔT_{ae}	temperature rise the of enclosure above the external ambient, K or °C
ΔT_{ext}	external temperature rise of the package surface above the local air temperature, K or °C
ΔT_{int}	internal temperature rise of the die above the package surface, K or °C
ΔT_n	average temperature rise above the ambient of the nth component, K or °C

Roman Letter Symbols (Cont'd)

ΔT_{TW}	temperature rise of thermal wake, K or °C
U	overal heat transfer coefficient, W/m^2-K
V	air velocity, m/s
$\hat{V}$	volumetric flow rate, m^3/s
w	an exponent [see eq (2.28)], dimensionless
x	an exponent [see eq (2.28)], dimensionless and length coordinate, m
y	an exponent [see eq (2.28)], dimensionless
z	an exponent [see eq (2.28)], dimensionless

Greek Symbols

α	fin parameter in eq (2.21) , dimensionless
β	fin parameter in eq (2.21), dimensionless
ϵ	emissivity, dimensionless
η	fin efficiency, dimensionless
ν	kinematic viscosity of liquid, m^2/s
ρ	density, kg/m^3
Φ	combination of Bessel Functions [see eq (2.23)

Roman Letter Subscripts

caps	pertains to encapsulant
copper	pertains to copper layers
d/a	pertains to the die attach layer
FR-4	pertains to FR-4 layers
h/s	pertains to the heat spreader layer
s	pertains to silicon layer

2.10 REFERENCES

Anderson, A (1994). *Decoupling Convective and Conductive Heat Transfer Using the Adiabatic Heat Transfer Coefficient*, J Elec Packaging, 116, 310-316.

Ashiwake, N, Nakayama, W, Daikoku, T and Kobayashi, F (1986). *Forced Convection Heat Transfer from LSI Packages in an Air Cooled Wiring Card Array*, ASME-HTD-28, 35-42.

Arvizu, D E (1981). *Experimental Heat Transfer from an Array of Heated Cubical Elements on an Adiabatic Channelm Wall*, Thermosciences Division Report HMT-33, Mechanical Engineering Department, Stanford University, Stanford, CA.

Arvizu, D E and Moffat, R J (1982). *The Use of Superposition in Calculating Cooling Requirements of Circuit Board Mounted Electronic Components*, Proc IEEE 32nd Electronic Components Conference, San Diega, Ca, 133-144.

Aung, W (1972). *Fully Developed Laminar Free Convection Between Vertical Plates Heatd Asymetrically*, Int J Heat and Mass Transfer, 15, 1577-1580.

Aung, W (1988). Cooling Technology for Electronic Equipment, Hemisphere Publishing Corp., New York, NY.

Bar-Cohen A, and Rohsenow, W (1980). *Thermally Optimal Spacing of Vertical Natural Convection Cooled Parallel Plates*, J Heat Transfer, 102, 221-227.

Buller M and Kilburn R (1981). *Evaluation of Surface Heat Transfer Coefficients for Electronic Module Packages*, ASME HTD-20, 15-20.

Davalath, J and Y Bayazitoglu (1987) *Forced Convection Cooling Across Rectangular Blocks.* J Heat Transfer, 109. 321-328.

Furkay, S (1984). *Convective Heat Transfer in Electronic Equipment: An Overview*, Int J Hybrid Microlectronics, 7 (3), 27-35.

Graham, K and Witzman S (1988). *Analytical Correlation of Thermal Design of Electronic Packages*, Cooling Technology for Electronic Equipment, Hemisphere Publishing Corporation, New York, 249-264.

Godfrey W (1990). *Interative Thermal Modeling of Electronic Circuit Boards*, ASME HTD-153.

Hollworth, B R and Fuller, H A (1987). *Heat Transfer and Pressure Drop in a staggered Array of Aircooled Components*, Proc Int

Symposium on Cooling Technology for Electronic Equipment, 742-748.

Landis, F (1986) *W. Elenbaas' Paper on Heat Dissipation of Parallel Plates by Free Convection*, ASME HTD-57, Boston, MA, 11-22.

Luikov A (1973). *Conjugate Convective Heat Transfer Problems*, Int J of Heat and Mass Transfer, 17, 257-265.

Mihan, K, Stacey, B, Nakhla M and Zhang, Q (1995). *Concurrent Thermal and Electrical Optimisation of High-Speed Packages and Systems.* Proc INTERPAK'95, Advances in Electronic Packaging, EEP-10-1, 221-228.

Miyatake, O and Fujii, T (1974). *Natural Convection Heat Transfer Between Vertical Parallel Plates with Unequal Heat Fluxes*, Heat Transfer- Japanese Research, 3 (3), 29-33.

Moffat R (1990). *Experimental Heat Transfer*, Proc Ninth Intl Heat Transfer Conference, 187-206.

Moffat R J and Anderson A (1990). *Applying Heat Transfer Coefficient Data To Electronics Cooling*, J Heat Transfer, 112, 882-890.

Moffat R J and Ortega, A (1988). *Direct Air Cooling of Electronic Components*, In Advances in Thermal Modeling of Electronic Components and Systems, A Bar-Cohen and A D Krausd (eds). Hemisphere Publishing Corp., New York, NY.

Nakayama, W and Park, S H (1996). *Conjugate Heat Transfer From a Single Surface-Mounted Block to Forced Convective Air Flow in a Channel*, J Heat Transfer, 118, 301-309.

Ortega, A and Moffat, R (1985). *Heat Transfer from an Array of Simulated Electronic Components: Experimental Results for Free Convection With and Without a Shrouding Wall*, Proc 23rd AICHE/ASME National Heat Transfer Conference, Denver, CO, 5-15.

Wirtz, R A and Stutzman R J (1982). *Experiments on Free Convection Between Vertical Plates with Symmetric Heating*, J Heat Transfer, 104, 501-507.

Witzman, S, Mix, D and Bar-Cohen, A (1994). *Thermal Stress in Convectively-Cooled Plastic-Encapsulated Chip Packages*, in Thermal Management of Electronic Systems, C. J. Hoogendoorn (ed), Kluwer Academic Publishers, Dordrecht, the Netherlands, 309-323

Witzman, S, Newport, D and Nicoletta, T (1989). *Free Convection Alr Cooling of the Electronic Equipment Still a Miracle Waiting to be Explored and Explained*, Proc SEMITHERM, San Diego, CA, 98-103.

Witzman, S, Newport, D and Nicoletta, T (1990). *An Alternative Methodology for Computing the Temperature Rise Or an Electronic Device Inside a Direct Air-Cooled Electronic Cabinet*, Proc Ninth International Heat Transfer Conference, 2, 313-319,

Yovanovich, M (1988). *General Expression for Forced Convection Heat and Mass Transfer From Isopotential Spheroids*, AIAA 26th Aerospace Science Meeting, AIAA 88-0743, Reno, NV.

Yovanovich, M and Vanoverbeke, C (1988). *Combined Natural and Forced Convection from Isothermal Spheres*, AIAA Thermophysics Plasma Dynamics and Lasers Conference, AIAA-88-2618, San Antonio, TX.

Zebib A and Wo, Y (1989). *A Two-Dimensional Conjugate Heat Transfer Model for Forced Air Cooling of an Electronic Device*, J Electronic Packaging, 111, 41-45.

CHAPTER 3

GOVERNING RELATIONS AND PERFORMANCE LIMITS IN AIR-COOLED HEAT SINKS

Allan D. Kraus
Allan D. Kraus Associates
Pacific Grove, CA, 93950
and
Avram Bar-Cohen
Mechanical Engineering Department
University of Minnesota
Minneapolis, MN, 55455

3.1 INTRODUCTION

Finned arrays and *arrays of fins* are often employed to augment the heat transfer from the surface of electronic components. Examples of these are shown in Fig 3.1 where it may be observed that the finned array (Fig 3.1a) can be composed of many individual fins which may be *cascaded* or *clustered* together or placed in *parallel* with one another. Moreover, the heat sink composed of arrays of fins (Fig 3.1b) is composed of many individual fins that are arranged in parallel.

A most useful measure of the effectiveness of augmentation of the heat flow from component to environment is the *input admittance* of the fin structure which was first proposed by Kraus et al (1978). Its use as a design parameter essentially replaces the more conventional *fin efficiency* which was proposed by Gardner (1945) and is capable of characterizing an entire array rather than just an individual fin. The input admittance, defined as the ratio of the heat entering the base of an individual fin or a finned array, q_b, to the temperature excess at the base of the fin or array, θ_b

$$Y_{\mathrm{IN}} \equiv \frac{q_b}{\theta_b} \quad (\mathrm{W/^\circ C \ or \ W/K})$$

The word *admittance*, borrowed from electric circuitry, is used to imply that a finned array is a *distributed heat flow configuration* [1] The form of Y_{IN} suggests that

- if q_b is known and Y_{IN} has been computed, the temperature excess, θ_b, required to induce the heat flow, q_b through the base of the array can be calculated from

$$\theta_b = \frac{q_b}{Y_{\mathrm{IN}}} \quad (^\circ\mathrm{C \ or \ K})$$

- If θ_b is known and Y_{IN} has been computed, the heat dissipated by the array for a prescribed temperature excess can also be calculated

$$q_b = Y_{\mathrm{IN}}\theta_b \qquad (\mathrm{W})$$

The *input impedance* is employed in place of the input resistance and is

$$Z_{\mathrm{IN}} = \frac{1}{Y_{\mathrm{IN}}}$$

The governing relations that are cited here will be those that lead to the input admittance for the entire finned array or arrays of fins consisting of the longitudinal fin of rectangular profile. It seems logical to make a division on the basis of the two categories of Fig 3.1. First, however, a summary of some governing relations for individual fins is in order.

[1] Following this reasoning, the terminolgy *conductance* or *resistance* would apply to a lumped parameter such as a contact conductance or resistance.

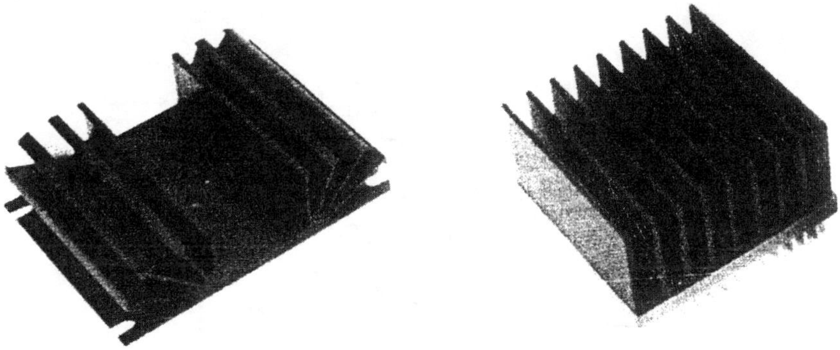

(a) (b)

Figure 3.1: Two forms of heat sink: (a) the finned array and (b) the array of fins. Photographs courtesy of the Wakefield Engineering Co., Wakefield, MA.

3.2 GOVERNING RELATIONS

3.2.1 THE LONGITUDINAL FIN OF RECTANGULAR PROFILE

For the longitudinal fin of rectangular profile displayed in Fig 3.2, the origin of the height coordinate, x, is taken at the fin tip. The Murray-Gardner assumptions [Murray (1938), Gardner (1945)] are then considered in the formulation of the differential equation for the temperature excess, which is based on the energy balance over the differential element of height, dx

$$\frac{d^2\theta}{dx^2} - m^2\theta = 0$$

where

$$m = \left(\frac{2h}{k\delta}\right)^{1/2}$$

and where θ is the temperature excess between the fin and the environment,

$$\theta = T - T_s$$

h is the heat transfer coefficient between the faces of the fin and the environment and k is the thermal conductivity of the fin material. When this differential equation is solved as an initial value problem

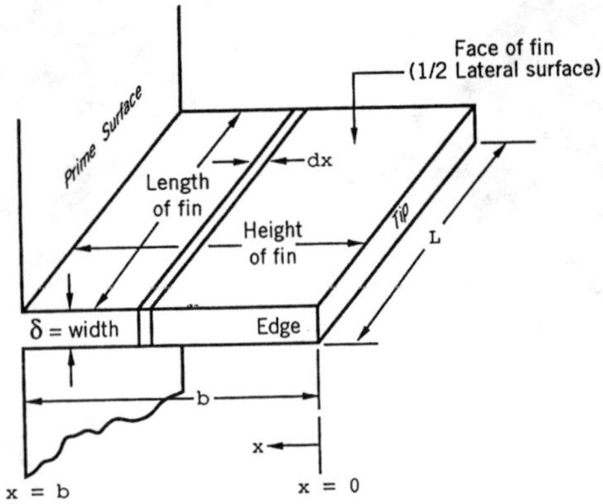

Figure 3.2: The longitudinal fin of rectangular profile.

subject to the initial conditions taken at the fin base

$$\theta(x = b) = \theta_b \qquad \text{and} \qquad q(x = b) = q_b$$

the solution is a linear transformation that maps conditions at the fin base to conditions at the any point, x

$$\theta(x) = \theta_b \cosh m(b - x) - \frac{q_b}{Y_o} \sinh m(b - x)$$

and

$$q(x) = -\,\theta_b Y_o \sinh m(b - x) + q_b \cosh m(b - x)$$

where Y_o is designated as the *characteristic admittance* of the fin

$$Y_o = (2hk\delta)^{1/2} L$$

Because x is taken from the tip of the fin where $\theta(x = 0) = \theta_a$ and $q(x = 0) = q_a$, the temperature excess at the fin tip, θ_a, is given by

$$\theta_a = \theta_b \cosh mb - \frac{q_b}{Y_o} \sinh mb$$

and

$$q_a = -\,\theta_b Y_o \sinh mb + q_b \cosh mb$$

These may be put into matrix form

$$\begin{bmatrix} \theta_a \\ q_a \end{bmatrix} = \mathbf{\Gamma} \begin{bmatrix} \theta_b \\ q_b \end{bmatrix} = \begin{bmatrix} \gamma_{11} & \gamma_{12} \\ \gamma_{21} & \gamma_{22} \end{bmatrix} \begin{bmatrix} \theta_b \\ q_b \end{bmatrix} \tag{3.1}$$

where

$$
\begin{aligned}
\gamma_{11} &= \cosh mb \\
\gamma_{12} &= -\frac{1}{Y_o} \sinh mb \\
\gamma_{21} &= -Y_o \sinh mb
\end{aligned}
$$

and

$$\gamma_{22} = \cosh mb$$

The matrix $\mathbf{\Gamma}$ represents a linear transformation that maps base conditions to tip conditions. It is an important parameterization for the fin. Not only does it demonstrate that the tip heat flow and temperature excess are obtained from a superposition of effects of heat flow and temperature excess at the fin base, to be expected in this linear *system*, but it satisfies the intuitive feeling that conditions at the fin base induce conditions at the fin tip.

When dealing with finned arrays, it is useful to have a linear transformation that maps fin tip conditions into base conditions

$$\begin{bmatrix} \theta_b \\ q_b \end{bmatrix} = \mathbf{T} \begin{bmatrix} \theta_a \\ q_a \end{bmatrix} = \begin{bmatrix} \tau_{11} & \tau_{12} \\ \tau_{21} & \tau_{22} \end{bmatrix} \begin{bmatrix} \theta_a \\ q_a \end{bmatrix} \tag{3.2}$$

where

$$
\begin{aligned}
\tau_{11} &= \cosh mb \\
\tau_{12} &= \frac{1}{Y_o} \sinh mb \\
\tau_{21} &= Y_o \sinh mb
\end{aligned}
$$

and

$$\tau_{22} = \cosh mb$$

Along with the *input admittance*, defined as the ratio of the heat flow into the fin to the temperature excess at the base of the fin,

$$Y_{\mathrm{IN}} = \frac{q_b}{\theta_b} \tag{3.3}$$

eqs (3.1) and (3.2) represent two of the linear transformations used in the design of heat sinks. A third is the *thermal admittance matrix* that maps the temperature excess at base and tip, as independent variables to the heat flows at base and tip, as dependent variables

$$\begin{bmatrix} q_b \\ q_a \end{bmatrix} = \mathbf{Y} \begin{bmatrix} \theta_b \\ \theta_a \end{bmatrix} = \begin{bmatrix} y_{11} & y_{12} \\ y_{21} & y_{22} \end{bmatrix} \begin{bmatrix} \theta_b \\ \theta_a \end{bmatrix} \tag{3.4}$$

where

$$\begin{aligned} y_{11} &= Y_o \coth mb \\ y_{12} &= -Y_o \operatorname{csch} mb \\ y_{21} &= Y_o \operatorname{csch} mb \end{aligned}$$

and

$$y_{22} = -Y_o \coth mb$$

3.2.2 OTHER PROFILES

Observe that $\mathbf{T}$ is the inverse of $\mathbf{\Gamma}$ and that, because the inverse exists, the longitudinal fin of rectangular profile is referred to as a *regular fin.* In certain fins, the thickness vanishes at the tip (at $x = 0$). The longitudinal fins of triangular profile and concave parabolic profile and the conical and concave parabolic spine) are such fins and are called *singular fins.* For singular fins, the parameterization is the *thermal transmission ratio*

$$\mu = \frac{q_b}{\theta_b}$$

This parameterization is identical in form to the input admittance. There are, however, subtle differences in how these two parameterizations are employed.

All regular fins, regardless of their profile, can be parameterized by any or all of six linear transformations. These are developed, in excruciating detail, by Kraus and Bar-Cohen (1995) but, because of space limitations, only the $\mathbf{T}, \mathbf{\Gamma}$ and $\mathbf{Y}$ parameterizations will be considered here.

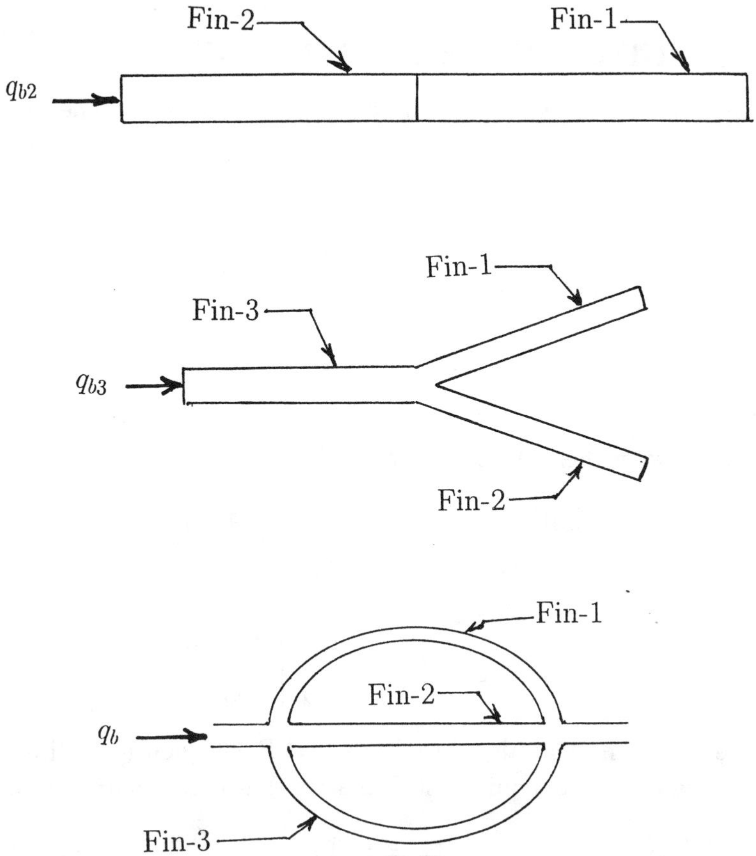

Figure 3.3: Three arrangements of longitudinal fins (a) Cascade, (b) Cluster and (c) Parallel.

3.3 THE ANALYSIS OF FINNED ARRAYS

There is no formal design method for the finned array shown in Fig 3.1a. However, trial-and-error designs can be achieved quite expeditiously by using the concept of the input admittance. Moreover, there is no set procedure for the selection of the design heat transfer coefficient, although designs resulting from estimated coefficients can be surprisingly accurate. The analysis of a finned array which has been assembled from several individual fins is expedited by three algorithms for array assembly.

3.3.1 THE CASCADE ALGORITHM

Figure 3.3a shows two fins in the *cascade arrangement.* The two-fin entity is the array and the base of the array is taken at the base of fin-2. Continuity and compatability at the interface between the two fins requires that the heat leaving fin-2 enter fin-1 and that the temperature excesses at the tip of fin-2 and the base of fin-1 must match. These conditions can be represented in matrix form by

$$\begin{bmatrix} q_{a2} \\ \theta_{a2} \end{bmatrix} = \begin{bmatrix} q_{b1} \\ \theta_{b1} \end{bmatrix}$$

and along with the **T** representations for fins-1 and -2

$$\begin{bmatrix} \theta_{b1} \\ q_{b1} \end{bmatrix} = \begin{bmatrix} \tau_{11,1} & \tau_{12,1} \\ \tau_{21,1} & \tau_{22,1} \end{bmatrix} \begin{bmatrix} \theta_{a1} \\ q_{a1} \end{bmatrix}$$

and

$$\begin{bmatrix} \theta_{b2} \\ q_{b2} \end{bmatrix} = \begin{bmatrix} \tau_{11,2} & \tau_{12,2} \\ \tau_{21,2} & \tau_{22,2} \end{bmatrix} \begin{bmatrix} \theta_{a2} \\ q_{a2} \end{bmatrix}$$

an exercise in matrix algebra [Kraus and Bar-Cohen (1995)] shows that there will be an equivalent **T**-matrix that is the matrix product

$$\mathbf{T_e} = \mathbf{T_2 T_1}$$

and if n-fins are involved in the cascade connection

$$\mathbf{T_e} = \mathbf{T_n T_{n-1} T_{n-2}} \cdots \mathbf{T_2 T_1} \tag{3.5}$$

which is the *cascade algorithm.*

The input admittance derives from a consideration of the equivalent **T** representation of the entire array composed of n fins.

$$\begin{bmatrix} \theta_{bn} \\ q_{bn} \end{bmatrix} = \begin{bmatrix} \tau_{11,e} & \tau_{12,e} \\ \tau_{21,e} & \tau_{22,e} \end{bmatrix} \begin{bmatrix} \theta_{a1} \\ q_{a1} \end{bmatrix}$$

A matrix expansion provides

$$\theta_{bn} = \theta_{a1}\tau_{11,e} + q_{a1}\tau_{12,e}$$

and

$$q_{bn} = \theta_{a1}\tau_{21,e} + q_{a1}\tau_{22,e}$$

If the second of these is divided by the first, then

$$Y_{\text{IN,n}} = \frac{\theta_{a1}\tau_{21,e} + q_{a1}\tau_{22,e}}{\theta_{a1}\tau_{11,e} + q_{a1}\tau_{12,e}}$$

or

$$Y_{\text{IN,n}} = \frac{\tau_{21,e} + Y_{a1}\tau_{22,e}}{\tau_{11,e} + Y_{a1}\tau_{12,e}} \tag{3.6}$$

where

$$Y_{a1} = \frac{q_{a1}}{\theta_{a1}}$$

is the admittance at the tip of fin-1 and is sometimes referred to as the *tip admittance.* If the tip of fin-1 is dissipating heat, $q_{a1} = h_a A_a \theta_{a1}$ and $Y_{a1} = h_a A_a$. If the tip is not dissipating, then $Y_{a1} = 0$ and eq (3.6) reduces to

$$Y_{\text{IN,n}} = \frac{\tau_{21,e}}{\tau_{11,e}}$$

3.3.2 THE CLUSTER ALGORITHM

Figure 3.3b shows three fins. Fins-1 and-2 form a *cluster* at the tip of fin-3 and the point where fins-1, -2 and -3 meet is called the *cluster point.* Again, the entity is the entire array composed of all three fins and the base of the array is taken at the base of fin-3. Continuity at the cluster point requires that the heat leaving fin-3 enter fins-1 and -2

$$q_{a3} = q_{b1} + q_{b2}$$

and that the temperature excesses for all fins at the cluster point must match

$$\theta_{a3} = \theta_{b1} = \theta_{b2}$$

The tip admittance at fin-3 is the *cluster admittance*

$$Y_{a3} = Y_C = \frac{q_{a3}}{\theta_{a3}}$$

and use of the continuitity and compatability conditions for the cluster point provides

$$Y_{a3} = Y_C = \frac{q_{a1} + q_{a2}}{\theta_{a3}} = \frac{q_{a1}}{\theta_{a1}} + \frac{q_{a2}}{\theta_{a2}} = Y_{\text{IN,1}} + Y_{\text{IN,2}}$$

This may be extended to n-fins in cluster to provide the *cluster algorithm*

$$Y_C = \sum_{i=1}^{n} Y_{\mathrm{IN,i}} \tag{3.7}$$

and it is noted that if the cluster is composed of j-regular fins and k-singular fins

$$Y_C = \sum_{i=1}^{j} Y_{\mathrm{IN,i}} + \sum_{i=1}^{k} \mu_i$$

3.3.3 THE PARALLEL ALGORITHM

Three fins in parallel are shown in Figure 3.3c. The base of this array is taken at the left and continuity and compatability demand that

$$\begin{aligned} q_b &= q_{b1} + q_{b2} + q_{b3} \\ q_a &= q_{a1} + q_{a2} + q_{a3} \\ \theta_b &= \theta_{b1} = \theta_{b2} = \theta_{b3} \end{aligned}$$

and

$$\theta_a = \theta_{a1} = \theta_{a2} = \theta_{a3}$$

Because the heat flows at the base and tip of any regular fin can be related to the temperature excesses at base and tip by the admittance representation of eq (3.4)

$$\begin{bmatrix} q_b \\ q_a \end{bmatrix} = \mathbf{Y} \begin{bmatrix} \theta_b \\ \theta_a \end{bmatrix} = \begin{bmatrix} y_{11} & y_{12} \\ y_{21} & y_{22} \end{bmatrix} \begin{bmatrix} \theta_b \\ \theta_a \end{bmatrix} \tag{2.4}$$

an exercise in matrix algebra shows that

$$\begin{bmatrix} q_b \\ q_a \end{bmatrix} = \mathbf{Y_1} \begin{bmatrix} \theta_{b1} \\ \theta_{a1} \end{bmatrix} + \mathbf{Y_2} \begin{bmatrix} \theta_{b2} \\ \theta_{a2} \end{bmatrix} + \mathbf{Y_3} \begin{bmatrix} \theta_{b3} \\ \theta_{a3} \end{bmatrix}$$

or

$$\begin{bmatrix} q_b \\ q_a \end{bmatrix} = [\mathbf{Y_1} + \mathbf{Y_2} + \mathbf{Y_3}] \begin{bmatrix} \theta_b \\ \theta_a \end{bmatrix}$$

This can be extended to n-fins in parallel, and such extension shows that an equivalent $\mathbf{Y}$-matrix is obtained

$$\begin{bmatrix} q_b \\ q_a \end{bmatrix} = \mathbf{Y_e} \begin{bmatrix} \theta_b \\ \theta_a \end{bmatrix}$$

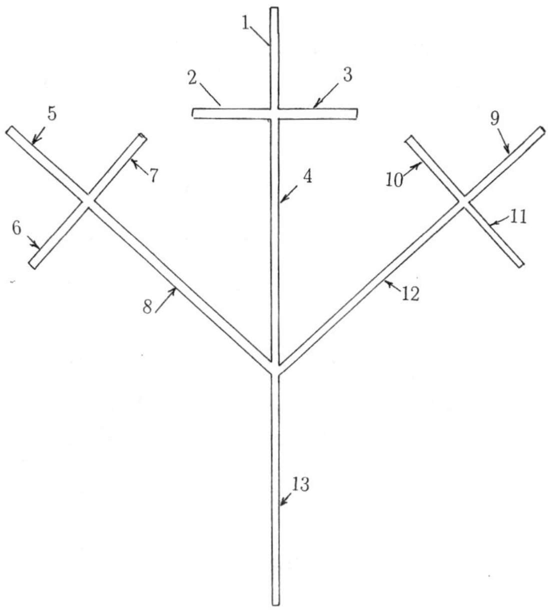

Figure 3.4: Finned array composed of thirteen longitudinal fins of rectangular profile.

where

$$\mathbf{Y_e} = \mathbf{Y_1} + \mathbf{Y_2} + \mathbf{Y_3} \cdots \mathbf{Y_n} \tag{3.8}$$

and this is the *parallel algorithm.*

3.3.4 EXAMPLE

Example 3.1, which now follows, is intended to illustrate the use of the cascade and cluster algorithms in an organized computational procedure. At the end of the example, the limitations of the finned array are noted.

Example 3.1: The finned array displayed in Fig 3.4 is composed of thirteen longitudinal fins of rectangular profile. The thermal conductivity is $k = 180\,\mathrm{W/m\text{-}K}$ (aluminum) and the array is to operate in a forced convection environment where $h = 60\,\mathrm{W/m^2\text{-}K}$. All fins have a thickness of 0.2286 cm and a length of 2 cm. Fins-1, -5 and -9 are 1.25 cm high, fins-4, -8, -12 and -13 are 3 cm high and all other fins are 1 cm high. The objective for the array in Fig 3.4 is to provide a dissipation of 18.75 W at a base temperature excess of 50°C.

Table 3.1: Summary of values of mb and Y_o for the fins in the finned array of Fig. 3.4.

Fin	mb, dimensionless	Y_o, W/°C
1, 5, 9	0.14054	0.21346
2, 6, 10	0.14054	0.17077
3, 7, 11	0.14054	0.17077
4, 8, 12	0.14054	0.51232
13	0.14054	0.51232

Solution

The first step is to determine the τ-parameters for each fin. These parameters are based on Y_o which is the same for all fins[2]

$$\begin{aligned} Y_o &= (2hk\delta)^{1/2}L \\ &= [2(60)(180)(0.002286)]^{1/2}(0.02) \\ &= (49.37760)^{1/2}(0.02) \\ &= 0.14054\,\mathrm{W/^\circ C} \end{aligned}$$

and mb with

$$\begin{aligned} m &= \left(\frac{2h}{k\delta}\right)^{1/2} \\ &= \left[\frac{2(60)}{180(0.002286)}\right]^{1/2} \\ &= (291.6302)^{1/2} \\ &= 17.07718\,\mathrm{m}^{-1} \end{aligned}$$

Values of mb and Y_o are listed for the thirteen fins in Table 3.1. After all τ-parameters for each of the fins are determined, the cluster

[2]Five decimal places are carried in these computations to avoid the effects of two phenemona in matrix computation known as round-off and catastrophic cancellation.

and cascade algorithms are employed to determine the input admittance at the base of fin-13. The procedure for establishing the input admittance for the entire array is summarized in Table 3.2.

Table 3.2 shows that the array will not accommodate the specified heat dissipation at a base temperature excess of 50°C. The input admittance at the base of fin-13 is $Y_{\text{IN}} = 0.1900\,\text{W}/^\circ\text{C}$ and this yields a heat dissipation of $q_b = 50(0.1900) = 9.50\,\text{W}$. The footprint of the finned array on its base surface is

$$S_b = \delta L = (0.002286)(2.00) = 0.4572\,\text{cm}^2$$

and the thermal resistance for this strip of base surface is, thus,

$$R_b = \frac{1}{hS_b} = \frac{1}{60(4.572 \times 10^{-5})} = 364.5^\circ\text{C/W}$$

This may be compared to the input impedance of the array

$$Z_{\text{IN}} = \frac{1}{0.1900} = 5.26^\circ\ \text{C/W}$$

There is no question that the use of the finned array provides an improvement over the use of just the bare surface. However, the admittance at the tip of fin-13 is higher than the admittance at its base and indicates that fin-13 is *choking* the heat flow to the rest of the array composed of fins-1 through -12.

The *heat flow choke* is the major limitation in the design of finned arrays and its elimination, based on an investigation by Kraus and Snider (1985), will be considered in section 3.4.2. The elimination of the heat flow choke involves the replacement of fin-13 with three fins in parallel.

3.4 FINNED ARRAY LIMITATIONS

Example 3.1 points to some limitations with regard to the finned array.

Table 3.2: Summary of input admittances.

Fin	Y_{IN}	Y_{IN}, Total	Remarks
1	0.02955	0.02955	
2	0.02337	0.05332	after cluster with (1)
3	0.02337	0.07709	after cluster with (1) and (2)
4		0.11391	after cascade to base of (4)
5	0.02955	0.02955	
6	0.02337	0.05332	after cluster with (5)
7	0.02337	0.07709	after cluster with (5) and (6)
8		0.11391	after cascade to base of (8)
		0.22782	after cluster of (4) and (8)
9	0.02955	0.02955	
10	0.02337	0.05332	after cluster with (9)
11	0.02337	0.07709	after cluster with (9) and (10)
12		0.11391	after cascade to base of (12)
		0.34174	after cluster of (4) and (8) and (12)
13		0.19004	after cascade to base of (13)

3.4.1 AIR AS A COOLANT

Air, while having the advantage of being inexpensive, does not possess the thermal properties that are requisite in an ideal coolant. Thus, finned arrays that employ air to dissipate heat to the environment can be bulky and, because of this, have limited input admittance. Moreover, if the efficiency of these arrays is to be considered, it can be obtained from

$$\eta = \frac{Y_{\mathrm{IN}}}{hS} = \frac{q_b}{hS\theta_b} \tag{3.9}$$

where S is the total surface in the finned array. For the foregoing example,

$$S = 2[3(1.25) + 6(1.00) + 4(3.00)](2) = 4(21.75) = 87\,\text{cm}^2$$

so that

$$\eta = \frac{0.1900}{60(0.0087)} = 0.364$$

This low value of efficiency is a limitation that goes with the addition of individual fins in the array in attempt to increase the wetted surface area so as to obtain a higher hS product. Part of the problem can be attributed to the heat flow choke in fin-13.

3.4.2 ELIMINATION OF THE HEAT FLOW CHOKE

A fin that accommodates a cluster at its tip may exhibit a heat flow choke because the supporting fin dimensions may not permit the transfer of a sufficient amount of heat to the cluster at its tip with a required $Y_a = q_a/\theta_a$. Table 3.1 shows that the input admittance at the base of fin-13 is $Y_{\text{IN}} = 0.1900\,\text{W/°C}$ while the cluster admittance at its tip is almost twice as high, $Y_a = 0.3417\,\text{W/°C}$.

In the case of the longitudinal fin of rectangular profile, eq (3.6) can be written with $Y_a = q_a/\theta_a$ and its matrix elements as

$$Y_{\text{IN}} = \frac{Y_o \sinh mb + Y_a \cosh mb}{\cosh mb + Y_a/Y_o \sinh mb} \tag{3.10}$$

If numerator and denominator are divided by $\cosh mb$, the result is

$$Y_{\text{IN}} - \frac{Y_o \tanh mb + Y_a}{1 + (Y_a/Y_o) \tanh mb}$$

and it may be observed that

$$\lim_{b \to 0} Y_{\text{IN}} = Y_a$$

which is the tip admittance and

$$\lim_{b \to \infty} Y_{\text{IN}} = Y_o$$

which is the characteristic admittance. These provide the extremes for Y_{IN} over the range of values of the fin height b.

In order to find the fin height that makes Y_{IN} a maximum, the customary method of determining the value for which $\partial Y_{\text{IN}}/\partial b$ vanishes, is employed. In eq (3.10), once again observe that neither Y_o nor Y_a is a function of b, and find the point where $\partial Y_{\text{IN}}/\partial b = 0$. After some algebraic tedium, the result shows that

$$Y_o = Y_a \tag{3.11}$$

Because this equation does not involve b, it implies that the slope of $Y_{\text{IN}} = f(b)$ is never zero unless $Y_o = Y_a$ in which case it is always zero. Thus, there is no question of optimizing with respect to fin height and three cases are evident.

- **Case I** $Y_o < Y_a$: When the characteristic admittance is less than the tip admittance, the fin is choking the heat flow regardless of its height. In this case, no improvement can be achieved by varying the fin height, b. To save metal, the fin responsible for the choke should be removed but, if a support structure is required as a standoff for the rest of the array, b should be as low as possible ($b \to 0$).

- **Case II** $Y_o > Y_a$: In this case, no choking occurs and b should be made as high as possible because a benefit exists for high $b\,(b \to \infty)$.

- **Case III** $Y_o = Y_a$: In this case, the value of b has no effect on the heat transfer process and the fin is neither contributing to nor interfering with the performance of the array.

These conclusions may appear somewhat surprising but they are certainly useful as a design guideline. For the longitudinal fin of rectangular profile, $Y_o = (2hk\delta)^{1/2}L$, so that to achieve the optimum, the fin thickness, δ, must be equal to or greater than

$$\delta \geq \frac{Y_a^2}{2hkL^2} \tag{3.12}$$

For fin-13 in Example 3.1, the fin thickness can be increased so that Y_{IN} for the array is equal to Y_a at the tip of fin-13. Use of eq

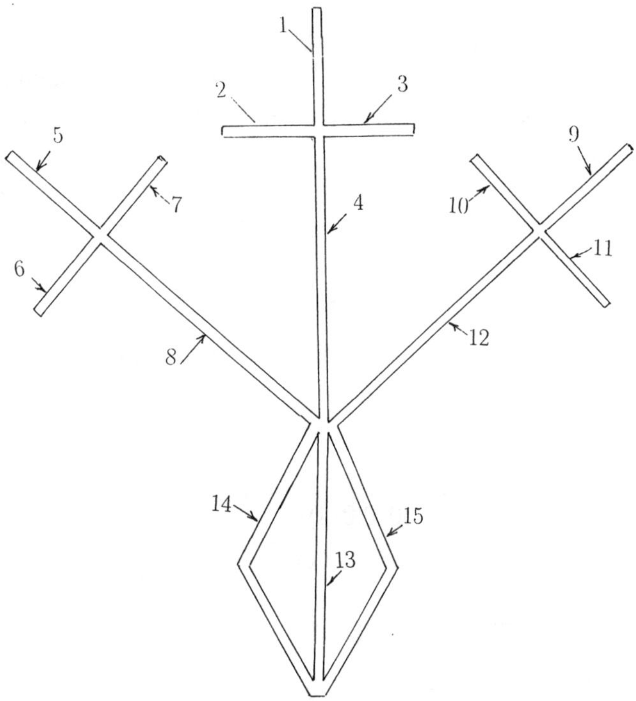

Figure 3.5: Finned array composed of fifteen longitudinal fins of rectangular profile. In this array, fin-13 in Fig 4 has been replaced by three fins.

(3.12) gives

$$\begin{aligned}\delta &= \frac{Y_a^2}{2hkL^2} \\ &= \frac{(0.34174)^2}{2(60)(180)(0.02)^2} \\ &= 0.01352\,\mathrm{m} \quad (1.352\,\mathrm{cm})\end{aligned}$$

A fin with this value of fin thickness would violate the Murray-Gardner assumption that the fin must be thin with respect to its height. Such a fin thickness is not desireable and the alternative is to modify fin-13.

Another way to modify the finned array of the example is to add two fins in parallel to fin-13 as indicated in Fig 3.5. Here fin-13 is not altered but fins-14 and -15 have a height of 4 cm and a width of 0.2286 cm. This arrangement adds about three times as much surface at the base of the array. An analysis of this arrangement, which relies heavily on the parallel algorithm of eq (3.8), shows that the input admittance at the base of the array (at the point where fins-13, -14 and -15 intersect) is $Y_{\text{IN}} = 0.3918\,\text{W}/^\circ\text{C}$. This yields a heat dissipation of $q_b = 50(0.3918) = 19.59\,\text{W}$. Thus, the specification of $q_b = 18.75\,\text{W}$ is not only satisfied, but the design is slightly conservative.

Now the surface area of the array is

$$\begin{aligned} S &= 2[3(1.25) + 6(1.00) + 4(3.00) + 2(4.00)](2) \\ &= 4(29.75) = 119\,\text{cm}^2 \end{aligned}$$

so that

$$\eta = \frac{0.3918}{60(0.0119)} = 0.549$$

which is still quite low but a distinct improvement.

3.4.3 FIN OPTIMIZATION

Kern and Kraus (1972) provided a modest treatment of the optimization of longitudinal and radial fins. Later, Aziz (1992), in a comprehensive article, summarized the optimum dimensions for several fin profiles. For the optimum longitudinal fin of rectangular profile, where the profile area, A_p, for a single fin is specified, Aziz gives

$$b_{\text{opt}} = 1.0023\left(\frac{A_p k}{h}\right)^{1/3} \tag{3.13a}$$

and

$$\delta_{\text{opt}} = \frac{A_p}{b_{\text{opt}}} = 0.9977\left(\frac{A_p^2 h}{k}\right)^{1/3} \tag{3.13b}$$

If the desired value of Y_{IN} is specified, then Aziz provides

$$b_{\text{opt}} = \frac{0.7978}{h}\left(\frac{Y_{\text{IN}}}{L}\right) \tag{3.14a}$$

and

$$\delta_{\text{opt}} = \frac{0.6321}{hk}\left(\frac{Y_{\text{IN}}}{L}\right)^2 \tag{3.14b}$$

Notice that, in eqs (3.13a) and (3.14a), the optimum fin height varies inversely with the magnitude of the heat transfer coefficient and that, in natural convection applications, the individual optimized fin tends to be high. Observe from eq (3.13a) that

$$b \sim \left(\frac{A_p k}{h}\right)^{1/3}$$

and because $A_p = b\delta$

$$\begin{aligned} b &\sim \left(\frac{b\delta k}{h}\right)^{1/3} \\ b^{2/3} &\sim \left(\frac{k}{h}\right)^{1/3} \delta^{1/3} \\ \left(\frac{b}{\delta}\right)^{2/3} &\sim \left(\frac{k}{h}\right)^{1/3} \delta^{-1/3} \end{aligned}$$

and

$$\left(\frac{b}{\delta}\right) \sim \left(\frac{k}{h\delta}\right)^{1/3}$$

Thus, a minimization of the aspect ratio, δ/b, is seen to depend on a low value of the material thermal conductivity, k, or a high value of the heat transfer coefficient, h, or a high value of the fin thickness, δ, or any or all of these acting in concert.

On the other hand, eq (3.13a) also gives

$$\begin{aligned} b &\sim \left(\frac{A_p k}{h}\right)^{1/3} \\ &\sim \left(\frac{b\delta k}{h}\right)^{1/3} \\ b^{2/3} &\sim \left(\frac{k\delta}{h}\right)^{1/3} \end{aligned}$$

and

$$b \sim \left(\frac{k\delta}{h}\right)^{1/2}$$

Here, a minimization of the fin height, b, depends upon a low value of the material thermal conductivity, k, or a high value of the heat transfer coefficient, h, or a low value of the fin thickness, δ, or any or all of these acting together.

The profile area for the 15-fin array under consideration is

$$\begin{aligned} A_p &= 0.2286[3(1.25) + 6(1.00) + 4(3.00) + 2(4.00)] \\ &= 0.2286(29.75) = 6.8009\,\mathrm{cm}^2 \end{aligned}$$

and the optimum dimensions for a single fin of this profile area is obtained from eqs (3.13)

$$\begin{aligned} b_{\mathrm{opt}} &= 1.0023\left(\frac{A_p k}{h}\right)^{1/3} \\ &= 1.0023\left[\frac{(6.8009\times 10^{-4})(180)}{60}\right]^{1/3} \\ &= 1.0023(2.0403\times 10^{-3})^{1/3} \\ &= 1.0023(0.1268) \\ &= 0.1271\,\mathrm{m} \end{aligned}$$

or

$$b = 12.71\,\mathrm{cm}$$

and

$$\begin{aligned} \delta_{\mathrm{opt}} &= \frac{A_p}{b_{\mathrm{opt}}} \\ &= \frac{6.8009\times 10^{-4}}{0.1271} \\ &= 0.00535\,\mathrm{m} \end{aligned}$$

or

$$\delta = 0.535\,\mathrm{cm}$$

These computations show the inherent limitation in the use of optimum fins. If it is desired to employ a single optimum fin, the aspect ratio is approximately 24:1. The height of the non-optimum array is approximately 2(3.00) + 1.25 = 7.25 cm. This may be compared to the single fin $b_{\text{opt}} = 12.71\,\text{cm}$. The difference is about 5.50 cm and may prevent the use of the single optimum fin because of "space envelope" considerations.

Of course, every case must stand on its own and after a great deal of trial-and-error, it is often possible to find a more optimum arrangement of the individual fins in the array. And, while it is noted that the aspect ratios for the optimum single fins do, in general, preclude their use, it has been found that a step-by-step method which optimizes individual fins in the array has possibilities. This will be explored in the next section.

3.5 A STEP-BY-STEP OPTIMIZATION METHOD

In a complex finned array, each of the remote fins can be optimized through the use of equations (3.13) and (3.14).

$$b_{\text{opt}} = 1.0023\left(\frac{A_p k}{h}\right)^{1/3} \tag{3.13a}$$

$$\delta_{\text{opt}} = \frac{A_p}{b_{\text{opt}}} = 0.9977\left(\frac{A_p^2 h}{k}\right)^{1/3} \tag{3.13b}$$

$$b_{\text{opt}} = \frac{0.7978}{h}\left(\frac{Y_{\text{IN}}}{L}\right) \tag{3.14a}$$

and

$$\delta_{\text{opt}} = \frac{0.6321}{hk}\left(\frac{Y_{\text{IN}}}{L}\right)^2 \tag{3.14b}$$

These optimum dimensions are based on assumption that no heat flows from the tip of a remote fin.

Finned arrays also include fins in cascade and fins in cluster at the tips of non-remote connecting fins. In such fins, heat flows out of the tip and expressions for the optimum thickness and height need development. Aziz (1978) has addressed this problem. His expression for the heat dissipated is

$$q = (2hk\delta)^{1/2}\theta_b\left[\frac{\sinh mb + H\cosh mb}{\cosh mb + H\sinh mb}\right]$$

where $H = (\delta/2hk)^{1/2}h_a$ and where h_a is the tip heat transfer coefficient. For a fixed profile area, $A_p = \delta b$, the height, b, can be replaced by A_p/δ and the point where $dq/d\delta$ vanishes may be found. This procedure leads to a relationship between $\beta = mb$ and H [Aziz (1978)]

$$6\beta(1 - H^2) = (1 + H^2)\sinh 2\beta + 2h(1 + \cosh 2\beta)$$

and a trial-and-error procedure is necessary to fix the optimum fin dimensions, b_{opt} and δ_{opt}.

The foregoing discussion is based on the optimization of a single fin with tip heat loss. This optimization does not allow the user to consider the temperature excesses at the base or tip. Because of this, the discussion is difficult to apply to a finned array. Thus, attention turns to the "*reflection relationship*" of eq (3.10) for the longitudinal fin of rectangular profile

$$Y_{\text{IN}} = \frac{Y_o \sinh mb + Y_a \cosh mb}{\cosh mb + Y_a/Y_o \sinh mb} \tag{3.10}$$

or with $\Upsilon = Y_a/Y_o$

$$Y_{\text{IN}} = Y_o \left[\frac{\sinh mb + \Upsilon \cosh mb}{\cosh mb + \Upsilon \sinh mb}\right]$$

where the optimization will involve finding the point where $dY_{\text{IN}}/d\delta$ vanishes.

The first step is to eliminate b in favor of δ via the profile area ($b = A_p/\delta$) so that

$$\beta = mb = \left(\frac{2h}{k\delta}\right)^{1/2}\frac{A_p}{\delta} = \left(\frac{2hA_p^2}{k}\right)^{1/2}\delta^{-3/2}$$

Then with

$$Y_o = (2hk\delta)^{1/2}L = [(2hk)^{1/2}L]\delta^{1/2}$$

Y_{IN} becomes

$$Y_{\text{IN}} = \frac{C_1\delta^{1/2}\sinh\beta + Y_a\cosh\beta}{\cosh\beta + C_2\delta^{-1/2}\sinh\beta}$$

where

$$C_1 = (2hk)^{1/2}L \qquad \text{and} \qquad C_2 = \frac{Y_a}{(2hk)^{1/2}L}$$

Let

$$u = Y_o \sinh mb + Y_a \cosh mb$$

and

$$v = \cosh mb + Y_a/Y_o \sinh mb$$

Then, because in

$$\frac{du}{dv} = \frac{vdu - udv}{v^2} = 0$$

the product, v^2, is not of any consequence,

$$\frac{dY_{\text{IN}}}{d\delta} = A + B + C + D = 0$$

where

$$A = \frac{1}{2}(C_1\delta^{-1/2} + C_2 Y_a \delta^{-3/2}) \sinh\beta \cosh\beta$$

$$B = \left\{\frac{3}{2}\left(\frac{2hA_p^2}{k}\right)^{1/2}\left[C_1\delta^{-2} - C_2 Y_a \delta^{-3}\right]\right\} \sinh^2\beta$$

$$C = C_1 C_1 \delta^{-1} \sinh^2\beta$$

$$D = \frac{3}{2}\left(\frac{2hA_p^2}{k}\right)^{1/2}(C_2 Y_a \delta^{-3} - C_1\delta^{-2}) \cosh^2\beta$$

After multiplication by 2δ, recognition of the hyperbolic function identity,

$$\sinh^2\beta = 2\sinh\beta\cosh\beta$$

and some algebraic simplification, the result, with $\Upsilon = Y_a/Y_o$, is

$$\sinh^2\beta + \left[\frac{1+\Upsilon^2}{4\Upsilon}\right]\sinh 2\beta = 3\left[\frac{1-\Upsilon^2}{2\Upsilon}\right]\beta \qquad (3.15)$$

Observe that if eq (3.15) is multiplied throughout by 4Υ, the result is

$$4\Upsilon \sinh^2\beta + (1+\Upsilon^2)\sinh 2\beta = 6(1-\Upsilon^2)\beta$$

and if $\Upsilon = 0$, that is $Y_a/Y_o = 0$, this reduces to

$$\sinh 2\beta = 6\beta$$

which is the transcendental equation used by Kern and Kraus (1972) and Aziz (1992) in the development of the optimum dimensions of the longitudinal fin of rectangular profile when the profile area is specified and the tip is adiabatic.

3.5.1 THE OPTIMIZATION PROCEDURE

Equation (3.15) is the key to the optimization. But, the value of the tip admittance is all that is specified and a solution must be obtained by trial-and-error. Some preliminaries are necessary before the trial-and-error procedure to determine suitable values of b and δ can proceed.

1. Because the parameter, β, involves both the fin height, b, and the fin thickness, δ, one of these must be selected to begin the procedure. However, because Y_o involves δ and not b, select a value for the fin thickness. This will force the value of Y_o

   $$Y_o = (2hk\delta)^{1/2}L$$

   The value of fin thickness will be the optimum fin thickness, not only because it will help dictate the value of the optimum fin height, $b_{\rm opt}$, but it will form the fin profile area

   $$A_p = \delta_{\rm opt} b_{\rm opt}$$

   - Observe that Y_o must be greater than Y_a to avoid the heat flow choke considered in Section 3.4.2.
   - A selection of δ that is too small will provide a value of Y_o that will yield, after the laborious trial-and-error procedure needed to solve eq (3.15), a value of β that will cause the fin height, b to be very large. A large value of b may cause the array to exceed the envelope permitted for the heat sink.

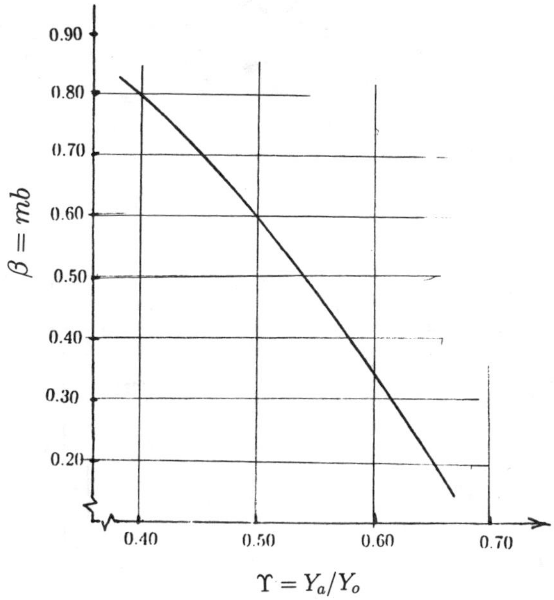

Figure 3.6: The optimization parameter, β, as a function of the ratio $\Upsilon = Y_a/Y_o$.

- On the other hand, too large a value of δ will yield a value of Y_o that will provide a value of β that will cause the value of the fin height to be quite low. A low value of b coupled with a high value of δ yields a low aspect ratio, b/δ, and low aspect ratios mean that the fin height is comparable in magnitude to the fin thickness. This leads to a violation of one of the Murray-Gardner assumptions that the heat flow in the fin must be one-dimensional. Figure 6 shows the variation of β with Υ and may be used to make ball-park estimates for use in eq (3.15). Observe that the value of Υ in the trial and error procedure should be restricted to values between 0.4 and 0.65. Values outside this range will yield values of β that provide unrealistic aspect ratios. b/δ.

2. Find the value of β that will satisfy eq (3.15).

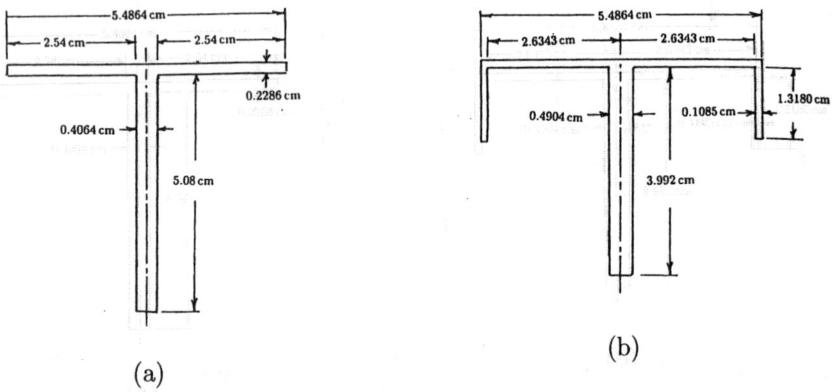

Figure 3.7: (a) Edge view of a heat sink to be used in a forced convection application and (b) a possible optimization.

3. Then find the optimum fin height. If the selection of δ is called δ_{opt}, then

$$b_{\mathrm{opt}} = \frac{\beta}{m} = \beta \left(\frac{k\delta_{\mathrm{opt}}}{2h} \right)^{1/2}$$

4. A check of the aspect ratio should then be made. Aspect ratios should be at least 6:1 in order to not violate the Murray Gardner assumption that the heat flow from base to tip is one-dimensional.

3.5.2 A FORCED CONVECTION EXAMPLE

Finned arrays that are cooled by forced convection, operate at higher values of the heat transfer coefficient. Because this mode provides more moderate values of the input admittance, it is deemed improper to attempt to cluster more than two are three remote fins at the tip of a supporting fin. A cluster admittance (equal to the tip admittance,

$Y_a = q_a/\theta_a$) that possesses a high magnitude will require a high value of input admittance to maintain Υ within the feasible range. All things being equal, a high value of the input admittance can only derive from a high value of fin thickness. This, in turn, may lead to a violation of the Murray-Gardner assumption that the fin height must be substantially higher than the fin thickness to assure that the heat flow is one-dimensional (from base to tip). Thus, when dealing with the forced convection mode, there are limitations to the number of fins that can be put into a cluster.

Example 3.2: Figure 3.7a shows a finned array that is to be employed in a forced convection application where the heat transfer coefficient is 60 W/m^2-K.[3] The fins are fabricated from a material having a thermal conductivity of 180 W/m-K and the array is 0.10 m long (perpendicular to the plane of the paper). Pertinent computed values along with the length, L, height, b and thickness, δ, for the fins are:

	Fins-*A*	**Fin-*B***
δ, m	2.286×10^{-3}	4.064×10^{-3}
b, m	0.0254	0.0508
L, m	0.100	0.100
AR	11.11	12.50
Y_o, W/°C	0.7027	0.9369
m, m^{-1}	17.0772	12.8079
mb	0.4378	0.6506

Fins-A have negligible tip heat dissipation. Under this circumstance, $Y_a = q_a/\theta_a = 0$ and eq (3.10) reduces to

$$Y_{\mathrm{IN}} = Y_o \left[\frac{\sinh mb}{\cosh mb}\right]$$

or

$$Y_{\mathrm{IN}} = Y_o \tanh mb$$

Use of this relationship with $Y_o = 0.7027$, W/°C and $mb = 0.4378$ gives $Y_{\mathrm{IN}} = 0.2870$ W/°C and the admittance at the tip of fin-B is $Y_a = 2Y_{\mathrm{IN}} = 2(0.2870) = 0.5740$ W/°C.

[3]This is the same value as used in Example 3.1 and is merely a coincidence.

Equation (3.15) can now be employed using mb and Y_o for fin-B with $\Upsilon = Y_a/Y_o = 0.5740/0.9369 = 0.6127\,\mathrm{W/°C}$. This gives $Y_{\mathrm{IN}} = 0.8220\,\mathrm{W/°C}$. An optimization can then be sought to maintain Y_{IN} at this value.

Fins-A can be optimized using eqs (3.14)

$$b_{\mathrm{opt}} = \frac{0.7978}{h}\left(\frac{Y_{\mathrm{IN}}}{L}\right) \tag{3.14a}$$

and

$$\delta_{\mathrm{opt}} = \frac{0.6321}{hk}\left(\frac{Y_{\mathrm{IN}}}{L}\right)^2 \tag{3.14b}$$

This straightforward computation yields

$$b_{\mathrm{opt}} = 3.816\,\mathrm{cm}$$

and

$$\delta_{\mathrm{opt}} = 0.1085\,\mathrm{cm}$$

with an aspect ratio of

$$\mathrm{AR} = 3.816/0.1085 = 35.17$$

Fin-B is optimized using the trial and error procedure which is based on eq (3.15). The procedure can be summarized by

$$\begin{aligned}
\delta_{\mathrm{opt}} &= 0.4904\,\mathrm{cm} \\
Y_o &= 1.0293\,\mathrm{W/°C} \\
\Upsilon &= Y_a/Y_o = 0.5740/1.0293 = 0.5577 \\
\beta &= 0.4655 \\
m &= 11.6595\,\mathrm{m}^{-1} \\
b_{\mathrm{opt}} &= 3.992\,\mathrm{cm}
\end{aligned}$$

with an aspect ratio of

$$\mathrm{AR} = 3.992/0.4904 = 8.14$$

The original design can be compared to the optimum design in a side-by-side listing and this optimization is shown in Fig 3.7b.

	Original Design	Optimum Design
Fin-A		
δ, cm	0.2286	0.1085
b, cm	2.54	3.816
A_p, cm^2	0.5806	0.4140
AR	11.11	35.17
Fin-B		
δ, cm	0.4064	0.4904
b, cm	5.08	3.992
A_p, cm^2	2.0645	1.9557
AR	12.5	8.14
Total A_p	3.2258	2.7857

On a profile area basis, the optimum design, uses only 2.7857/3.2258 = 0.864 or 86.4 percent as much metal.

3.5.2.1 WHAT IF?

What if an attempt is made to design Fins-A for the specified profile area of 0.5806 cm^2? In this case eqs (3.14) are employed and the result is

$$b_{\text{opt}} = 5.80\,\text{cm} \qquad \text{and} \qquad \delta_{\text{opt}} = 0.100\,\text{cm}$$

These dimensions provide an aspect ratio of 58 and the fin is somewhat unwieldly. Moreover, when compared with the original design, no metal is saved and the fin does not transfer as much heat as in the design where q_b or Y_{IN} is specified.

Finned arrays that operate in the free or natural convection mode operate at lower values of the heat transfer coefficient. Because of this, attempts to optimize individual fins contained in a cluster at the tip of a supporting fin can meet with rather dismal results. However, because the input admittance for these individual fins will be rather low in a natural convection application, the use of a cluster conataining many fins is encouraged. In such cases, the supporting fin can be optimized because the value of $\Upsilon = Y_a/Y_o$, due to the low value of Y_a, can still be made to fall in the range $0.40 \leq \Upsilon \leq 0.65$.

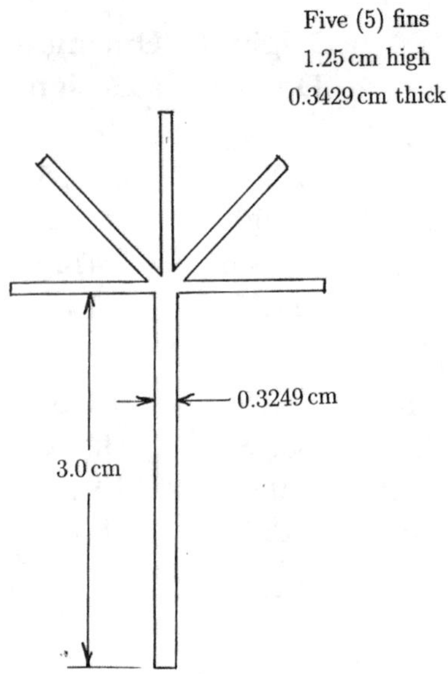

Figure 3.8: Edge view of a heat sink to be used in a natural convection application.

Example 3.3: Consider the array shown in Fig 3.8 which is to be employed in a natural convection application where the heat transfer coefficient is 6 W/m^2-K. The fins are fabricated from a material having a thermal conductivity of 180 W/m-K and the array is 0.10 m long (into the plane of the paper). Pertinent computed values along with the length, L, height, b and thickness, δ, for the fins are:

	Fins-A	**Fin-B**
number, n	5	1
δ, m	2.286×10^{-3}	3.429×10^{-3}
b, m	0.0125	0.03
L, m	0.10	0.10
AR	5.47	8.75
Y_o, W/°C	0.2222	0.2722
m, m^{-1}	5.4003	4.0931
mb	0.0675	0.1329
A_p m^2	2.858×10^{-5}	1.0287×10^{-4}

Use of eq (3.10) for fins-A with $Y_a = q_a/\theta_a = 0$ gives $Y_{\rm IN} =$ 0.0150 W/°C. This makes the cluster admittance at the tip of fin-B,

$Y_a = 5(0.0150) = 0.0750\,\mathrm{W/^\circ C}$. Equation (3.10) can then be used to show that $Y_{\mathrm{IN}} = 0.10682\,\mathrm{W/^\circ C}$ at the base of the array (the base of fin-B).

An attempt can be made to optimize fins-A using eqs (3.14)

$$b_{\mathrm{opt}} = \frac{0.7978}{h}\left(\frac{Y_{\mathrm{IN}}}{L}\right) \tag{3.14a}$$

and

$$\delta_{\mathrm{opt}} = \frac{0.6321}{hk}\left(\frac{Y_{\mathrm{IN}}}{L}\right)^2 \tag{3.14b}$$

but the results of such an attempt shows an outlandish result.

$$b_{\mathrm{opt}} = 1.99\,\mathrm{cm} \quad \text{and} \quad \delta_{\mathrm{opt}} = 0.0013\,\mathrm{cm}$$

This shows that, under natural convection conditions, not all fins can be optimized or a thermal optimization is not practical. An optimization can then be sought to maintain Y_{IN} at this value.

Fin-B is optimized using the trial and error procedure which is based on eq (3.15). Noting that fins-A remain untouched, the procedure can be summarized by

$$\begin{aligned}
\delta_{\mathrm{opt}} &= 0.0829\,\mathrm{cm} \\
Y_o &= 01338\,\mathrm{W/^\circ C} \\
\Upsilon &= Y_a/Y_o = 0.0750/0.1338 = 0.5598 \\
\beta &= 0.4604 \\
m &= 8.9700\,\mathrm{m}^{-1} \\
b_{\mathrm{opt}} &= 5.13\,\mathrm{cm} \\
A_p &= 0.452\,\mathrm{cm}^2
\end{aligned}$$

with an aspect ratio of

$$\mathrm{AR} = 5.13/0.0813 = 61.88$$

The original design can be compared to the optimum design by comparing the profile areas. The original design has a total profile area of $2.458\,\mathrm{cm}^2$ of which $1.209\,\mathrm{cm}^2$ is taken by fin-B. The final design has a total profile area of $1.854\,\mathrm{cm}^2$ of which $0.425\,\mathrm{cm}^2$ is taken by fin-B. The optimization shows that the original design has 1.33 times as much total metal and 2.85 times more metal in fin-B.

3.6 THE ANALYSIS OF ARRAYS OF FINS

The array of fins is composed of many individual fins in parallel which may be analyzed on an individual fin basis when care is taken to include heat transfer from the adjacent unfinned "prime surface area" of the base. In forced convection applications, the fins are usually shrouded so that all of the cooling air is confined to pass through the channel formed by the fins, the base plate and the shroud. Once the heat transfer coefficient is determined, the input admittance of the repeating section composed of the fin and that part of the shroud that is assigned to the repeating section is evaluated using the cluster and cascade algorithms. The base surface works in parallel with the fins and its input admittance is determined from its own heat transfer coefficient, h_b, and its surface area, S_b. The total input admittance is

$$Y_{\mathrm{IN}} = Y_{\mathrm{IN,b}} + nY_{\mathrm{IN,f}}$$

where n is the number of repeating sections in the array of fins which are $s + z$ wide.

Natural convection applications are considered in this section. The heat transfer coefficient can be obtained from the correlations of Starner and McManus (1963), Welling and Wooldridge (1965), Van de Pol and Tierney (1973), Bar-Cohen and Rohsenow (1984), Bilitsky (1986) and Yovanovich (1987a) and (1987b). The Bar-Cohen and Rohsenow (1984) relationships, based on the earlier work of Elenbaas (1942) and their use in designing the array of fins are discussed here.

In what follows, it is useful to define the fin efficiency in terms of the actual heat flow

$$q = hS\bar{\theta}$$

where $\bar{\theta}$ is the average or mean fin temperature excess. In the case of the longitudinal fin of rectangular profile where $x = a = 0$, the average or mean value of the temperature excess derives from the area under the temperature excess curve divided by the width of the curve. Hence

$$\bar{\theta} \quad = \quad \frac{\theta_b}{b}\int_{x=a}^{x=b}\frac{\cosh mx}{\cosh mb}dx$$

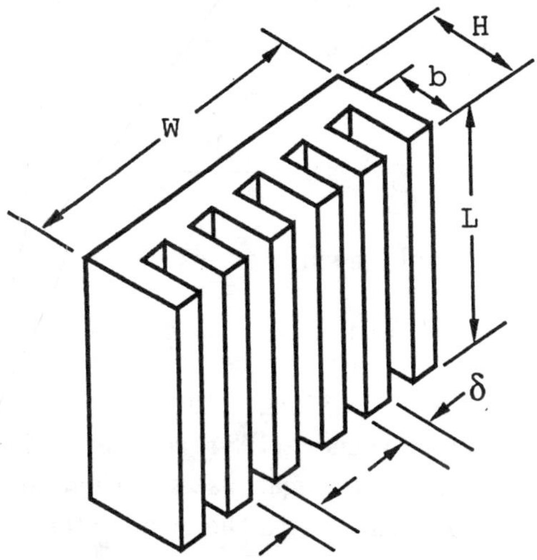

Figure 3.9: An array of fins.

$$= \frac{\theta_b}{mb}\left[\frac{\sinh mx}{\cosh mb}\right]_{x=0}^{x=b}$$

$$= \frac{\theta_b}{mb}\frac{\sinh mb}{\cosh mb}$$

$$= \theta_b \frac{\tanh mb}{mb}$$

or

$$\eta = \frac{\bar{\theta}}{\theta_b} \tag{3.16}$$

For the array of fins displayed in Fig 3.9, Bar-Cohen and Rohsenow give for the optimum fin spacing, $z_{\rm opt}$

$$z_{\rm opt} = 2.714P \tag{3.17}$$

where P is called the *plate-air parameter* with the thermal properties of the air evaluated at the average temperature between the plate and the surroundings

$$P = \left(\frac{\mu k L}{c_p \rho^2 g \beta \bar{\theta}}\right)^{1/4}$$

It is to be noted that this relationship offers a slight modification of the former Bar-Cohen and Rohsenow (1984) result for isothermal plates, by introducing the use of the average surface temperature.

With z_{opt} known, the number of fins can be determined. With W taken as the width of the array

$$n = \frac{W}{z_{opt} + \delta}$$

and this leads to the base surface area

$$S_b = WL - nL\delta = L(W - n\delta)$$

where, as indicated in Fig 3.9, L is the height of the array.

The Bar-Cohen-Rohsenow (1984) correlation for natural convection between vertical isothermal plates is then employed. First the Elenbaas number is formed with the average temperature, $\bar{\theta}$

$$\mathrm{El} = \frac{\rho^2 \beta g c_p z^4 \bar{\theta}}{\mu k L} \tag{3.18}$$

and the heat transfer coefficient is then determined from the Nusselt number

$$\mathrm{Nu} = \frac{hz}{k} = \left[\frac{576}{(\mathrm{El})^2} + \frac{2.873}{(\mathrm{El})^{1/2}}\right]^{-1/2}$$

as

$$h = \frac{k}{z}\left[\frac{576}{(\mathrm{El})^2} + \frac{2.873}{(\mathrm{El})^{1/2}}\right]^{-1/2} \tag{3.19}$$

The input admittance of each fin, $Y_{IN,f}$ is then obtained (perhaps through the use of the reflection relationship of eq (3.10)). Using these relations, it is possible to determine whether a chosen heat sink offers a solution that meets the specification.

Example 3.4: Suppose that an area of 4 cm × 4 cm must dissipate 4.5 W at 75 °C by natural convection to air at an environmental temperature of 25°C with a heat sink containing fins no higher than 3 cm. Can a heat sink employing fins with a length of 4 cm and a thickness of 0.2286 cm and a thermal conductivity of 180 W/m-K be used for this application?

Here

$$\begin{aligned} W &= 4\,\text{cm} \quad (0.04\,\text{m}) \\ L &= 4\,\text{cm} \quad (0.04\,\text{m}) \\ b &= 3\,\text{cm} \quad (0.03\,\text{m}) \\ \delta &= 0.2286\,\text{cm} \quad (0.002286\,\text{m}) \\ \theta_b &= 75 - 25 = 50°\text{C} \end{aligned}$$

and

$$T_{\text{avg}} = 50°\text{C}$$

The thermal properties of air at 50°C are

$$\begin{aligned} c_p &= 1007\,\text{J/kg-K} \\ \rho &= 1.095\,\text{kg/m}^3 \\ k &= 0.0274\,\text{W/m-K} \\ \mu &= 1.95 \times 10^{-5}\,\text{kg/s-m} \end{aligned}$$

and

$$\beta = \frac{1}{T_{av}} = \frac{1}{323} = 0.003096\,\text{K}^{-1}$$

Now, assume that the fin efficiency for this natural convection application will be very high and, indeed, close to unity. Hence, assume that $\eta = 0.990$ so that

$$\bar{\theta} = \eta\theta_b = 0.990(50) = 49.5°\text{C}$$

This must be verified at the end of the process.

The plate air parameter, P will be

$$\begin{aligned} P &= \left(\frac{\mu k L}{c_p \rho^2 g \beta \bar{\theta}}\right)^{1/4} \\ &= \left[\frac{(1.95 \times 10^{-5})(0.0274)(0.04)}{1007(1.095)^2(9.81)(0.003096)(49.5)}\right]^{1/4} \\ &= (1.177 \times 10^{-11})^{1/4} \\ &= 0.001853\,\text{m} \quad (0.1853\,\text{cm}) \end{aligned}$$

The optimum clear space between the fins is then

$$\begin{aligned} z &= 2.714P \\ &= 2.714(0.001853\,\mathrm{m}) \\ &= 0.005028\,\mathrm{m} \quad (0.5028\,\mathrm{cm}) \end{aligned}$$

The centerline spacing of the fins is $z + \delta$ and allowing n spaces for n fins gives

$$n = \frac{W}{z+\delta} = \frac{4}{0.5028 + 0.2286} = \frac{4}{0.7314} \approx 5\,\text{fins}$$

The Bar-Cohen and Rohsenow (1984) correlation for natural convection between vertical isothermal plates may be used for this configuration if $\bar{\theta} = \eta\theta_b$. First the Elenbaas number is formed:

$$\begin{aligned} \mathrm{El} &= \frac{\rho^2 \beta g c_p z^4 \bar{\theta}}{\mu k L} \\ &= \frac{(1.095)^2(0.003096)(9.81)(1007)(0.005015)^4(49.5)}{(1.95 \times 10^{-5})(0.0274)(0.04)} \\ &= 53.72 \end{aligned}$$

and then eq (3.19) may be employed

$$\begin{aligned} h &= \frac{k}{z}\left[\frac{576}{(\mathrm{El})^2} + \frac{2.873}{(\mathrm{El})^{1/2}}\right]^{-1/2} \\ &= \frac{0.0274}{0.005015}\left[\frac{576}{(53.72)^2} + \frac{2.873}{(53.72)^{1/2}}\right]^{-1/2} \\ &= (5.464)(0.592)^{-1/2} \\ &= (5.464)(1.300) \\ &= 7.10\,\mathrm{W/m^2\text{-}K} \end{aligned}$$

Because the fin temperature gradient can substantially reduce the fin tip temperature relative to the fin base, it is prudent to neglect heat transfer from the fin tip in the initial evaluation of heat sink performance. Thus, the input admittance of the fins can be obtained from eq (3.10) with $Y_a = 0$. Thus with

$$\begin{aligned} m &= \left(\frac{2h}{k\delta}\right)^{1/2} \\ &= \left[\frac{2(7.10)}{180(0.002286)}\right]^{1/2} \\ &= 5.876\,\mathrm{m}^{-1} \\ mb &= 5.876(0.03) = 0.1763 \end{aligned}$$

and

$$\begin{aligned} Y_o &= (2hk\delta)^{1/2}L \\ &= [2(7.10)(180)(0.002286)]^{1/2}(0.04) \\ &= (5.843)^{1/2}(0.04) \\ &= 0.0967\,\mathrm{W/^\circ C} \end{aligned}$$

the input admittance can be computed as

$$\begin{aligned} Y_{\mathrm{IN,f}} &= Y_o\frac{\sinh mb}{\cosh mb} = Y_o \tanh mb \\ &= (0.0970)\tanh 0.1763 \\ &= (0.0970)(0.1745) \\ &= 0.0169\,\mathrm{W/^\circ C} \end{aligned}$$

The surface area of the base is the total area, WL, minus the *footprints* of the 5 fins

$$\begin{aligned} S_b &= WL - nL\delta = L(W - n\delta) \\ &= 4.00[4.00 - 5(0.2286)] \\ &= 4.00(2.8570) = 11.428\,\mathrm{cm}^2 \\ &= 11.428 \times 10^{-4}\,\mathrm{m}^2 \end{aligned}$$

Hence, for the base surface

$$\begin{aligned} Y_{\mathrm{IN,b}} &= h_b S_b \\ &= 7.10(11.428 \times 10^{-4}) \\ &= 0.0081\,\mathrm{W/^\circ C} \end{aligned}$$

Then, the total input admittance is

$$\begin{aligned} Y_{\mathrm{IN}} &= Y_{\mathrm{IN},b} + nY_{\mathrm{IN},f} \\ &= 0.0081 + 5(0.0169) \\ &= 0.0081 + 0.0845 \\ &= 0.0926\,\mathrm{W/^\circ C} \end{aligned}$$

and the heat sink can dissipate

$$q_b = Y_{\mathrm{IN}}\theta_b = 0.0926(50) = 4.63\,\mathrm{W}$$

Because the dissipation exceeds the required dissipation ($4.66 > 4.50\,\mathrm{W}$), this heat sink can be accepted as a satisfactory design if the fin efficiency of $\eta = 0.990$ can be verified. Here, with mb already computed as $mb = 0.1767$

$$\eta = \frac{\tanh mb}{mb} = \frac{\tanh 0.1763}{0.1763} = \frac{0.1745}{0.1763} = 0.9898$$

Thus, the assumed value of η has been verified and the design can be accepted.

Observe that there will be a small dissipation from the fin tip but this dissipation is negligible. With the temperature excess profile given by

$$\theta(x) = \theta_b \frac{\cosh mx}{\cosh mb}$$

at $x = a = 0$

$$\theta(x=0) = (50)\frac{\cosh 0}{\cosh 0.1763} = (50)\frac{1.0000}{1.0156} = 49.2^\circ\mathrm{C}$$

The heat flow from the tip assuming that the tip heat transfer coefficient is $h_a = h = 7.10\,\mathrm{W/m^2\text{-}^\circ C}$ will be

$$\begin{aligned} q_a &= nh_aS_a\theta_a \\ &= n\delta L\theta_a \\ &= (5)(7.10)(0.002286)(0.04)(49.2) \\ &= 0.160\,\mathrm{W} \end{aligned}$$

The heat flow from the fin tips is indeed negligible.

3.6.1 THE LIMITATION IN THE ARRAY OF FINS

The quest for a reasonable heat sink volume envelope usually prevents the use of n-optimum fins. The profile area of each fin in the foregoing design example is

$$A_p = 3(0.2286) = 0.6858\,\mathrm{cm}^2 \quad (6.858 \times 10^{-5}\,\mathrm{m}^2)$$

and equations (3.13)

$$b_{\mathrm{opt}} = 1.0023\left(\frac{A_p k}{h}\right)^{1/3} \tag{3.13a}$$

and

$$\delta_{\mathrm{opt}} = \frac{A_p}{b_{\mathrm{opt}}} = 0.9977\left(\frac{A_p^2 h}{k}\right)^{1/3} \tag{3.13b}$$

can be used to show that

$$\begin{aligned}
b_{\mathrm{opt}} &= 1.0023\left(\frac{A_p k}{h}\right)^{1/3} \\
&= 1.0023\left[\frac{(6.858 \times 10^{-5})(180)}{7.14}\right]^{1/3} \\
&= 1.0023(1.729 \times 10^{-3}) \\
&= 1.0023(0.1200) \\
&= 0.1203\,\mathrm{m}
\end{aligned}$$

or

$$b_{\mathrm{opt}} = 12.03\,\mathrm{cm}$$

and

$$\begin{aligned}
\delta_{\mathrm{opt}} &= \frac{A_p}{b_{\mathrm{opt}}} \\
&= \frac{6.858 \times 10^{-5}}{0.1203} \\
&= 0.000507\,\mathrm{m}
\end{aligned}$$

or

$$\delta_{\text{opt}} = 0.057\,\text{cm}$$

The use of fins with an optimum fin height of $b_{\text{opt}} = 12.03\,\text{cm}$ may conceivably lead to space envelope problems. Thus, attention now turns to a procedure for the optimization of the entire array of fins proposed by Bar-Cohen and Jelinek (1986).

3.7 OPTIMUM (LEAST MATERIAL) HEAT SINKS

3.7.1 INTRODUCTION

The optimization of single fins, discussed in Section 3.4 provides the starting point for the optimization of the array of fins (the heat sink). For a specified convective heat transfer coefficient, it is possible to determine the optimum dimensions of the rectangular, longitudinal fin, as well as several other classes of fin geometries. However, because the convective heat transfer coefficient varies with the spacing between the fins, heat sink optimization involves an understanding of convection in the interfin channels as well as conduction in the fins.

3.7.2 THEORETICAL BACKGROUND

Following the method outlined by Kern and Kraus (1972) and Aziz (1992) employing the Murray-Gardner assumptions, it can be shown that the temperature excess profile for the longitudinal fin of rectangular profile, with x taken as the height coordinate with origin at the fin tip, is

$$\theta = \theta_b \frac{\cosh mx}{\cosh mb} \tag{3.20a}$$

and the heat entering the base, per unit fin length, is

$$\frac{q_b}{L} = \frac{Y_o}{L}\theta_b \tanh mb \tag{3.20b}$$

where the reader is reminded that

$$m = \left(\frac{2h}{k\delta}\right)^{1/2} \qquad \text{and} \qquad Y_o = (2hk\delta)^{1/2}L$$

and where it must be emphasized that these expressions pertain to a single fin with an adiabatic tip.

For convenience in later calculations, eq (3.20b) can be recast in terms of the fin profile area, $A_p \equiv \delta b$, as

$$\frac{q}{L} = \left(\frac{2hkA_p}{b}\right)^{1/2} \theta_b \tanh\left[b^{3/2}\left(\frac{2h}{kA_p}\right)^{1/2}\right] \tag{3.21}$$

Examination of eq (3.21) suggests that for a fixed fin profile area, the fin heat flow will depend on the fin height, b, approaching zero for both very narrow and very thick fins and peaking at an intermediate value of b. This "optimum" value of b can be found by differentiating eq (3.21) with respect to b and determining where this derivative vanushes. The result of this operation with

$$\beta \equiv mb = b^{3/2}(2h/kA_p)^{1/2}$$

is a transcendental equation

$$(\tanh\beta - 3\beta\,\mathrm{sech}^2\beta)_{s,\mathrm{opt}} = 0 \tag{3.22}$$

where the subscript, s, opt, is intended to mean "*single fin optimum*". This equation corresponds to that developed by Kern and Kraus (1972) and, after some algebraic adjustment, matches the result of Aziz (1992)

$$(\sinh 2\beta - 6\beta)_{s,\mathrm{opt}} = 0$$

The solution of eq (3.22) can be expected to yield the rectangular fin geometry that will maximize the fin heat dissipation for a given profile area, volume, or mass. Alternatively, this fin geometry can be expected to yield the least material fin for a specified heat dissipation. The single-fin, optimum value of $\beta_{s,\mathrm{opt}}$, is found to equal $\beta_{s,\mathrm{opt}} = 1.4192$ [Kern and Kraus (1972) and Aziz (1992)].

With $\beta_{s,\mathrm{opt}} = 1.4192$, the aspect ratio, δ/b, of the single, optimum fin can be derived from the expression for mb

$$mb = b\left(\frac{2h}{k\delta}\right)^{1/2} b = 1.4192$$

$$\left(\frac{2hb^2}{k\delta}\right)^{1/2} = 1.4192$$

$$\left(\frac{2hb^2}{k\delta}\right) = (1.4192)^2$$

and hence

$$\left(\frac{\delta}{b}\right)_{s,\mathrm{opt}} = \frac{2}{(1.4192)^2}\frac{hb}{k} = 0.993\frac{hb}{k} \tag{3.23}$$

The profile area of the optimum fin can be found from eqs (3.14)

$$b_{\mathrm{opt}} = \frac{0.7978}{h}\left(\frac{Y_{\mathrm{IN}}}{L}\right) \tag{3.14a}$$

and

$$\delta_{\mathrm{opt}} = \frac{0.6321}{hk}\left(\frac{Y_{\mathrm{IN}}}{L}\right)^2 \tag{3.14b}$$

to be

$$A_{p,\,\mathrm{opt}} = b_{\mathrm{opt}}\delta_{\mathrm{opt}} = \frac{(0.6321)(0.7978)}{h^2k}\left(\frac{Y_{\mathrm{IN}}}{L}\right)^3$$

Then, because $Y_{\mathrm{IN}} = q/\theta_b$

$$\left(\frac{q}{\theta_b L}\right)_{s,\mathrm{opt}} = \left(\frac{h^2 k A_p}{0.5043}\right)^{1/3}$$

and the optimum single fin heat dissipation per unit length is given by

$$\left(\frac{q}{L}\right)_{s,\mathrm{opt}} = 1.258\theta_b(h^2A_pk)^{1/3} \tag{3.24}$$

3.7.3 OPTIMUM MULTIPLE FIN ARRAY

3.7.3.1 INDIVIDUALLY OPTIMUM FINS

The results of Section 3.7.2 can be applied to the optimization of an array of fins by assuming that the array is to be composed of n individually optimum fins. These fins are to possess an aspect ratio determined by eq (3.23) and are capable of transferring heat according to eq (3.24). In such an array, the required number of fins, n, is approximately equal to the desired or design point total heat dissipation divided by the heat transfer rate of each fin, Q/q. In examining eq (3.24), it may be observed that the optimum fin profile area, A_p, is proportional to q^3/h^2. Consequently, the profile area of the entire array of fins, nA_p, is proportional to Qq^2/h^2.

As a consequence of this proportionality, minimization of the total array profile area and mass would appear to require either an increase in the heat transfer coefficient (an obvious approach) or a decrease in the heat dissipation of each fin by reducing the fin dimensions. Either approach, carried to its logical conclusion, would yield a thermally optimum array of a near-infinite number of vanishingly thin fins, which would, together, have a near-zero profile area and a near-zero mass!

Unfortunately, manufacturing constraints and available pumping power (or the buoyant driving force) generally limit the minimum fin thickness and minimum interfin spacing respectively and thus make the realization of a near-zero added mass, optimum array impossible. Instead, the thermal optimization must seek to find the minimum viable array profile area (or mass) consistent with the appropriate system constraints. When the fin material, the prime mover, and the working fluid have been determined, the requisite analysis can often be cast in the form of minimizing the total fin profile area (or mass) for a specified number of fins and a given heat transfer coefficient. A least material array optimization, consistent with this definition, is presented in Section 3.8.

In optimizing an array of longitudinal fins, as may be done in the design of a heat sink, it must be recognized that the base temperature, the local fluid temperature, and the surface heat transfer coefficient may all vary in the flow direction. Consequently, in applying eqs (3.23) and (3.24) to a single fin in an array, the designer can elect to use average values for θ_b and h or to perform the calculations in discrete steps in the flow direction. Such discretization can, of course, follow naturally from the geometry of the array when an interrupted fin structure is used.

It must be noted that while the fin heat transfer rate is proportional to both θ_b and $h^{2/3}$, the optimum single-fin aspect ratio, as indicated by eq (3.23), is independent of θ_b and, in fact, becomes invariant in the flow direction as the fully developed condition ($h =$ constant) is attained. As a consequence, the use of average values can be expected to introduce only a minor error in the determination of the optimum aspect ratio for such a fin.

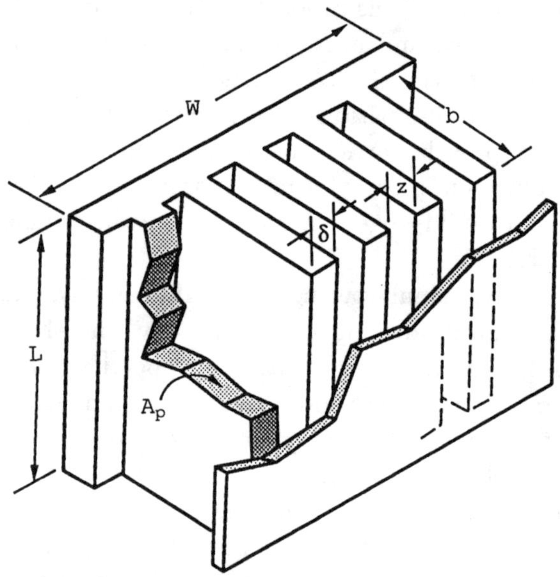

Figure 3.10: The geometric definition of an array of longitudinal fins.

3.7.3.2 THE OPTIMUM ARRAY

The rate of heat transfer from a shrouded array of longitudinal, rectangular, fins, as shown in Fig 3.10, can be found by summing the thermal contribution of the fins and the interfin area, according to

$$\frac{Q}{L} = n\left[\left(\frac{2hkA_p}{b}\right)^{1/2}\theta_b \tanh\beta + h_w\theta_b\left(\frac{W}{n} - \frac{A_p}{b}\right)\right] \tag{3.25}$$

where A_p, as before, is the profile area of the single fin and equal to δb, W is the array width, and h_w the heat transfer coefficient along the interfin surfaces (the prime or *wall* surfaces). While provision has been made in eq (3.25) for differences between the heat transfer coefficient prevailing along the fin surfaces and in the interfin spaces, it has been assumed that the primary surface area (including both the fin base and interfin surface area) is isothermal. The validity of this common assumption can be expected to improve as the fin spacing decreases and the thermal conductivity, as well as thickness of the fin support plate, increases.

To find the maximum array dissipation at a fixed profile area, nA_p, or (with the number of fins predetermined by system considerations) for a fixed A_p, it is possible to differentiate $Q/Ln\theta_b$ with respect to b and to set the derivative equal to zero. Upon so doing, it is found that the array optimum is governed by

$$\left[\frac{1}{2}(2hkA_p)^{1/2}f(\beta,b)+\frac{h_wA_p}{b^2}\right]_{a,\mathrm{opt}}=0 \qquad (3.26)$$

where

$$f(\beta,b)=3\beta b^{-1/2}\mathrm{sech}^2\beta-b^{-3/2}\tanh\beta$$

and where the subscript, a,opt, is intended to mean "array optimum." After multiplying through by $2b^3/kA_p$ and cancelling common terms

$$(\beta\tanh\beta-3\beta^2\,\mathrm{sech}^2\beta)_{a,\mathrm{opt}}=\frac{2h_wb}{k} \qquad (3.27)$$

Comparison of eqs (3.22) and (3.27) reveals the relation for the optimum array value of β to differ from the single-fin optimization by a term, $2h_wb/k$, which pertains to the heat transfer in the interfin spaces. Indeed, for h_w equal to zero, eq (3.27) reduces to eq (3.22) and $\beta_{a,\mathrm{opt}}=\beta_{s,\mathrm{opt}}=1.4192$. The iteratively obtained solution of eq (3.27) for nonzero values of h_w is shown in Fig 3.11 and may be approximated by the linear relation

$$\beta_{a,\mathrm{opt}}=1.4192+1.125\frac{h_wb}{k} \qquad (3.28)$$

in the range $1.4192<\beta_{a,\mathrm{opt}}<1.60$.

Example 3.5: Determine the array-optimum fin parameter, $\beta_{a,\mathrm{opt}}$, for typical heat sink configurations and compare its value to the single-fin optimum value. The cases to be considered are

(a) A 3 cm high aluminum fin cooled by forced convection in air,

(b) A 20 cm high cast iron fin cooled by natural convection in air and

(c) A 3 cm high aluminum fin cooled by natural convection in air.

Solution

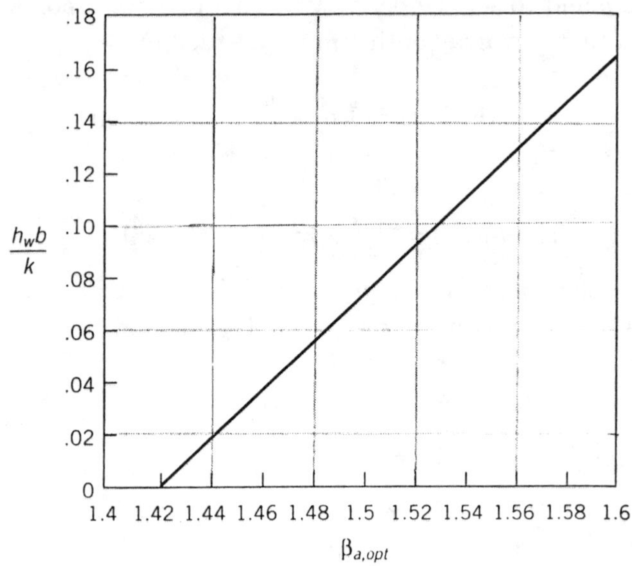

Figure 3.11: Optimizing value of β for an array of fins.

(a) For the 3 cm high aluminum fin cooled by forced convection in air, $b = 0.03$ m. Suppose that $k = 175$ W/m-K and $h = 100$ W/m^2-K. Then

$$\begin{aligned}\beta &= 1.4192 + 1.125\left(\frac{hb}{k}\right)\\ &= 1.4192 + 1.125\left[\frac{100(0.03)}{175}\right]\\ &= 1.4192 + 0.0193\\ &= 1.4385\end{aligned}$$

and the ratio of the array-optimum β to the single-fin optimum is $1.4385/1.4192 = 1.0136$. The array-optimum value is 1.4 percent higher.

(b) For the 20 cm high cast iron fin cooled by natural convection in air, $b = 0.20$ m. Suppose that $k = 30$ W/m-K and $h = 6$ W/m^2-K. Then

$$\beta = 1.4192 + 1.125\left(\frac{hb}{k}\right)$$

$$
\begin{aligned}
&= 1.4192 + 1.125 \left[\frac{6(0.20)}{30} \right] \\
&= 1.4192 + 0.0450 \\
&= 1.4642
\end{aligned}
$$

and the ratio of the array-optimum β to the single-fin optimum is 1.4642/1.4192 = 1.0317. The array-optimum value is 3.2 percent higher.

(c) For the 3 cm high aluminum fin cooled by natural convection in air, $b = 0.03\,\text{m}$. Ssuppose that $k = 175$ W/m-K and $h = 6$ W/m^2-K. Then

$$
\begin{aligned}
\beta &= 1.4192 + 1.125 \left(\frac{hb}{k} \right) \\
&= 1.4192 + 1.125 \left[\frac{6(0.03)}{175} \right] \\
&= 1.4192 + 0.0012 \\
&= 1.4204
\end{aligned}
$$

and the ratio of the array-optimum β to the single-fin optimum is 1.4204/1.4192 = 1.0008. The array-optimum value is essentially the same as the individual fin optimum.

Because the fin aspect ratio, δ/b, is, as shown in eq (3.23), inversely proportional to β, the marginally higher values for the optimum array imply that the array-optimum fins can be expected to be somewhat thinner than those based on the individual fin optimization. However, judging by the values, shown in the foregoing example, this difference may often be sufficiently small to justify the use, in commercial practice, of the individual-optimum fin values. The implications of this conclusion are explored in the next two sections, dealing with natural and forced convection optimum fin arrays in air, respectively.

As noted earlier, the validity of the foregoing analysis is limited to configurations in which the primary surface area is (or can be assumed) isothermal, as is often the case in the thermal control of electronic components. Furthermore, this analysis is limited to situations in which the heat transfer coefficient along the finned surface can be assumed constant.

3.8 OPTIMUM NATURAL CONVECTION HEAT SINKS

3.8.1 INTRODUCTION

The design of natural convection fin arrays or heat sinks is complicated by the coupling between the flow and temperature fields. This is a characteristic of all natural convection heat transfer situations. Thermal optimization of such an array of fins requires selection of both the optimum fin spacing and optimum fin thickness or thickness, each determined to maximize total heat dissipation for a given added mass or profile area.

In natural convection, a balance between viscous dissipation and buoyant potential, rather than pump or blower characteristics, establishes the desired interfin channel width. It may be anticipated that as the channel width or plate spacing is reduced, the heat transfer coefficient would decrease below the value associated with an isolated plate in an infinite medium. Based on his experimental results, and his subsequent correlation for relatively short plates in air

$$\mathrm{Nu} = \frac{hz}{k} = \frac{\mathrm{El}}{24}\left(1 - e^{-35/\mathrm{El}}\right)^{3/4} \tag{3.29}$$

Elenbaas (1942) determined that the optimum spacing between parallel, isothermal, thin, vertical plates in air could be obtained by setting the channel Rayleigh number (based on plate spacing and channel aspect ratio and here referred to as the Elenbaas number) equal to 50. At this spacing, the product of the total plate surface area and the heat transfer coefficient is maximized, yielding the peak volumetric heat dissipation rate for the entire array. The optimum Nusselt number was found by Elenbaas (1942) to equal 1.25 thus providing an optimum average heat transfer coefficient along the plate surfaces of

$$h_{\mathrm{opt}} = 1.25\frac{k_g}{z_{\mathrm{opt}}} \tag{3.30}$$

In Bar-Cohen (1979), the Elenbaas' optimum spacing was extended to optimum fin arrays by basing the plate air paramater, P, on an average surface temperature. Thus

$$P = \left(\frac{L\nu^2}{g\beta\hat{\theta}\mathrm{Pr}}\right)^{1/4}$$

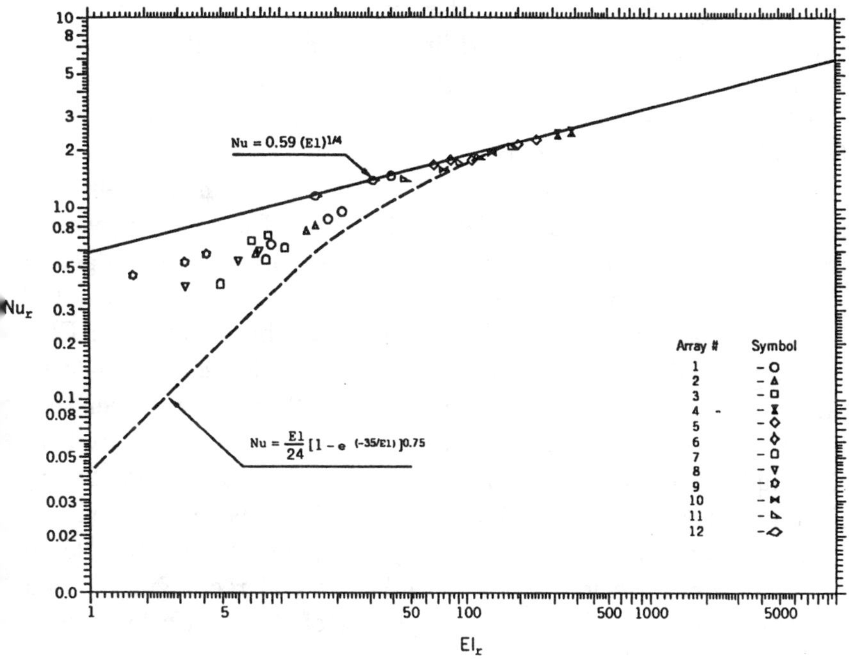

Figure 3.12: Nusselt numbers for vertical arrays (Bilitsky (1986)].

and

$$z_{\text{opt}} = \left(\frac{50L\nu^2}{\eta g\beta\bar{\theta}\text{Pr}}\right)^{1/4} = 2.66P \tag{3.31}$$

As previously noted in eq (3.17), the average temperature excess, $\bar{\theta}$, can be expressed as the product of the fin efficiency, η, and the base temperature excess, θ_b. The fin efficiency for the single optimum fin, for which $mb = 1.4192$ can be found to equal 0.627.

$$\eta = \frac{\tanh mb}{mb} = \frac{\tanh 1.4192}{1.4192} = \frac{0.8894}{1.4192} = 0.627$$

The extension of such an optimization approach to fins where the thickness is not negligible is facilitated by the use of the composite Nusselt number equation of eq (3.19) and is discussed by Bar-Cohen and Rohsenow (1983).

It is interesting to recall that the Elenbaas results were obtained with relatively tall fins where inflow from the open edges had a negligible impact on the heat transfer rate. However, studies on actual heat sinks, most recently by Bilitzky (1986), as well as of narrow vertical plates by Sparrow and Bahrami (1980), have shown that the Elenbaas correlation, eq (3.29), underpredicts the heat transfer coefficients for such configurations at low Elenbaas numbers. As displayed in Fig 3.12 for closely-spaced narrow fins, the Nusselt numbers lie between the isolated plate limit and the Elenbaas correlation, thereby reflecting the contribution of inflow from the open edge. As the Elenbaas number increases, the Elenbaas correlation asymptotically approaches the isolated plate limit and in the range of El = 50, corresponding to the optimum fin spacing, the isolated plate relation yields Nusselt numbers that are just 25 percent higher than eq (3.29). In this same intermediate range of Elenbaas numbers, the empirical heat transfer coefficients, for the heat sinks studied by Bilitzky (1986), are approximately 15 percent above the predicted values. Moreover, examination of eq (3.23) reveals that the "least-material" optimization of a longitudinal fin of rectangular profile yields relatively tall fins, (for example, a b of 10 cm for a 1 mm thick fin with a thermal conductivity of 50 W/m-K).

Consideration of these twin factors, intermediate Elenbaas numbers and relatively tall fins, suggests that the optimization of a multiple fin array can proceed on the basis of the Elenbaas correlation for the natural convection heat transfer coefficient. Furthermore, because the previous section has shown the optimizing parameter for the fin aspect ratio to be nearly independent of the fin array characteristics, it would now appear possible to use eqs (3.24), (3.30) and (3.31) to determine the geometric parameters of an optimum heat sink using vertical, natural convection air-cooled fins.

3.8.2 ARRAY DISSIPATION

To obtain the heat dissipation of an array of optimally spaced, least material fins, eq (3.24) with eq (3.30) can be used to quantify the heat flow through each fin, while eq (3.31) sets the interfin spacing. Moreover, Assuming, that h_{opt} can also be used to characterize the heat transfer coefficient in the area between the fins and that the number of fins is equal to the number of fin spaces (as is nearly the case for a large number of fins), the total heat dissipation of the array is expressible as

$$\frac{Q}{L} = \frac{\left(\frac{q_{\text{opt}}}{L} + h_{\text{opt}} z_{\text{opt}} \theta_b\right) W}{z_{\text{opt}} + \delta_{\text{opt}}} \tag{3.32}$$

Following substitution for $q_{\text{opt}}/L, h_{\text{opt}}$ and z_{opt} and using the fin efficiency of 0.627 as is appropriate for optimum rectangular fins, eq (3.32) can be modified to yield

$$\frac{Q}{LW\theta_b} = \frac{0.862 \frac{k_g k \delta_{\text{opt}}}{P^{1/2}} + 1.25 k_g}{2.66P + \delta_{\text{opt}}} \tag{3.33}$$

A graphical presentation of eq (3.33) for aluminum, brass and steel arrays of least material fins, dissipating heat to ambient air, is shown in Figs 3.13 through 3.15 respectively, for a typical parametric range. Array dissipation is seen to vary greatly with both the array parameter, P, and the thickness of the optimum fin, δ_{opt}. Furthermore, a maximizing value of δ_{opt} appears to exist for each value of P.

The locus of the array dissipation maxima, shown graphically in Figs 3.13 through 3.15, can be obtained by differentiation of eq (3.33) or, more simply, by differentiation of an approximate form of this relation, in which heat transfer from the area between the fins is neglected. With this additional assumption, clearly justified by the large surface area of the optimum natural convection fin, eq (3.32) yields

$$\frac{Q}{LW\theta_b} \cong 1.258 \frac{(hk\delta_{\text{opt}})^{1/2}}{z_{\text{opt}} + \delta_{\text{opt}}} \tag{3.34}$$

Following Bar-Cohen (1979), differentiating eq (3.34) with respect to δ_{opt} and setting the derivative to zero, the maximizing value of δ_{opt} is found to equal z_{opt}.

This particular value of fin thickness defines an array geometry that maximizes heat dissipation from a prime area of given length, L, and specified temperature excess, θ_b, while using least-material fins of rectangular profile. Substitution of $\delta_{\text{opt}} = z_{\text{opt}}$ into eq (3.33) yields the heat transfer rate for the maximum dissipation, optimum

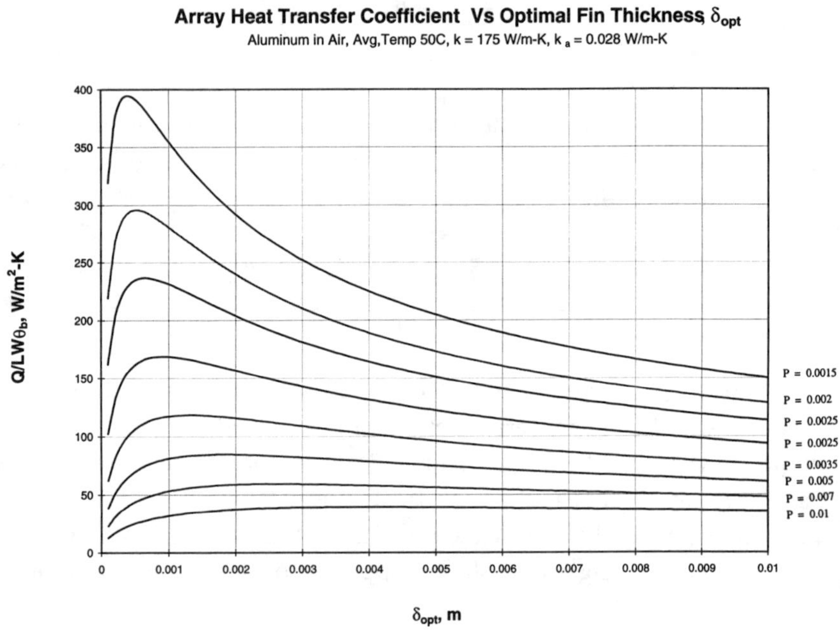

Figure 3.13: Natural convection heat dissipation of an array of optimally spaced, vertical, rectangular profile least material aluminum fins.

array as

$$\left(\frac{Q}{LW\theta_b}\right)_{\text{opt}} = 0.264\left(\frac{kk_g}{P^2}\right)^{1/2} \tag{3.35}$$

where k is the thermal conductivity of the fin material and k_g is the thermal conductivity of the fluid.

3.8.3 EXAMPLE

Example 3.6: To evaluate the efficacy of this approach, consider an example of an air-cooled natural convection fin array. A single side of an aluminum ($k = 175\,\text{W/m-K}$) module containing a laboratory data aquisition system dissipates 80 W and is to be cooled to a temperature of 65°C by natural convection in air. The module wall

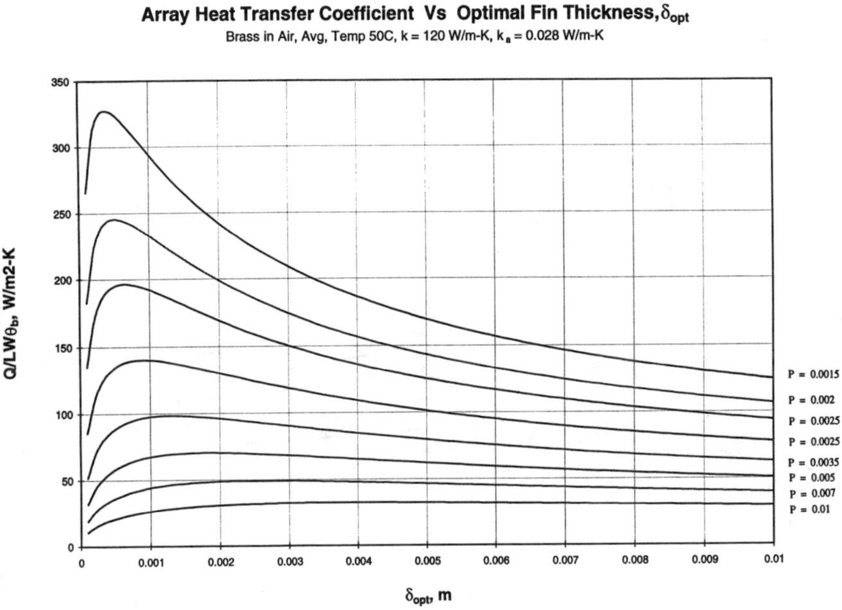

Figure 3.14: Natural convection heat dissipation of an array of optimally spaced, vertical, rectangular profile least material brass fins.

is 35.6 cm high and 30.5 cm wide and is exposed to ambient air at 35°C. Find the maximum "least material" heat transfer rate from this module wall and determine the added mass and geometry (fin height, thickness and fin spacing) of the finned array.

Solution

For the stated conditions

$$\theta_b = 65°\text{C} - 35°\text{C} = 30°\text{C}$$

and

$$T_{\text{av}} = \frac{65°\text{C} + 35°\text{C}}{2} = 50°\text{C}$$

the thermal properties of air are

$$\text{Pr} = 0.701$$

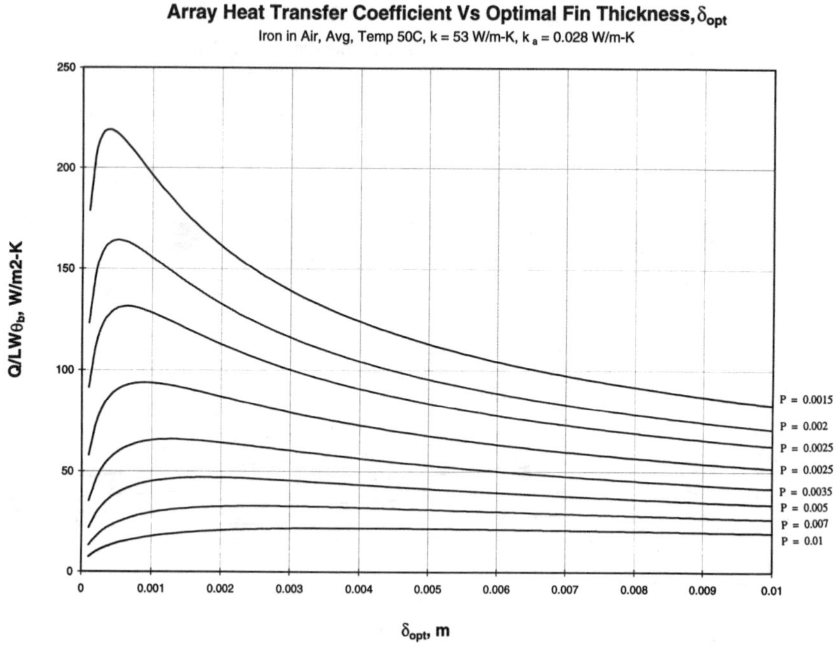

Figure 3.15: Natural convection heat dissipation of an array of optimally spaced, vertical, rectangular profile least material steel fins.

$$\nu = 1.79 \times 10^{-5}\,\mathrm{m^2/s}$$
$$k = 0.0274\,\mathrm{W/m\text{-}K}$$

and

$$\beta = \frac{1}{T_{av}} = \frac{1}{323} = 0.003096\,\mathrm{K^{-1}}$$

The fin array parameter, P is evaluated as

$$P = \left(\frac{L\nu^2}{g\beta\bar{\theta}\mathrm{Pr}}\right)^{1/4} = \left(\frac{L\nu^2}{g\beta\eta\theta_b\mathrm{Pr}}\right)^{1/4}$$

or

$$P = \left[\frac{(0.356)(1.79\times10^{-5})^2}{(9.81)(0.003096)(0.627)(30)(0.701)}\right]^{1/4}$$

$$
\begin{aligned}
&= (2.853 \times 10^{-10})^{1/4} \\
&= 0.004110\,\mathrm{m}
\end{aligned}
$$

The optimum fin spacing is found from eq (3.31)

$$
z_{\mathrm{opt}} = 2.66P = 2.66(0.004110) = 0.01093\,\mathrm{m} \quad (10.93\,\mathrm{mm})
$$

and the optimum heat transfer coefficient is found using eq (3.30)

$$
h_{\mathrm{opt}} = 1.25\frac{k_g}{z} = 1.25\frac{0.0274}{0.01903} = 3.13\,\mathrm{W/m^2\text{-}K}
$$

In accordance with the findings of this section, the peak performance of such a natural convection array is achieved by setting the fin thickness equal to the fin spacing (10.93 mm). Consequently, the fin thickness is set equal to $\delta = 0.01093\,\mathrm{m}$ and the fin height may determined from eq (3.23)

$$
\left(\frac{\delta}{b}\right)_{s,\mathrm{opt}} = \frac{2}{(1.4192)^2}\frac{hb}{k} = 0.993\frac{hb}{k} \tag{3.23}
$$

or

$$
\begin{aligned}
b &= \left[\frac{k\delta}{0.993h}\right]^{1/2} \\
&= \left[\frac{(175)(0.01093)}{(0.993)(3.13)}\right]^{1/2} \\
&= (0.615)^{1/2} \\
&= 0.784\,\mathrm{m}
\end{aligned}
$$

The dimensions of $\delta = 1.093\,\mathrm{cm}$ and $b = 0.784\,\mathrm{m}$ define an array which maximizes heat transfer yielding an effective cooling heat transfer coefficient in accordance with eq (3.35).

$$
\begin{aligned}
\left(\frac{Q}{LW\theta_b}\right)_{\mathrm{opt}} &= 0.264\left(\frac{kk_g}{P^2}\right)^{1/2} \\
&= 0.264\left[\frac{(175)(0.0274)}{(0.004110)^2}\right]^{1/2} \\
&= 0.264(2.839 \times 10^5)^{1/2} \\
&= 140.7\,\mathrm{W/m^2\text{-}K}
\end{aligned}
$$

This rather remarkable heat transfer coefficient, obtained as the combined contribution of fins on a base surface reference is rather ideal. While it leads to a heat disspiation capability of

$$Q = 140.7LW\theta_b = 141.4(0.356)(0.305)(30) = 458.2\,\mathrm{W}$$

it is to be noted that that this array is based on 14 fins, each 78.4 cm high, 1.1 cm thick and weighing a total of 116.5 kg. This is certainly a more massive and more expensive fin array than appropriate to the task. However, the "optimum" heat dissipation is also substantially more than the required 80 W dissipation. Consequently, it is clear that a natural convection array of far shorter fins will suffice. This can be achieved by

1. Maintaining the optimum fin spacing at $z_{\mathrm{opt}} = 1.093\,\mathrm{cm}$,

2. Keeping the local heat transfer coefficient unchanged at $h = 3.13\,\mathrm{W/m^2\text{-}K}$, and

3. The required effective heat transfer coefficient is now just

$$h = \frac{80}{458.2}(140.7) = 24.6\,\mathrm{W/m^2\text{-}K}$$

3.8.3.1 DESIGN CONSIDERATIONS

Examination of Figs 3.13 through 3.15 and eq (3.33) shows $Q/LW\theta_b$ (constituting an effective array heat transfer coefficient referenced to the primary area) to be governed by two key parameters, P, embodying the influence of geometric, thermal, and air property variations and δ_{opt}, the individual optimum fin thickness. At fixed values of P, a maximizing value of δ_{opt} exists, although at larger P values, the array dissipation is only slightly dependent on δ_{opt} and a maximizing "*plateau*" is encountered.

Examining the array configuration at which the peak heat dissipation rates are encountered, it is apparent that by proper design it is possible to attain effective heat sink heat transfer coefficients in natural convection, $Q/LW\theta_b$, that are comparable to values normally associated with forced convection transfer (in air) from the primary surface alone. In particular, for aluminum fins (k= 175 W/m-K), in air, the maximum effective heat transfer coefficient, as presented in

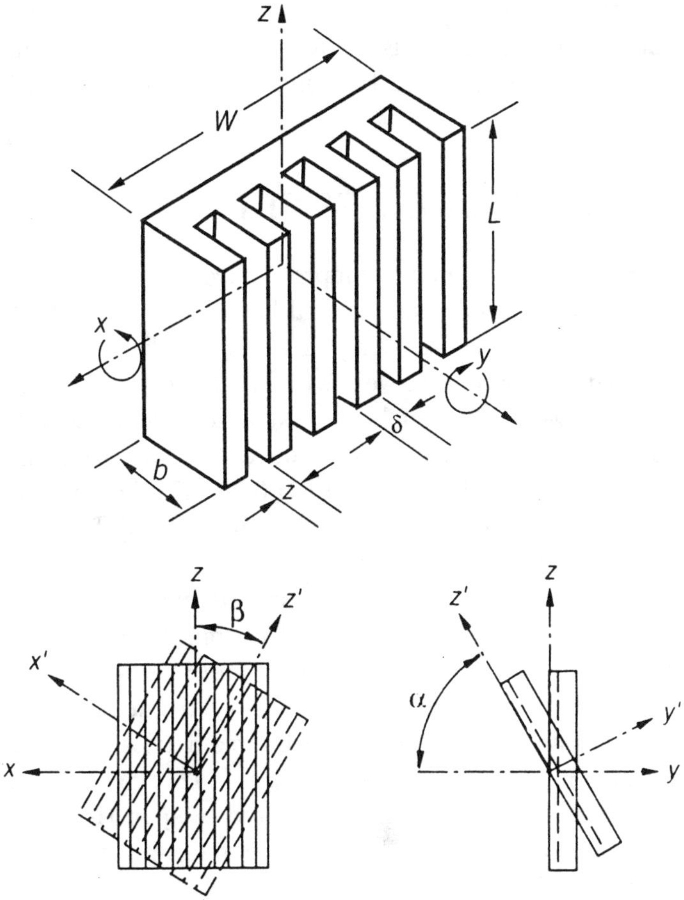

Figure 3.16: Geometry and orientation of Belitsky (1986) finned arrays.

Fig 3.13, is seen to range from 100 to 350 W/m^2-K in the parametric range of interest, providing a 15- to 45-fold improvement over natural convection heat transfer from an unfinned surface. The transfer rates attainable with an optimum array of brass (k = 120 W/m-K) and steel fins (k = 53 W/m-K) are plotted in Figs 3.14 and 3.15. While the improvement over the unfinned surface is seen to decrease, the margin nevertheless remains most significant even for the low conductivity steel array. These transfer rates are attained when the fins are optimally spaced (based on the modified Elenbaas criteria), the thickness of each fin closely approximates the optimum spacing, and the thickness of the fins is determined by the least-material criterion.

Array design that compromises any one of these three requirements cannot be expected to yield the foregoing maximum heat transfer rates.

3.9 EMPIRICAL RESULTS

In 1986, Bilitzky completed a comprehensive investigation of natural convection heat transfer from several heat sinks which differed, primarily, in fin length and spacing. The heat sinks were operated at different heat dissipations as well as different angles of inclination and orientation. Twelve distinct heat sinks and a flat plate were tested in a room within which extraneous convection had been suppressed. The range of angles is indicated in Fig 3.16. The base was first kept vertical, while the fins were rotated through four different positions (90°, 60°, 30°, and 0°). Then, the base was tilted backwards towards the horizontal orientation through four different positions (90°, 60°, 30° and 0°). Six of the heat sinks used 144 mm long by 115 mm wide bases to support plate fins, nominally 2 mm in thickness and 6 mm to 13.8 mm apart, ranging in length from 8.6 mm to 25.5 mm. Six additional heat sinks, with identical fin geometries, were supported on 280 mm long and 115 mm wide bases. The geometric parameters of the twelve heat sinks were selected to span the base and fin dimensions encountered in electronic cooling applications and are summarized in Table 3.3.

In all twelve heat sinks studied, the vertical-vertical orientation, that is, a vertically-oriented base with vertical fins and channels, yielded the highest heat transfer coefficients most often. However, in a relatively large number of situations, the thermal performance of the vertical-vertical array was indistinguishable from that attained by a vertical base plate with fins rotated 30° from the axis, a horizontal base plate, and a base plate inclined 60° from the horizontal with unrotated fins. On the other hand, the vertical-horizontal orientation, that is, the base plate vertical and the fins rotated 90° from the axis, led to the lowest heat dissipation rates. For the unrotated fins, the lowest heat transfer coefficients were almost always found to occur at a base plate angle of 30° from the horizontal. The use of smoke revealed a relatively complex three-dimensional flow pattern around the heat sinks, with very substantial inflow from the direction

Table 3.3: **Geometric Parameters for Fin Arrays [Bilitzky (1986)].**

Array	L		W		b		z		δ	
	in	mm	in	mm	in	mm	in	mm	in	mm
1	5.67	144	4.53	115	1.004	25.5	.236	6.0	.075	1.9
2	5.67	144	4.53	115	.677	17.2	.232	5.9	.079	2.0
3	5.67	144	4.53	115	.339	8.6	.228	5.8	.083	2.1
4	5.67	144	4.53	115	1.004	25.5	.547	13.9	.075	1.9
5	5.67	144	4.53	115	.677	17.2	.543	13.8	.079	2.0
6	5.67	144	4.53	115	.339	8.6	.539	13.7	.083	2.1
7	11.02	280	4.53	115	1.004	25.5	.236	6.0	.075	1.9
8	11.02	280	4.53	115	.669	17.0	.232	5.9	.079	2.0
9	11.02	280	4.53	115	.335	8.5	.228	5.8	.083	2.1
10	11.02	280	4.53	115	1.004	25.5	.547	13.9	.075	1.9
11	11.02	280	4.53	115	.669	17.0	.543	13.8	.079	2.0
12	11.02	280	4.53	115	.335	8.5	.539	13.7	.083	2.1

of the fin tips when the base plate was strongly inclined and when the heat sinks were in the vertical base-horizontal fins orientation.

The influence of the spacing between the fins, z, for short and long base plates, was examined by comparing pairs of heat sinks which differed only in the geometric parameter, z (fin arrays 1 and 4, 2 and 5, 3 and 6, 7 and 10, 8 and 11 and 9 and 12 in Table 3.3). Bilitzky observed that, in nearly all of the configurations and operating conditions examined, the highest heat transfer coefficients were attained with the larger fin spacing. However, the improvement in the heat transfer coefficient was not always sufficient to compensate for the loss of wetted fin surface area. Moreover, for the horizontal base plate configuration, as well as for the vertical base with horizontal fins, the total array dissipation appeared to not depend on this parameter.

A similar exploration of the fin height effect, by comparing the performance of four sets of three fins revealed that, in most cases, the heat transfer coefficient decreases with fin height. Nevertheless, for the shorter heat sinks (with length equal to 144 mm) with closely-spaced fins, the heat transfer coefficient attained a minimum value

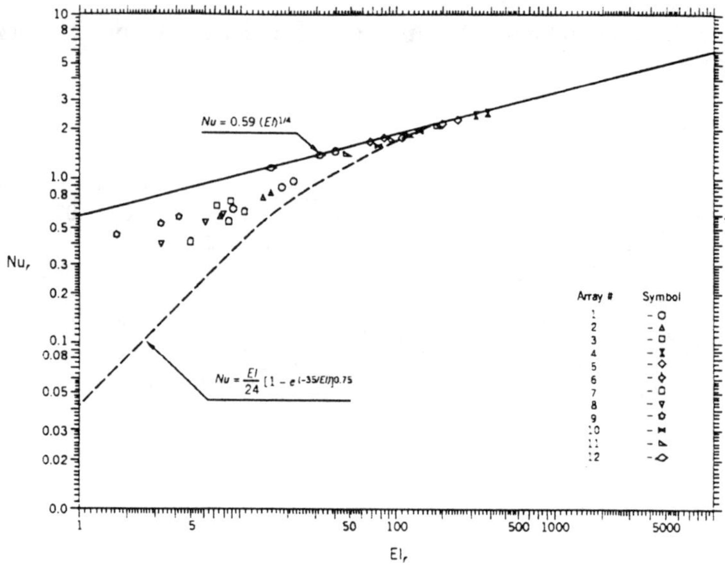

Figure 3.17: Nusselt numbers for Belitsky (1986) vertical finned arrays.

at the intermediate fin thickness (at 17 mm). An increase in the heat sink length was found to monotonically increase the total heat dissipation at identical base-to-air temperature differences and angles of orientation, but to decrease the actual convective heat transfer coefficient along the wetted surfaces.

After reviewing his extensive data base of empirical results, Bilitzky 1986) concluded that, for all of the geometries examined, the heat transfer coefficients along the fin surfaces, for the most common vertical base-vertical fin orientation, lie between the values observed for flat plates and those associated with flow in parallel-plate channels. However, as may be seen in Fig 3.17, for this orientation, the experimental Nusselt numbers are substantially higher than the fully developed limit and those obtained from the Elenbaas correlation. This holds true, even for Elenbaas numbers below 10, which are normally associated with the fully-developed flow limit. This behavior would appear to reflect the very strong impact of three-dimensional flow on the heat transfer from the relatively short fins typical of the actual heat sink configurations examined in this study. A similar result can be seen in Fig 3.18 where the relatively poor

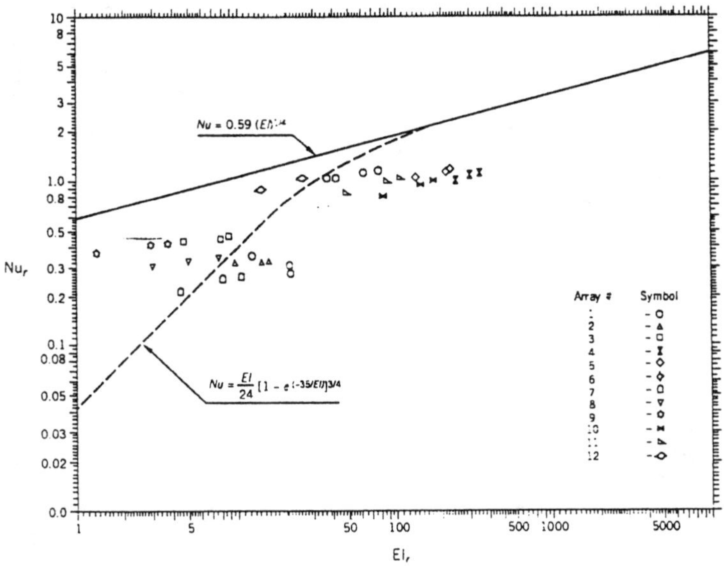

Figure 3.18: Nusselt numbers for Belitsky (1986) vertical base-horizontal finned arrays.

vertical base-horizontal fins orientation shows remarkably high Nu values, approaching 85 percent of the vertical flat plate value at high Elenbaas numbers and still 20 percent of the vertical flat plate value at Elenbaas numbers as low as 5. For the horizontal base plate orientations, shown in Fig 3.19, the heat transfer coefficients appear to straddle the horizontal flat plate correlation, displaying 10 percent higher values at large El and 50 percent lower values at small El numbers.

Bilitsky (1986) also used his experimental results to examine the effectiveness with which the added-mass was utilized in the various heat sink geometries. Defining a utility function, M, as

$$M \equiv \frac{Q_{\text{Heat Sink}}/Q_{\text{Base Plate}}}{\text{added mass of fins}} \tag{3.36}$$

he found significant variations among the twelve heat sinks. The values of M were higher for the short heat sinks than for the long heat sinks (144 mm vs 280 mm) and peaked for the heat sink arrangement (number 6 in Table 3.3) with narrow, widely-spaced fins.

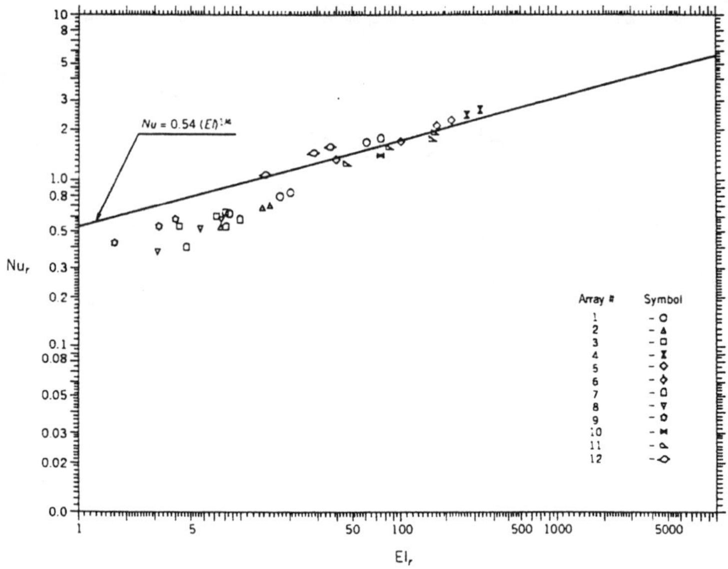

Figure 3.19: Nusselt numbers for Belitsky (1986) horizontal finned arrays.

3.10 CLOSURE

The finned array and the plate fin heat sink (the array of fins), perhaps the most commonly recognized fin stuctures, have found wide application in the electronic industry for the thermal management of microprocessors, logic and memory devices, power transistors, telecommunication modules and laboratory measurement equipment.

The design of such heat sinks requires an understanding of both natural and forced convection flow across the surfaces and through the channels, a knowledge of the convection-conduction interplay in the individual extended surfaces themselves and a gut feel for optimization and the avoidance of heat flow choking.

3.11 NOMENCLATURE

Roman Letter Symbols

A Cross-sectional area, m^2

a Coordinate of fin tip, m

Roman Letter Symbols (Cont'd)

b	Coordinate of fin base, m
	Fin height, m
c_p	Specific heat at constant pressure, J/kg-K
El	Elenbaas number, dimensionless
g	Acceleration of gravity, m/s^2
h	Heat transfer coefficient, W/m^2-K
k	Thermal conductivity, W/m-K
L	Fin length, m
	Heat sink length, m
m	Fin performance perameter, m^{-1}
n	Number of fins
P	Plate air parameter, m^{-4}
q	Heat flow, W
S	Surface area, M^2
W	Heat sink width, m
x	Height coordinate, m
$\mathbf{Y}$	Admittance matrix, W/° C
Y	Admittance, W/° C
y	Elements of the admittance matrix, W/° C
z	Clear space between fins, m

Greek Letter Symbols

β	Volumetric coefficient of expansion, k−1
$\mathbf{\Gamma}$	Thermal transmission matrix
γ	Elements of thermal transmission matrix
δ	Fin thickness, m
η	Fin efficiency, dimensionless
θ	Temperature excess, K or ° C
μ	Dynamic viscosity, N-s/m^2
	Thermal transmission ratio, W/°C
ρ	kg/m^3
L	Fin length, m
$\mathbf{T}$	Inverse transmission matrix
τ	Elements of the inverse transmission matrix

Subscripts

a	Fin or array tip
b	Fin or array base
C	Cluster
e	Equivalent
IN	Input
o	Characteristic
opt	Optimum
p	Profile

3.12 REFERENCES

Aziz, A (1978). Optimum rectangular convection fin with tip heat loss and having a temperature dependent thermal conductivity, J Eng Sci (University of Riyadh), Saudi Arabia, 4, 133.

Aziz, A (1992). Optimum Dimensions of Extended Surfaces Operating in a Convective Environment, ASME Appl Mech Rev, 45 (5), 155.

Bar-Cohen, A (1979). Fin Thickness for an Optimized Natural Convection Array of Rectangular Fins, J. Heat Trans, 101, 564-566.

Bar-Cohen, A and Jelinek M (1990). Optimization of Longitudinal Finned Arrays - London's 25 KW Power Tube Revisited, Compact Heat Exchangers, R. Shah, A. D. Kraus and D. Metzger (eds), Hemisphere Publishing Corp, New York, NY, 105.

Bar-Cohen, A and Rohsenow, W M (1984). Thermally Optimum Spacing of Vertical, Natural Convection Cooled, Parallel Plates, J Heat Trans, 106 (5), 116.

Bilitsky, A (1986). The Effect of Geometry on Heat Transfer by Free Convection From a Fin Array. MS Thesis, Department of Mechanical Engineering, Ben-Gurion University of the Negev, Beer Sheva, Israel.

Elenbaas, W (1942). Heat Distribution of Parallel Plates by Free Convection. Physica, 9 (1), 665.

Gardner, K A (1945). Efficiency of Extended Surface. Trans ASME, 67, 621.

Kraus, A D, Snider, A D, and Doty L F (1978). An Efficient Algorithm for Evaluating Arrays of Extended Surface. J Heat Trans, 100, 288.

Kraus, A D and Snider, A D (1985). The Choking and Optimization of Extended Surface Arrays. J Heat Trans, 107, 746.

Kraus, A. D and Bar-Cohen, A (1995). Design and Analysis of Heat Sinks, John Wiley and Sons, New York, NY.

Lee, S, Yovanovich, M M, and Jafarpur K (1991). Effects of Geometry and Orientation on Laminar Natural Convection from Isothermal Bodies. AIAA J of Thermophysics and Heat Trans, 5, 208.

Murray, W M (1938). Heat Transfer Through an Annular Diak or Fin of Uniform Thickness. Trans ASME, 60, A78.

Sparrow, E M and Bahrami, P A (1978). Experiments on Natural Convection from Vertical Plates with Either Open or Closed Edges, J. Heat Trans, 102, 221-227

Starner, K E and McManus, H N (1963). An Experimental Investigation of Free Convection Heat Transfer From Rectangular Fin Arrays. J Heat Trans, 85, 273.

Van de Pol, D W and Tierney, J K (1973). Free Convection Nusselt Number for Vertical U-Shaped Channels. J Heat Trans, 95, 542.

Welling, J R and Wooldridge, C B (1965). Free Convection Heat Transfer Coefficients from Rectangular Vertical Fins. J Heat Trans, 87, 439.

Yovanovich, M M (1987a). Natural Convection From Isothermal Spheroids in the Conductive to Laminar Flow Regimes. AIAA Paper AIAA-87-1587.

Yovanovich, M M (1987b). On the Effect of Shape, Aspect Ration and Orientation Upon Natural Convection From Isothermal Bodies. ASME HTD, vol 82, 79-85.

CHAPTER 4
BIBLIOGRAPHY OF HEAT TRANSFER IN ELECTRONIC EQUIPMENT: 1990–1994

Robert E. Simons
Consultant
Poughkeepsie, New York

4.1 INTRODUCTION

For more than 50 years, the thermal control of electronic components and equipment has been one of the primary areas for the application of advanced heat transfer techniques. Improvements in thermal analysis and design have been major contributors to the advances in reliability, increased power density and physical miniaturization associated with modern day microelectronics. Now, more than ever before, the demand for improved thermal control of electronic components and equipment, coupled with the need for more efficient methods of heat removal, has brought increased attention to the important task of thermal management.

The level of importance that thermal management or heat transfer in electronic equipment has attained, is attested to by the number of papers published each year on the various aspects of the subject. A bibliography prepared by Antonetti and Simons (1985) covering the period 1970 to 1984, listed 237 publications relating to various aspects of heat transfer in electronic equipment. More recent annotated bibliographies by Simons (1988, 1990 and 1993) covering the period from 1986 to 1989 reported on nearly 600 publications on the same subject. An annotated bibliography including more than 400 selected papers, articles, and patents published during the period 1990 and 1994 pertaining to heat transfer in electronic equipment is provided here. The material presented here was obtained from technical journals, trade magazines, conference proceedings, and U.S. patents.

4.2 STATE-OF-THE-ART

Included here are eight publications which addressed the state-of-the-art of heat transfer in electronic equipment, providing an overview of both past and recent developments and applications. A paper by Nakayama and Bergles (1990) reviewed the evolution of cooling technology for electronic equipment with special emphasis on the thermal control of solid-state devices. Topics discussed included: the rising levels of overall power dissipation and chip level heat fluxes, techniques for calculating and improving conduction heat flow, natural and forced convection air cooling and liquid cooling. In the conclusion it was noted "that electronic equipment today is no longer confined to performing tasks for specialists, but is also making inroads into the vital control centers of society." It was further noted that reliability is playing an increasingly important part "in setting the design goals of new equipment, for which heat transfer research is expected to play an even greater role." A paper by Chu and Simons (1990) provided an overview of thermal design for multi-chip modules (MCM's). Noting that thermal design played a key role in the successful introduction of multi-chip modules, the techniques used to cool MCMs in the IBM 3081, NEC SX-2, Fujitsu FACOM M-780, Cray-2, and the IBM 4381 computers were discussed.

Another paper by Chu and Simons (1993) reviewed the development of cooling technology used in computer systems; covering air, air-liquid hybrid and water cooling. The development of liquid immersion technology was also discussed with examples of cooling applications. A paper by Mudawar (1992) provided a review of several direct liquid immersion cooling techniques investigated at Purdue University. The techniques considered included pool boiling, falling film cooling, channel flow boiling and jet impingement.

A trade press article by Robinson (1990) discussed the increasing importance of thermal management for electronic applications across the board and some of the approaches being undertaken to address these needs. Another article by Remsburg (1991) addressed the need for new or advanced cooling techniques created by increasing electronic packaging density and power dissipation. The cooling techniques discussed included micro heat pipes, thermoelectrics, microchannels, liquid heat sinks, jet impingement cooling, direct liquid immersion and liquid cold plates. In addition, a book edited by Bergles (1990) constituted the archival proceedings of the International Symposium on Heat Transfer in Electronic and Microelectronic Equipment held in Dubrovnik, Yugoslavia, in 1988, under the sponsorship of the International Centre for Heat and Mass Transfer. A total of 63 papers were included in chapters titled: Overview, Air Cooling, Liquid Cooling, Conduction Aspects and Thermal Analysis.

Another book, edited by Kakac et al (1994) contained 36 lecture papers presented at the NATO Advanced Study Institute on Cooling of Electronic Systems, held in Cesme-Izmir, Turkey in 1993. A paper by Simons (1994) on the occasion of the 10th anniversary of the Semiconductor Thermal Measurement and Management Symposium provided a look back at some of the cooling developments that took place since its founding. The topics covered included thermal measurement, thermal characterization, thermal analysis and modeling, air cooling, water cooling and immersion cooling.

4.3 MATERIALS

The choice of materials is an important consideration in both the thermal design and analysis of electronic packages. Before heat can be removed from any external surface it must be transported to the surface by the process of thermal conduction. In many electronic modules containing integrated circuit (IC) chips, thermal conduction through the substrate is the principal path for heat flow from the chip to the external module surface. Consequently, the thermal conductivity of the substrate material is a property of principal interest in thermal design. Of course, other material characteristics such as the thermal coefficient of expansion (TCE), electrical insulation capacity, low dielectric constant and mechanical strength are equally important in terms of other design considerations.

Although, alumina (Al_2O_3) is still the most widely used substrate material, efforts to develop improved substrate designs using materials such as aluminum nitride (AlN) and silicon carbide (SiC) are continuing. A paper by Enck et al (1991) reported measurements of thermal conductivity for a variety of commercial substrates made of AlN, SiC, BeO, or Al_2O_3. The thermal conductivity of each specimen was measured over a range of temperatures from 22°C to 400°C and fitted to a simple model of the form

$$k = \frac{1}{a + bT}$$

Figure 4.1 shows the temperature dependence of thermal conductivity of the substrate materials for a range of temperatures which might be encountered in chip packaging applications. This figure is based upon the a and b coefficients reported in the paper. The ac and dc electrical properties of commercially available AlN substrates were also reported. Another paper by Goosey et al (1991) reviewed the various materials available for use in high thermal conductivity packages. The authors concluded that aluminum nitride is the preferred material because of its high thermal conductivity and thermal coefficient of expansion which is closely matched to that of silicon. Processing technology was described, along with thick- and thin-film package technology demonstrators. A paper by Copeland and Powell (1990) reported the results of a study to evaluate the feasibility of using AlN as a substrate material for metallized ceramic modules and a paper by Kuhlman and Sehitoglu (1992) also considered the effects of various substrate materials on the thermal performance of an MCM package.

In many component packages (such as plastic dual-in-line packages) the thermal conductivity of the mold compound can have a significant effect on the ability to conduct heat away from the chip heat source. Three papers reported in this section considered the effects of molding compound thermal conductivity. A paper by Proctor and Solc (1991) reported on an investigation of the practical and achievable improvements in mold compound thermal conductivity using several different filler materials (including boron nitride, silicon nitride, and diamond). It was concluded that there are practical limits to the thermal conductivity of a composite as predicted theoretically and verified experimentally. It was suggested that "a balance needs to be sought which maximizes the thermal dissipation effect of the molding compound while still achieving superior device protection as evidenced by low stress, dimensional stability and moisture resistance."

Michael and Nguyen (1992) discussed the results of a study of the effects of varying the thermal conductivity of molding compounds on the thermal behavior of plastic IC packages and the combined effects of the thermally enhanced mold compounds with the enhanced lead-frame and heat spreader designs. The thermal performance of three types of small outline packages (SO-8L, SO-16W and SO-24L) was simulated using a finite element model. Among the conclusions, it was noted that "the highest thermal conductivity gains were achieved with mold compounds formulated to only 8 to 10 times higher thermal conductivities than unfilled compounds," and "a molded-in copper heat spreader lead frame design can provide an attractive alternative to formulating a component with 100 times higher thermal conductivity than the neat resin in order to achieve similar thermal performance."

Azar et al (1993) reported on a computational study to investigate the effect of molding compound thermal conductivity on the thermal resistance of a molded multi-chip module. Seven different molding compound thermal conductivities and external cooling conditions ranging from natural convection to high velocity forced convection were considered. It was found that an eight-fold increase in molding compound thermal conductivity reduced the junction-to-ambient thermal resistance by 20 percent in natural convection and by 54 percent in high velocity forced convection.

The printed circuit board (PCB) upon which electronic components are mounted can also provide additional thermal paths for heat flow to air, especially if the board is not fully populated. The ability of the board to conduct heat away from powered components depends upon both the thermal conductivity and thickness of its various layers. A paper by Terada et al (1990) described a new metal core multilayer PCB structure for surface mount technology (SMT) components. The thermal conduction characteristics of the board were enhanced by providing metal planes (e.g. aluminum) directly

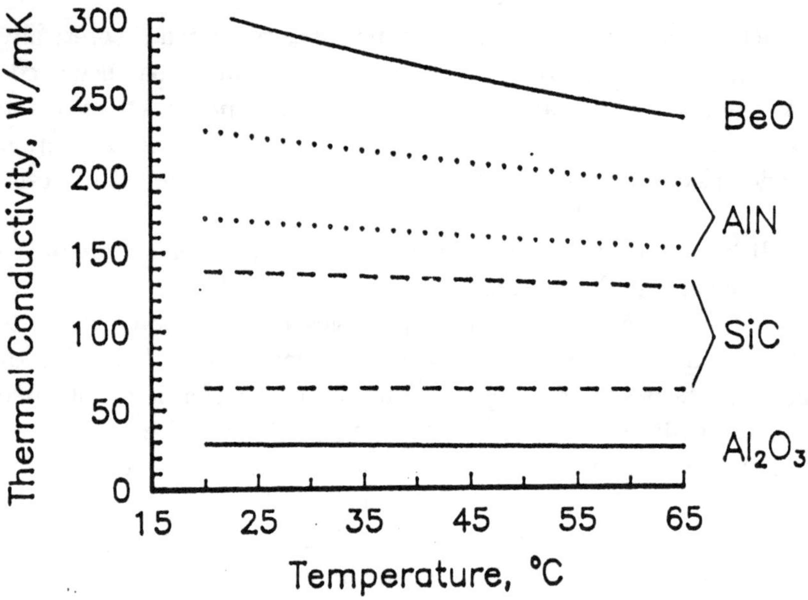

Figure 4.1: Dependence of substrate thermal conductivity on temperature [Adapted from Enck et al (1991)].

under the surface layers and by loading the epoxy resin with alumina powder. A paper by McKeown and LeVasseur (1991) discussed the use of a thermally conductive heat sink material between two SMT boards to provide a thermal path for heat removal in avionics applications. The paper focused on the advantages of a laminated copper clad molybdenum heat sink with a graphite core (CCMG). The CCMG structure was reported to provide improved performance for avionics applications in terms of thermal conductivity, density, and thermal expansion. Another paper by Johnson et al (1992) described the thermal properties of copper clad invar composites and discussed examples of applications at the chip, board, package and system levels.

The thermal conductivity of the leadframe material of the component plays an important role in determining the thermal resistance from the component to the board on which it is mounted. A paper by Sharan et al (1991) reported the results of a study to evaluate the use of high strength copper alloys in place of Alloy-42 as leadframe materials in plastic quad flat pack (PQFP) packages. It was reported that the use of any of the copper alloys would result in a minimum of a 30 perecent decrease in overall thermal resistance between junction and ambient, R_{JA}, compared to Alloy-42.

A paper by Zumbrunnen (1991) provided an example of material selection and design to improve the thermal performance of a large multi-chip module. The paper discussed the material change from helium to oil for the interstitial gap material within the IBM Thermal Conduction Module (TCM) used in the IBM ES/9000 series mainframe computer. The thermal impact of anodization barriers used for chip electrical isolation was also discussed.

4.4 THERMAL CONTACT RESISTANCE

In virtually all electronic packages and equipment, there are physical, mechanical, or metallurgical joints between different materials or parts of the structure across which heat is conducted. The thermal resistance across such interfaces is often referred to as a thermal contact resistance, and is generally a complex function of the geometric, physical, and thermal properties of the contacting solids and the interstitial substance. The ability to transport heat across two surfaces in contact may also be characterized in terms of the thermal contact conductance, h_c (W/cm^2-°C). A paper by Fletcher (1990) provided a review of recent thermal enhancement techniques for maximizing thermal contact conductance, by using greases, metallic foils and screens, composite materials and cements and surface treatments. A paper by Babus'Haq et al (1991) also addressed the enhancement of thermal contact conductance by employing interfacial fillers. Two papers by Maudgal et al (1990b, 1990c) discussed the experimental determination of the thermal contact conductance of mechanical joints in a diode heat sink assembly. Results were presented for:

1. bare mating surfaces,

2. aluminum foil between the mating surfaces and

3. high thermal conductivity grease applied to the mating surfaces.

As noted in the previous section, in a majority of integrated circuit packages today, the principal heat flow path is from the chip into

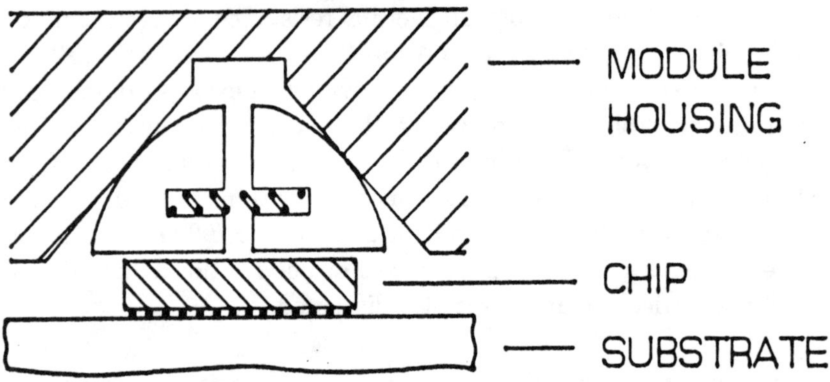

Figure 4.2: Cross-sectional view of line contact coolilng design [Morihara et al (1990)].

the substrate to which it is bonded. It is therefore, important that the thermal contact or "die bond" resistance between the chip and its substrate carrier be as small in magnitude as possible. A variety of methods have been used to bond the chip to the substrate including organic glues and hard or soft solders. A paper by Chung et al (1991) discussed die attachment to MCM substrates using low-stress, thermally conductive epoxies. Another paper by Bjorneklett et al (1993) reported on an investigation of the thermomechanical stress induced during the chip attachment process and the effect of temperature cycling on adhesive bonds. An increasing thermal resistance with number of thermal cycles which strongly depends on mismatch in thermal expansion between the silicon chip and substrate was found.

For some large multi-chip modules with many chips, the substrate can no longer be effectively used as the principal thermal conduction path for heat removal from the chips. It is shown in detail in Section 4.9, that for such situations, an approach is required which removes heat from the back of the chip. One such approach, termed "line contact cooling" was described in a paper by Morihara et al (1990). As shown in Fig 4.2, the back surface of an integrated circuit chip is in contact with the bottom surface of two conducting blocks of quasi-triangular cross-section, each with a curved side in contact with a slanted inner surface of the housing. As described in the paper, this geometry makes it possible to provide an effective thermal path while accommodating both chip tilting and vertical displacement of the chip. Minimization of the thermal contact resistance between the mating surfaces plays an important role in controlling the overall thermal resistance in such a scheme. Another paper by Naganuma et al (1990) discussed both analysis and experiments to study the effect of cylinder radius on thermal contact resistance for the "line contact cooling" approach.

Another area where thermal contact resistance is important is in the attachment of a heat sink to a module. This is especially true if a non-metallurgical joint is used. A paper by Gaynes and Shaukatullah (1994) describes a comparative evaluation of paste and tape adhesives for attaching heat sinks to plastic quad flatpacks (PQFP's). It was reported that only slight differences in thermal performance were measured between tapes and pastes for both metal and plastic flat packs, and that double sided tapes are a viable alternative to thermal conductive paste adhesives. A paper by Moy et al (1992) addressed the thermal, mechanical, material, process and cost considerations in selecting a heat sink attachment method for ceramic through-hole pin grid array packages.

In some instances a "dry" joint may be used to provide a thermal path from the module to an air-cooled heat sink or water cooled cold plate. In such cases, the appropriate level of thermal contact resistance may be achieved by control of the flatness and surface roughness of the mating surfaces, and the application of mechanical forces by means of bolts and clamps. A paper by Song et al (1993) discussed experiments performed to study contact radius and thermal/electrical contact resistance across two plates of unequal thickness. Models developed to predict the contact radius and thermal/electrical contact resistance across the bolted joint interface were also discussed. Another paper by Kunkel and Peck (1994) discussed development of a model to predict thermal contact resistance between a typical avionics electronics chassis and a PCB thermal core, and the chassis to the PCB clamp.

4.5 AIR COOLING

Although a number of cooling designs which utilize some form of liquid cooling have appeared, air continues to be the most widely used cooling medium in electronic equipment. The principal advantages of cooling with air are its ready availability and ease of application. For applications with low power and low packaging densities, natural convection cooling with air is often used. The advantage offered by natural convection cooling is that no fan or blower is required to move the air. Instead, air movement is induced by density differences resulting from the heat given off by the electronic components. Seven papers reported in this section are in the area of natural convection.

Chan et al (1991) reported on an experimental study of natural convection cooling of a simulated printed circuit board within a ventilated housing. It was found that the principal geometric factor influencing the heat transfer behavior is the separation between the PCB surface and the nearest wall of the enclosure. It was also found that the location of the vent holes had only a relatively marginal effect on the simulated PCB surface temperature. A paper by Chadwick et al (1991) reported the results of a theoretical and experimental study of natural convection in a discretely heated, sealed enclosure with single and multiple heater locations. The wall opposite the heater(s) was maintained at a constant temperature by means of a water-cooled plate. Another experimental investigation of an array of PCB's mounted vertically in a closed casing was reported by Wang and Mayinger (1993). The top and bottom walls were maintained at a constant temperature and the side walls of the enclosure were made of plexiglass. Heat transfer and air flow in the array of eight PCB's were measured using a holographic interferometer and a laser-doppler velocimeter. It was found that the heat transfer from the PCB's was "greatly influenced by the distance among the PCB's and relatively insensitive to the distance between the PCB array and the top and bottom wall." Empirical heat transfer correlations based on the experiments and an expression for the optimum PCB spacing were also reported. A paper by Eseteki et al (1993) reported the results of a theoretical study using a commercially available computational fluid dynamics (CFD) code to model laminar natural convection cooling of a vertical flat plate representing a PCB in a square sealed enclosure. The base of the enclosure was considered to be adiabatic and the other walls were considered isothermal. Results were presented for both isothermal and isoflux conditions at the surface of the plate. Another paper by Sefcik et al (1991) reported on a study of laminar natural convection in vertically vented two-dimensional enclosures.

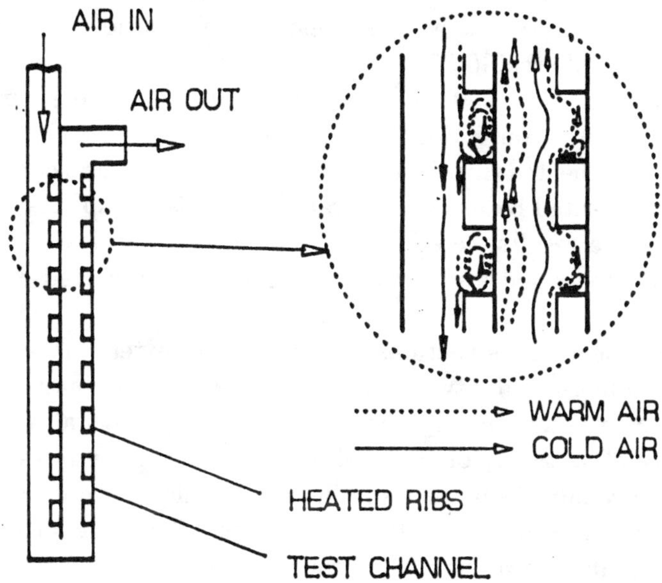

Figure 4.3: Schematic of test channels and typical flow patterns for opposing/assisting natural convection flows [Adapted from Lin and Hsieh (1990)].

The effects of vent gap spacing, enclosure aspect ratio, and Grashof number on the local flow structure were investigated. It was noted that the results showed asymptotic behavior to the classical vertical parallel plate result in the limit as the vent gap approaches the enclosure width.

Three papers provided the results of experimental studies conducted in natural convection and buoyancy-dominated mixed convection regimes. Tewari and Jaluria (1991) investigated the thermal interaction of wakes arising from two heat sources on a flat adiabatic plate in both horizontal and vertical orientations. Kang and Jaluria (1990) studied mixed convective transport from a protruding rib heat source mounted on a vertical surface in an externally induced low Reynolds number ($295 \leq \mathrm{Re} \leq 435$) flow. Shaukatulah et al (1994) reported the results of tests of 28 mm 208-leaded metal quad flat packs mounted on a thermal test card with air flows with velocities from 0 (natural convection) to 2 m/s. A non-dimensional correlation representing external thermal resistance ($\mathrm{Nu} = 1/(kDR_{\mathrm{ext}})$ in the mixed convection regime was developed using equations of the form suggested by Churchill(1977).

Lin and Hsieh (1990) reported the results of an experimental study of natural convection of opposing/assisting flows in vertical channels with heated ribs. The opposing/assisting natural convection flows were created by using two heated channels set-up as shown in Fig 4.3. The results were reported in terms of local and average Nusselt numbers and were compared with results reported by other investigators.

Finned heat sink structures may be used to extend natural convection cooling capability. Cengel and Ngai (1991) conducted experiments to determine the natural convection heat transfer characteristics of vertically oriented, shrouded, finned heat sinks. Commercially available heat sinks specifically designed to cool electronic components were used for the experiments. It was reported that "shrouding, in general, improves heat transfer from heat sinks significantly." As much as a 20 percent reduction in maximum heat sink temperature was obtained with shrouding at a given power level. It was found that the magnitude of temperature reduction first increases with increasing clearance, reaches a maximum reduction, and then decreases with further increases in clearance. It was also observed, that at zero or small clearances, shrouding can increase the maximum temperature of long heat sinks with closely spaced fins by as much as 40 percent. In another paper, Rodgers and Davis (1994) described an experimental investigation of three finned array heat sink designs cooled by natural convection. A hot wire measurement technique was used to measure air velocities and temperatures at the inlet and outlet of the passages between the fins. Non-dimensional temperature and velocity distributions at the inlet and outlet were presented for each design.

For packaging situations requiring more cooling capability or a greater degree of control, forced convection heat transfer with air is often the method of choice. Forced air cooling offers higher heat transfer coefficients than natural convection, is relatively insensitive to packaging orientation, is less affected by heating of the air by upstream components and is generally more easily controlled. The remaining papers in this section address forced convection cooling with air.

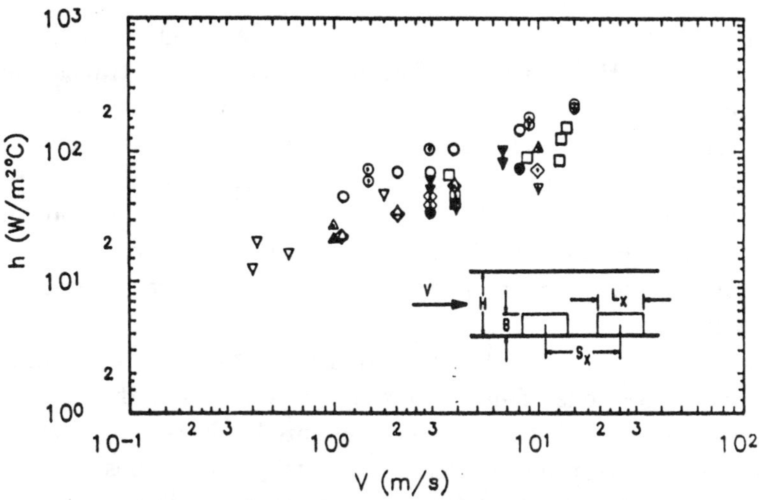

Figure 4.4: Fully developed heat transfer coefficient data from the literature [Adapted from Anderson and Moffat (1990)].

Forced air cooled packages or systems are often composed of arrays of vertically (or horizontally) stacked printed circuit cards. Mounted on these cards are arrays of electronic circuit modules. The efforts of a number of researchers continue to be directed at further understanding and characterizing heat transfer for such configurations. A paper by Moffat and Anderson (1990) discussed the adiabatic heat transfer coefficient (h_{ad}) and its application to the prediction of module surface temperature. It was noted that in most cases, "h_{ad} values are being used to calculate the temperature rise above the mean fluid temperature, rather than the adiabatic temperature, a procedure that may seriously underpredict the temperature rise of an element in service." An example of the wide range of fully developed heat transfer coefficient data reported in the literature was provided in another paper by Anderson and Moffat (1990) and is shown in Fig 4.4. In the paper, they described their approach to correlate a wide-range of module heat transfer data and improve the prediction of module temperatures. The results reported for an array of flat packs with variable channel heights and inlet velocity, showed that both, h_{ad}, and its superposition kernel functions can be calculated with good accuracy from the pressure drop characteristics of the channel.

In two other papers, Anderson and Moffat (1992a, 1992b) describe an investigation of the forced convection heat transfer and pressure drop characteristics of a regular in-line array of flatpacks for several channel heights and inlet velocities. Both adiabatic heat transfer coefficient and superposition kernel function data are presented. The first paper (1992a) presents heat transfer and pressure drop results and the second (1992b) presents a model for heat transfer that is based upon the maximum turbulence fluctuations in the channel.

Azar and Moffat (1991) reported the results of a study comparing three definitions of heat transfer coefficient, and the effects of nonuniform power distribution, on measured values using each definition of heat transfer coefficient. The three definitions examined differed in the choice of reference temperature as follows:

1. h_{in} using the channel inlet air temperature,

2. h_m using the mean air temperature based upon upstream power dissipation and

3. h_{ad} using the adiabatic component temperature.

In the conclusions it was noted that h_{in} and h_m are easier to measure, but change significantly when the surrounding power distribution changes. While h_{ad} is more difficult to measure and to use, it is independent of the power distribution.

A paper by Manno and Azar (1991) assessed the impact of the choice of heat transfer correlation on the prediction of component surface temperature, and explored the strengths and weaknesses of using the adiabatic heat transfer coefficient concept. Faghri et al (1991) reported the results of experiments to determine heat transfer coefficients in the entrance region of regular in-line arrays of rectangular blocks. Heat transfer coefficient data was correlated in terms of a composite geometric parameter termed the volume packing density, Reynolds number and a non-dimensional entry length. Wirtz and Chen (1991) reported on velocimetry, heat transfer and pressure drop experiments for laminar-transitional air flow in a channel. Rectangular transverse ribs located on one wall were used to simulate an array of low profile electronic modules. Heat transfer coefficients were correlated versus the maximum calculated velocity fluctuations as described by Anderson and Moffat (1990).

A paper by Wirtz and McAuliffe (1990) described experiments to investigate convective heat transfer immediately upstream of a row of closely spaced electronic modules which were simulated by a rib on the wall. It was found that a region of reduced heat transfer extends approximately 2.5 rib-heights upstream, with an average heat transfer coefficient approximately one-half of that which would occur in the absence of the rib. Two papers by Lehmann and Pembroke (1991) reported on the results of an experimental investigation of forced convection cooling of a simulated array of electronic circuit modules. The first paper reported the effects of channel spacing, row position, and flow rate ($3000 < \text{Re} < 20,000$) on component Nusselt numbers and thermal wake functions. The second paper reported on the effects due to the presence of an unheated longitudinally finned heat sink. Measurements of Nusselt numbers and thermal wake functions downstream of the heat sink were reported and compared with values obtained with no heat sink present. Further studies of the downstream effects of heat sinks were conducted and reported by Lehmann and Kosteva (1990).

In the design of electronic packages, it is often desireable to mount as much circuitry in a given volume by placing PCB's as close together as possible. Similarly, in the design of parallel plate heat sinks, it is desireable to maximize the heat transfer area by spacing fins as close together as possible. In both cases, however, if the spacing is too tight, flow reductions due to increased pressure drop may result in reduced transfer rates. A number of papers included here addressed the problem of optimum spacing to maximize heat transfer in forced convection. Bejan and Lee (1993) provided a review of fundamental results addressing the optimum spacing of heat generating boards arranged in parallel in a stack of fixed volume. Both natural and forced convection laminar flows were considered. The board geometry considered included smooth boards with uniform temperature or heat flux, flush mounted heat sources, protruding heat sources and a board cooled by its own insulated parallel plate channel. Mereu et al (1993) showed analytically how to optimize the spacing between heat generating boards cooled by laminar forced convection. Theoretical results for optimal spacing and maximum overall thermal conductance bewteen stack and coolant were validated by means of a numerical solution of the complete flow and temperature fields. Results were given for

1. fixed pressure drop,
2. fixed mass flow rate and
3. fixed pumping power

A paper by Morega and Bejan (1994) also showed how to select the optimal spacing between boards mounted in a specified volume so that the overall thermal conductance between the stack and the cooling fluid is a maximum. For laminar flow and fixed pressure drop, Δp, the optimum board-to-board spacing, D_{opt}, was shown to be

$$\frac{D_{\text{opt}}}{L} \cong 2.7 \left(\frac{\Delta p L^2}{\mu \alpha} \right)^{-1/4}$$

where L is the length of the plates in the flow direction and μ and α are the viscosity and the thermal diffusivity of the coolant. If the free stream velocity, U_∞, upstream of the stack is fixed, the optimal spacing was shown to be

$$\frac{D_{\text{opt}}}{L} \cong 3.2 \text{Pr}^{-1/4} \left(\frac{U_\infty L}{\nu} \right)^{-1/4}$$

where Pr is the Prandtl number and ν is the kinematic viscosity of the cooling fluid. Another paper by Bejan and Morega (1994) reported an analysis to determine the optimal plate-to-plate spacing for a stack of plates cooled by turbulent forced convection.

In order to provide adequate cooling, it is often necessary to achieve heat transfer rates higher than those normally attainable by forced convection with air. In such situations, cooling performance may be enhanced by reducing the cooling air temperature, increasing the surface area available for heat transfer or by increasing the heat transfer coefficient. The remaining papers in this section address some of the techniques for increasing forced convection air cooling capability.

Two papers addressed enhancement of module heat transfer by promoting thermal mixing in the coolant flow. Anderson and Moffat

(1991) discussed the use of small scoops behind leading edge modules to reduce the adiabatic temperature rise of downstream components by increasing thermal mixing in the flow. Experimental results were presented showing 10 to 50 percent reductions in adiabatic temperature rise resulting in up to a 19 percent reduction in overall module temperature rise. The greatest effect appeared for the two rows of modules directly behind the scoops. Lehmann and Huang (1991) reported on heat transfer enhancement resulting from the mixing action of longitudinal vortices trailing from delta wing shaped vortex generators (VG) mounted on a simulated component module. It was found that "the attachment of a VG can produce significant cooling enhancement throughout a column of components downstream of the VG."

Finned heat sink structures are often used to increase the convective heat transfer surface area of electronic circuit modules. Schenone and Tanda (1990) reported on heat transfer data obtained for five shrouded, staggered fin arrays over a range of Reynolds numbers from 2,000 to 12,000. The configurations investigated included fin heights of 20 and 35 mm, and spacing between fins of 6.3, 7.8 and 10 mm. The measured data was found to be substantially in agreement with standard correlations available in the literature for offset strip fin heat exchangers. Lee et al (1990) investigated the effects of air flow bypass on the heat transfer performance of straight fin heat sinks. The geometric parameters considered included clearance between fin tip(s) and the adjacent PCB, lateral space between heat sinks, fin height and spacing between fins. The results demonstrated that the effect of air flow bypass due to lateral spacing or tip clearance can be significant, with more than a 20 percent reduction in heat transfer performance.

Bazydola and Taslim (1993) reported heat transfer data obtained for a staggered array of heat sinks in the hydrodynamic and entrance region in a duct. Aluminum heat sinks with 13.8 mm tall fins and a gap between fins of 1.45 mm were attached to 40.6 mm × 40.6 mm simulated electronic modules. Three different channel heights were used to provide three different values of H/B (the ratio of duct height to module and heat sink height) from 1.0 to 5.0. Similarly, the array spacing was varied to provide five different values of L/D (the ratio of longitudinal to lateral heat sink spacing) from 1.0 to 3.0.

Heat transfer data was reported in the form of Nusselt number based upon heat sink length in the flow direction for Reynolds numbers of 20,250, 50,000 and 65,000 based upon the hydraulic diameter of the duct. Heat transfer correlations were obtained and reported for both upstream and downstream heat sinks.

Iwasaki et al (1994) reported the results of a numerical and experimental study to evaluate the air cooling characteristics of parallel plate (or straight fin) heat sinks. The heat sinks considered had base dimensions of 81 mm × 81 mm with 2 mm thick and 28 mm tall fins. Tests were conducted for designs with 3, 4 and 7 mm gaps between adjacent fins. A heat transfer correlation was developed based upon numerical flow modeling of the heat sink passages and used with a simple flow network model to predict the heat transfer performance with flow by-pass around the sides of the heat sink.

A paper by Kitajo et al (1992) discussed the development of a parallel flow trapezoidal fin heat sink with the minimum fin width at the base. Both theoretical and experimental results were presented demonstrating lower thermal resistance for the trapezoidal fins compared to conventional rectangular fins for the same air velocities. Vogel (1994) investigated the performance of hybrid heat sinks made of metallic and non-metallic materials. The materials considered included aluminum, copper, graphite and epoxy laminant. It was concluded that, through the use of alternate material hybrid designs, heat sink weight can be reduced to 50 percent of that of a copper design without sacrificing thermal performance.

Several papers addressed the improved heat sink thermal performance attainable by use of narrower seperation or microchannels between fins. Azar et al (1992) reported the results of experiments investigating the performance of an aluminum straight fin heat sink with base dimensions of 25 mm × 25 mm–12.5 mm tall fins and 1.1 mm gaps between adjacent fins. A paper by Ohara et al (1992) reported the results of an experimental investigation comparing the performance of a typical commercial heat sink with a gap between fins of 1.0 mm, and two narrow channel designs, one with a gap of 0.48 mm and the other with a gap of 0.24 mm. Even taking into account the reduced air velocity resulting from the pressure drop increase with the narrow channel design, the narrow channel

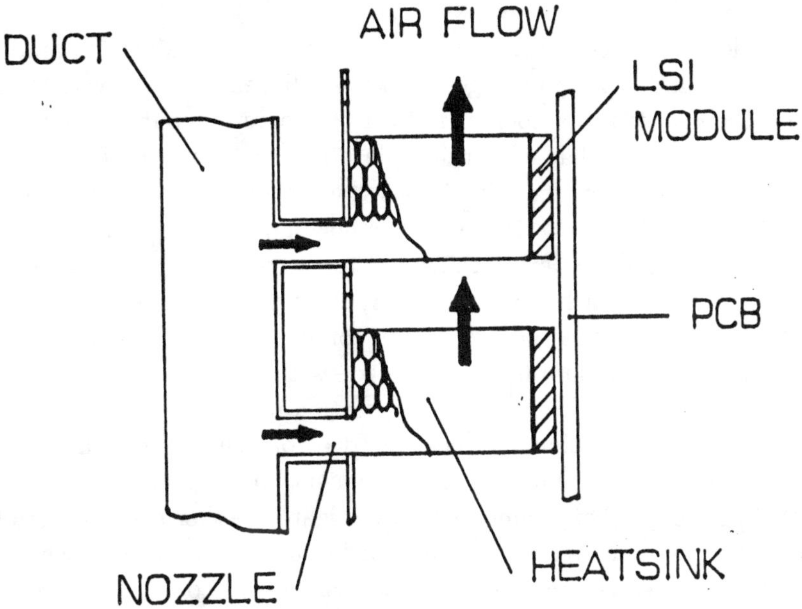

Figure 4.5: Parallel flow perforated fin heat sink cooling concept for LSI modules [Adapted from Takada et al (1990)]..

designs exhibited 3 to 5 times greater performance compared to the commercial heat sink in practical operation.

Three papers by Knight et al (1991, 1992b, 1992a) focused on the optimum design of microchannel heat sinks. The first two papers present the equations governing heat transfer and pressure drop in both laminar and turbulent flow through the channels between heat sink fins. A methodology is then described for solving these equations to determine the heat sink dimensions which will result in a minimal heat sink thermal resistance. The third paper discussed the use of the methodology to design three aluminum heat sinks with the same base dimensions and height, but with different fin thickness and gap. These designs were tested and the results were compared with the predicted values. Although the predicted values were significantly below the measured values, the design predicted to perform optimally did so.

A paper by Takada et al (1990) discussed a unique heat sink design for cooling electronic modules. As shown in Fig 4.5, the finned heat transfer surface consists of a stack of perforated copper plates, each bent to form alternating trapezoidal profiles. Each fin plate is soldered between two end-plates creating alternately expanding and contracting flow passages between adjacent fins. It was found that the heat transfer coefficient for this geometric configuration was about two times that of parallel plate fins for the same frontal velocity. The authors explained that the mechanism of heat transfer enhancement is due to the secondary flow induced by suction and injection through the perforations and to the frequent boundary layer interruptions at each contraction. Another paper by Chapman et al (1994) discussed heat transfer tests conducted to compare the performance of aluminum heat sinks made with extruded fins, cross-cut rectangular fins and elliptical cross-section fins.

One technique that has been used to obtain higher heat transfer coefficients with air is jet impingement cooling. Hollworth et al (1990) conducted experiments to investigate the thermo-fluid performance of a surface vented circuit board impingement cooling scheme. Each simulated component in a 5×8 array was cooled by a cluster of four jets issuing from orifices in a plenum opposite and facing the components. The space between the edges of the simulated PCB and the plenum were totally enclosed by a skirt on all four sides and spent air was exhausted through vent holes drilled through the test board adjacent to the edges of each module. In contrast to the results of a previous study in which spent coolant was exhausted out one edge of the channel formed by the board and plenum, the results of the current study showed no significant variation in heat transfer between elements at various sites in the array. Azar et al (1991a) conducted an experimental investigation of single jet impingement heat transfer using very small diameter, choked flow, air jets. Average heat transfer coefficients were reported for jet diameters of 0.254, 0.381 and 0.508 mm over a range of jet exit to surface spacing of $1 \leq Z/D \leq 30$.

Giammaruti et al (1991) reported the results of an experimental study investigating the effects of jet impingement cooling in combination with a cross flow. Chip heat sources were simulated using four ceramic heaters (2.54 cm by 2.29 cm) on two blank circuit boards mounted vertically in a horizontal wind tunnel. Impinging flow was provided by a jet nozzle facing each simulated chip. It was found that "for local heat transfer, jet impingement and cross flow can lead to increased heat transfer but certain combinations can lead to decreased local heat removal."

4.6 LIQUID COOLING

Chip or module cooling capability can be significantly improved when liquid cooling is used. Liquid cooling applications can generally be categorized as indirect or direct. An indirect scheme is usually considered to be one in which the liquid does not contact the electronics or the packages housing the electronics. Because there is no electrical contact, water can be used as the coolant, taking advantage of its superior thermophysical properties. Computers such as the IBM ES-9000 series, the NEC SX-2000, the FUJITSU VP-2000, and the HITACHI M-880 utilize indirect liquid cooling schemes. For direct liquid cooling the electronics or packages within which they are housed are in direct and intimate contact with the cooling liquid. Direct liquid cooling offers the opportunity of removing heat directly from the chip, or surface of the heat generating device, and transferring it into the liquid with virtually no intervening thermal conduction resistance paths. In recent years there has been considerable interest in direct liquid cooling as demonstrated by the many papers reported in this section.

A paper by Van San (1991) provided an overview of recent trends in microelectronics and focused on the direct immersion cooling approaches utilizing fluorocarbon liquids. In terms of electrical and chemical characteristics, the fluorocarbon family of liquids is generally considered to be the most suitable for direct immersion cooling of electronic devices and packages. Unlike the Freon liquids (such as R-113), the Fluorinert^{TM} liquids developed by the 3M Company do not contain any chlorine atoms and are not to be confused with the chlorofluorocarbon compounds (such as the CFC's) associated with the problem of atmospheric ozone depletion. Noting the need to replace CFC's in accordance with international agreements, a paper by Jomard et al (1991) discussed candidate replacement fluids and the constraints which they must satisfy. A paper by Joshi (1992) described Naval Postgraduate School research on passive liquid immersion cooling of electronic equipment. The discussion included experimental and computational studies of natural convection liquid cooling for three configurations of interest in applications along with recent studies on cooling by boiling.

Six papers reported here addressed natural convection liquid cooling. Prasad et al (1990) conducted an experimental and numerical investigation of natural convection in a vertical cavity with three heat sources spaced vertically on one wall and the opposing wall isothermally cooled. Numerical calculations were carried out for Prandtl numbers of 1, 25 (dielectric liquid) and 166 (ethylene glycol), and the experiments were carried out with ethylene glycol. The numerical predictions were reported to be in excellent agreement with flow visualization pictures and the measured heat transfer coefficients. Keyhani et al (1990) also conducted experiments with ethylene glycol in a vertical cavity to investigate the effect of aspect ratio on natural convection heat transfer. In these experiments the top surface of the test enclosure was an isothermal heat sink. One vertical wall was insulated and five heat sources were spaced vertically on the opposing wall. All of the data was well represented by a single correlation based upon local height measured from the bottom of the cavity to the middle of a heated section of each heater. This correlation includes an explicit aspect ratio effect and is independent of the number of heaters, protruding height and heater vertical height for the range of conditions investigated.

Another study with an isothermal heat sink at the top of a vertical cavity was reported by Chen et al (1991). This study utilized 10 vertically spaced heaters with ethylene glycol and water as the test fluids. Heat transfer correlations relating Nusselt number to the modified Rayleigh number (based on heater local height) were obtained for both fluids. Joshi and Knight (1990) investigated natural convection from a column of eight, in-line, rectangular heat sources flush mounted on one wall of a vertical channel in a tank of water. A single heat transfer correlation was obtained for all channel spacings larger than 3 mm. The effect of heater spacing was also investigated by making temperature measurements in the absence of a shroud with only selected heaters powered. Joshi et al (1991) conducted experiments using an array of protruding heat sources simulating 20 pin dual-in-line packages mounted on a vertical substrate within a liquid filled enclosure with heat exchangers at top and bottom. Heat transfer data was obtained for FC-43, FC-71, and FC-75 coolants. A single heat transfer correlation fitting 95 percent of the data above a power of 0.1 W within 15 percent was obtained. In another study Joshi and Paje (1991) investigated natural convection heat transfer from a 64 pin leadless chip carrier (LCC) mounted on a ceramic substrate vertically oriented in a bath of FC-75.

A paper by Mahaney et al (1990a) reported on studies investigating mixed convection heat transfer. Experiments were performed with FC-77 and water flowing over a 4×3 in-line array of 12.7 mm $\times$ 12.7 mm heaters, flush mounted to the lower wall of a horizontal rectangular channel. The results obtained included heat transfer regimes characterized by pure natural convection, mixed convection, laminar forced convection, and the initiation of transition to turbulence as shown in Fig 4.6. At the higher Reynolds numbers, the data is in the pure forced convection regime. However, as the Reynolds number decreases, enhancement of heat transfer occurs. This enhancement was attributed to the development of a buoyancy induced secondary flow. A comparison of experimental and numerical results in another paper by Mahaney et al (1990b) showed similar behavior. A third paper by Mahaney et al (1990c) included experimental data for augmentation of the heat sources with longitudinal fins. It was reported that, "in contrast to variations for the flush-mounted surfaces, row-average Nusselt numbers for the finned sources decrease monotonically with decreasing Reynolds number for rows 1 to 3 and are nearly independent of Reynolds number for row 4."

Three papers by Garimella and Eibeck (1990, 1991a, 1991b) reported the results of single phase forced convection experimental studies with water. The studies were conducted with an array of thirty simulated 2.54 cm $\times$2.54 cm modules, mounted in six rows on the bottom wall of a horizontal flow channel. The first paper (1990a) reported heat transfer and pressure drop measurements obtained at several channel heights and over a range of module streamwise and spanwise spacings. Results were included for both in-line and staggered array configurations. The second paper (1991a) also reported on the effect of spanwise spacing on heat transfer. The third paper (1991b) reported the heat transfer enhancement obtained by installing one vortex generator upstream of each column of modules. It was found that the maximum heat transfer enhancement (up to 40 percent) occurred at the second row of modules.

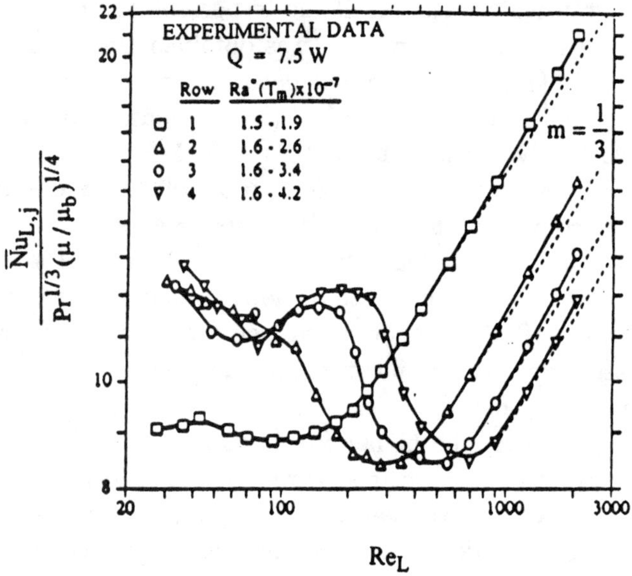

Figure 4.6: Effect of Reynolds number on the row averaged Nusselt number for water [Mahaney et al (1990a)].

A paper by Riddle et al (1991) discussed the design of microchannel heat sinks and reported the results of heat transfer experiments using water. A heat dissipation of 1 kW/cm^2 with a 100°C temperature difference between the active device surface and the cooling water was reported as an example of microchannel heat sink cooling capability. A paper by Harpole and Eninger (1991) also discussed calculations to optimize the design of a microchannel heat sink.

A paper by Chrysler and Simons (1990) discussed an experimental study which was conducted to investigate the use of FC-77 coolant flowing through a packed-bed as a possible method for cooling microelectronic circuit chips. Tests were performed for packed-beds with four different glass bead diameters. Heat transfer data for the four packed-beds were correlated by a single curve of Nusselt versus Reynolds number with the square root of the packed-bed permeability as the characteristic length.

Pool boiling has been of interest as a possible means for cooling electronic chips for many years. A number of papers in this section report additional studies of pool boiling for electronic cooling applications. Three papers focused on the problem of boiling incipience with highly wetting, dielectric liquids. As noted in a paper by Tong et al (1990), "the difficulty in predicting the boiling incipience of highly wetting liquids has slowed the application of immersion cooling technology in the electronics industry." In this paper the authors examined the influence of the dynamic solid/liquid contact angle and contact angle hysteresis on the incipience superheat. You et al (1990a) conducted boiling incipience experiments using chromel wires and a 0.51 mm diameter platinum thin film heater immersed in R-113. Wall incipience values up to 73°C and wide variations in incipience superheats (> 20°C) for nominally identical cases were reported. You et al (1990b) also conducted experiments using cylindrical heater surfaces of platinum, silicon, silicon dioxide and aluminum oxide immersed in FC-72 and R-113. Results were reported in terms of the probability of boiling incipience versus wall superheat. It was also reported that, although the difference in incipience wall superheat values between FC-72 and R-113 was significant, the surface material effect on boiling incipience was small. Further experiments investigating boiling incipience and heat flux using an array of flush mounted, 5 mm × 5 mm heaters on a 29 mm × 35 mm glass substrate immersed in FC-72 were reported by You et al (1992a).

Another paper by You et al (1992b) introduced particle layering as an effective technique for enhancing boiling nucleation on a surface immersed in FC-72. This technique provides an increased number of nucleation sites, resulting in a decreased wall superheat during boiling and an increased value of critical heat flux (CHF). You et al (1990c) reported the results of experiments to investigate the effects of pressure, subcooling, and dissolved gas on pool boiling heat transfer in FC-72 liquid. The results of a pool boiling study were also reported by Carvalho and Bergles (1990). This study investigated the enhancement on pool nucleate boiling and (CHF) in FC-72 with three commercially available enhanced boiling surfaces. Lee et al (1993) reported the results of an experimental investigation of the influence of mixtures of dielectric liquids in CHF in pool boiling.

Experiments were conducted with two different families of dielectric liquids; 3M perfluorocarbon liquids and Ausimont perfluoropolyether liquids. Two different liquids in each family were used in five different mixture ratios (0, 20, 50, 80 and 100 by volume). It was reported that for the perfluoropolyether liquids, CHF increased as the fraction of the higher boiling point liquid increased, while the perfluorocarbon liquids exhibited an almost constant CHF value over the entire range of mixtures. Another paper by Normington et al (1992) reported experiments to study the use of mixtures of perfluoropolyether fluids to control temperature overshoot associated with pool boiling of low surface tension dielectric coolants. Tests were conducted using a variety of mixtures of two perfluoropolyether coolants with boiling points of 81 and 110° C and sub cooling levels of 10 and 50°C. The effect of liquid mixtures in reducing temperature overshoot was substantial in the high subcooling case (50°C) with reductions in temperature overshoot of as much as 20 to 25°C. However, for the low subcooling cases (10°C), the effect was not significant.

A paper by Mudawar and Anderson (1990) reported the results of another study which investigated the effects of system pressure, subcooling, and surface augmentation, on pool boiling heat transfer. For this study a simulated 12.7 mm × 12.7 mm microelectronic chip was oriented vertically in a pool of FC-72 or FC-87. In another paper, Mudawar and Anderson (1993) discussed how experimental nucleate boiling data for a simulated flat chip, and chips with low profile microstructures, can be used as input boundary conditions for numerical prediction of boiling performance of smooth and microstructured cylindrical extended surfaces. Papers by Bhavnani et al (1991) and Goyal et al (1993) discussed pool boiling experiments in R-113 to investigate the performance of enhanced hybrid silicon surfaces. The enhanced surfaces were fabricated using crystallographic etching to produce pyramidal shaped cavities. Reductions were reported in values of both incipient wall superheat and temperature overshoot for the enhanced surfaces compared to untreated surfaces. Two papers by Phadke at al (1992a, 1992b) provided additional pool boiling results for these surfaces. Data was reported for boiling in both R-113 and FC-72. Surfaces with two different cavity sizes were tested at saturated and three different subcooling temperatures. A paper by Kristiansen et al (1990) reported the results of pool boiling experiments with test chips in FC-72 and FC-84. In another paper, Azimi-Rashti et al (1990) presented pool boiling results for a silicon test wafer immersed in Freon-12 and Freon-22.

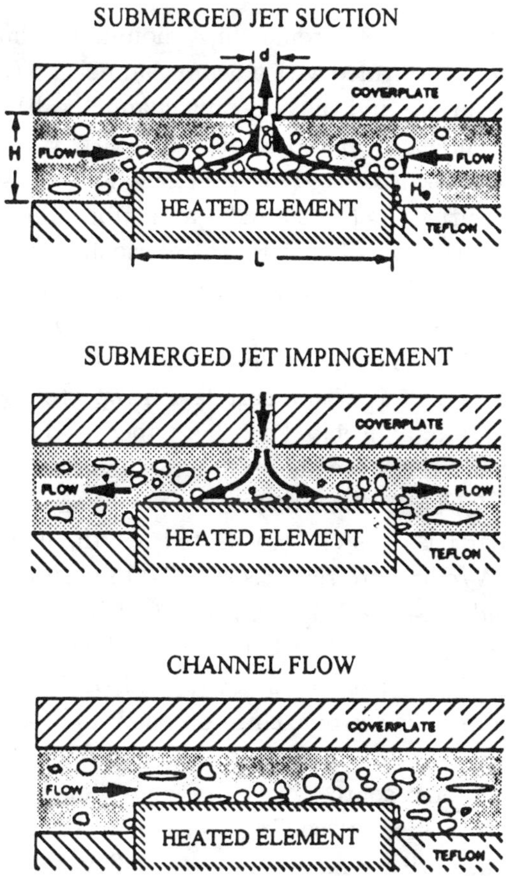

Figure 4.7: Flow configurations [McGillis and Carey (1990)].

To accommodate higher heat fluxes than are attainable in pool boiling, flow boiling may be used. A number of papers have derived from the laboratories at Purdue University. Mudawar and Maddox (1990) reported the results of a study investigating surface augmentation techniques to enhance the critical heat flux (CHF) from a simulated microelectronic chip. Comparisons to a smooth surface and to surfaces with low-profile microgrooves, low-profile microstuds and high-profile pin fins were presented. A critical heat flux as high as $361\,W/cm^2$ was achieved using a combination of higher flow velocity, high subcooling, and surface enhancement. Two papers by Gersey and Mudawar (1992, 1993) and one by Gersey et al (1992) discussed FC-72 forced convection boiling experiments with nine 10 mm

$\times$ 10 mm simulated microelectronic chips mounted in-line in a flow channel. Tests were performed with a 5 mm channel gap (the distance between the chip surface and the opposing channel wall) at eight different channel orientations spaced 45 degrees apart. Channel gaps of 2 mm and 10 mm were also tested for a vertical flow configuration. A range of flow velocities (13 to 400 cm/s) and subcooling levels (3 to 36°) were employed. It was found that the orientation changes had a major effect on CHF but did not effect single-phase or nucleate boiling characteristics. Upflow conditions were found to produce the highest values of CHF and the value for the most upstream chip in vertical upflow agreed well with a correlation reported by Mudawar and Maddox (1989) for an isolated chip. A paper by Bowers and Mudawar (1994) reported the results of an experimental study of CHF in mini-channel (2.54 mm diameter) and micro-channel (5 μm) heat sinks with circular passages of 1 cm heated length. Values of CHF in excess of 200 W/cm^2 were reported for R-113 with flow rates less than 65 mℓ/m and pressure drops less than 0.35 bar.

A paper by McGillis et al (1991b) reported the results of another study which investigated subcooled flow boiling of R-113 from a column of simulated microelectronic chips on one wall of a vertical flow channel. Measurements of CHF were obtained for both flush and slightly protruding heat sources over a range of flow velocities.

One of the most effective cooling methods in terms of supporting a high heat flux with a minimum coolant flow rate is liquid jet impingement. Accordingly, a number of researchers have conducted studies to investigate liquid jet impingement heat transfer with and without boiling. McGillis and Carey (1990) conducted an experimental study to investigate immersion jet impingement and immersion jet suction with boiling over flush and protruding heat sources. The test section for a channel flow boiling study by Strom et al (1989) was modified to accomodate the key flow configurations shown in Fig 4.7 and a comparison of the jet results and the earlier channel flow boiling results is shown in Fig 4.8. Schafer et al (1991) reported single phase heat transfer measurements made for discrete sources cooled by a water jet discharged from a rectangular slot in the opposite wall in the flow channel. Heat transfer data was reported in the form of Nusselt number vs. Reynolds number for heaters positioned

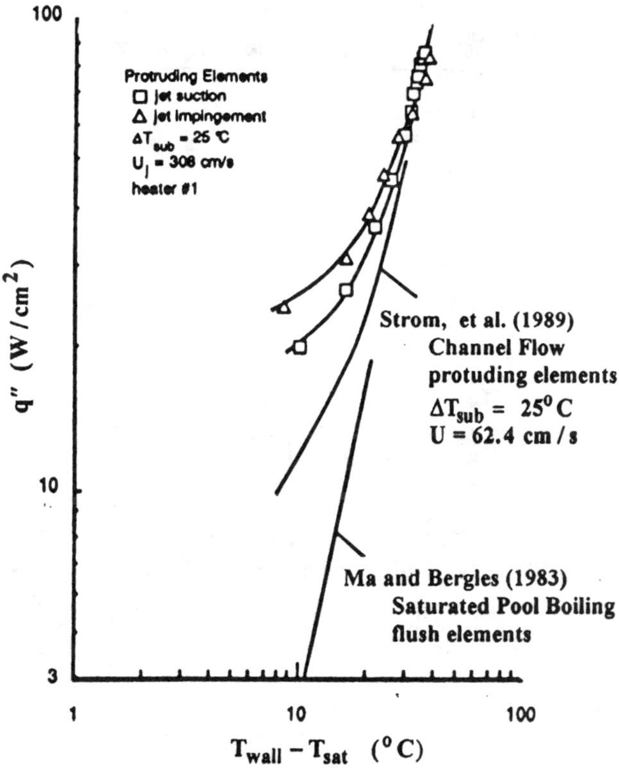

Figure 4.8: Boiling curve comparisons of immersion jet impingement and immersion jet suction with channel flow [McGillis and Carey (1990)].

directly below the jet as well as at locations offset from the midplane of the jet.

Wadsworth and Mudawar (1990) reported the results of experiments investigating single-phase heat transfer from a 3×3 array of simulated chips cooled by FC-72 slot jets in a liquid impingement cooling module. The effects of nozzle geometry, jet width, confinement channel height and impingement velocity were presented and discussed. Mudawar and Wadsworth (1991) reported further work using the same experimental apparatus to investigate jet impingement heat transfer under boiling conditions and to determine the CHF. Based upon the results, the authors concluded that the confined jet geometry tested "is well suited for uniformly cooling large arrays of heat sources dissipating heat fluxes as high as 250 W/cm^2." In another paper, Wadsworth and Mudawar (1992) described further

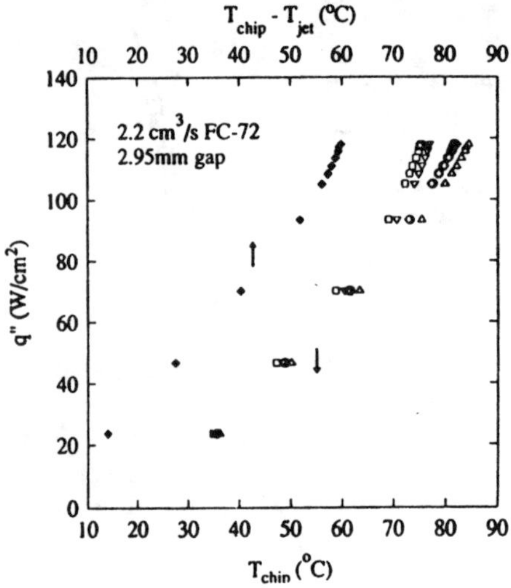

Figure 4.9: Typical liquid jet cooling data for a 0.50 mm diameter orifice and FC-72 at 20.5°C [Chrysler et al (1994)].

liquid impingement experiments to investigate single-phase and boiling heat transfer from smooth and microstructured surfaces with a planform area of 12.7 mm × 12.7 mm.

A paper by Chrysler et al (1994) reported the results of an experimental investigation into the effects of chip-to-orifice gap for jet impingement boiling of FC-72 at the back surface of a 6.5 mm × 6.5 mm chip. As shown in Fig 4.9, the results demonstrated the ability to support a chip heat flux of 120 W/cm^2. Copeland (1994) reported the results of an experimental study of single-phase and boiling heat transfer from pin fin array heat sinks. A flow of FC-72 liquid was directed to and from the pin fins by multiple slot nozzles, alternately providing impingement and suction flow.

Womac et al (1990) conducted experiments studying heat transfer with submerged and free-surface, single, circular jets impinging on a single 12.7 mm × 12.7 mm heat source. The experiments were conducted with water and FC-72 for nozzle discharge diameters from 0.457 mm to 6.55 mm. Besserman et al (1991) conducted a numerical simulation of axisymmetric, laminar jet impingement cooling of a circular heat source. The geometry studied was that of a jet discharged from a circular tube and redirected 180° after impingement to exit through an annular passage. Another paper by Besserman et al (1992) discussed the results of experiments to study heat transfer from a square source to an impinging water jet surrounded by an annular wall. A paper by Sullivan et al (1992) presented the results of experiments performed to study heat transfer enhancement for single-phase jet impingement cooling of simulated electronic chips. Enhancement was obtained using radial plate fins and arrays of square pin fins as extended surfaces. Sun et al (1993) reported the results of an experimental study to investigate heat transfer from a vertical simulated chip to an impinging, submerged, circular water jet.

A paper by Maddox and Bar-Cohen (1991) addressed thermofluid design and optimization of submerged jet impingement cooling for microelectronic chips. Utilizing the analogy between gas jet cooling and single-phase heat transfer to a submerged liquid jet, the authors used a modified form of the Martin correlation to examine the sensitivity of liquid jet cooling to parametric changes for both single and multiple jets. A case study of a possible high performance cooling application was used to illustrate the interplay of parameters and constraints on the thermofluid characteristics of a liquid jet impingement cooling design. Wang et al (1990) applied a previously developed analytic solution to the conjugate heat transfer problem of jet impingement cooling of a microelectronic chip. The analysis was used to predict the surface temperature and heat flux distribution on a uniformly heated chip cooled by a laminar impinging jet of FC-77 liquid or water.

Ghodbane and Holman (1991) conducted a study to investigate spray cooling with subcooled Freon-113 on a vertically oriented heat test surface. Two different size (15.24 cm × 15.24 cm and 7.62 cm × 7.62 cm) test surfaces were used to study the effects of test surface size. The effects of mass flux, spray droplet velocity, droplet diameter and distance between nozzle and test surface were investigated. The test results demonstrated that this cooling technique can support heat fluxes up to 50 W/cm^2 while maintaining a surface temperature below 80°C. A non-dimensional spray cooling correlation was developed for Freon-113 over a wide range of test conditions and parameters.

The last three papers in this section addressed boiling heat transfer within thermosyphon loops. Babin et al (1990a) conducted a study to investigate direct immersion cooling of a 6 by 8 array of

12 mm × 20 mm integrated circuit packages mounted on a PCB. The PCB was located in the evaporator section of a thermosyphon system using R-114 as the working fluid. Tests were also performed with a thin copper foil separating the IC packages from the working fluid in the evaporator, and with the IC packages on the PCB cooled by free air convection. The results showed that the thermosyphon system allowed ten times higher power with an 80 percent reduction in component surface temperature rise compared to free convection cooling. McGillis et al (1991a) studied pool boiling of water at low pressures from a heated 12.7 mm × 12.7 mm horizontal test surface. The test surface was mounted in the evaporator section at the bottom of a single pipe thermosyphon with an air-cooled finned-tube condenser at the top. Tests were conducted to investigate the pool boiling enhancement effects of rectangular finned geometries, particulate beds of copper or Teflon beads and different surface finishes. Cho et al (1990) conducted experiments to study boiling heat transfer from inclined channels in a thermosyphon using distilled water as the working fluid. Tests were conducted for heat fluxes of 9.09, 13.82 and 16.10 W/cm^2, and for channel spacings from 1 to 20 mm. It was found that a channel spacing above approximately 10 mm no longer influenced the heat transfer rate from the immersed heater surface.

4.7 THERMAL MEASUREMENTS AND SENSING

Thermal measurements play an important role in the successful design of cooling schemes for electronic components and systems. Accurate measurements of temperature and power are necessary to characterize the thermal resistance of components from the chip heat source to the cooling sink. This also includes accurate measurement of coolant flow and temperature conditions. In addition, temperature measurements may be utilized to determine the individual thermal resistances in one or more heat flow paths within a component assembly. Temperature measurement of prototype hardware has been a long-standing method to confirm the adequacy of a thermal design. The continuing trend towards higher power and packaging densities only serves to increase the need for thermal verification and accurate temperature measurements. Thermal measurements can also play an important role in assessing the quality of incoming vendor components, and outgoing manufactured electronic equipment.

For many years, thermocouples have proved a very useful device for measuring temperature, and they are widely used to measure both fluid and component temperatures. In some cases, however, it is desirable to measure temperatures at many points on the surface of an electronic package, or even obtain a complete temperature "map." Infrared (IR) surface temperature measurements may be used to obtain such temperature maps. Two articles by Walcutt (1991a, 1991b) discussed the operation and application of IR thermal imaging systems to obtain comprehensive and accurate thermal maps of circuit boards and component packages. Hussein et al (1991a) discussed the use of infrared thermal imaging to obtain experimental peak temperature measurements to validate results predicted by a computer model. A paper by Crall (1990) discussed the benefits of using IR thermal imaging as a thermal management tool in evaluating SMT device applications. A paper by Maudgal et al (1990a) described a thermal measurement method using a thermal imaging system to test circuit boards in a production environment. An article by Linnander (1993) discussed the application of infrared technology for non-destruvtive (or non-intrusive) temperature measurements of PCB's and other electronic components where contact with the measurement device would be impractical or would affect the results.

Another technique that may be used for thermal mapping in electronic applications is liquid crystal thermography. A paper by Parsley (1991) provided an introduction to thermochromic liquid crystals (TLC's) and their properties. An extensive list of references pertaining to the application of TLC's was also provided. A paper by Azar et al (1991b) discussed an experimental study conducted to verify the measurement capability and accuracy of the liquid crystal imaging technique.

Although thermocouples may be used to measure the temperature at some point on the external surface of a package, the so-called junction or chip temperature within the package is usually measured using temperature sensitive electrical parameters (TSEP) of the chip itself. Lu et al (1991) described an electrical technique which can be used to perform a fast measurement of the peak junction temperature of power transistors. The authors described their technique as involving three modifications to the standard approach using emitter-base voltage (V_{EB}) for the temperature sensitive parameter. It was claimed that in comparison with an infrared measured peak junction temperature, the measurement error of the reported technique is within 8 percent, compared to an error of up to 50 percent for the standard technique under the same conditions. Cripps (1990) described a technique which utilizes the inherent variation of RF gain with temperature for the measurement of channel temperature of microwave semiconductor devices.

Nishiguchi et al (1991) provided a discussion and comparison of three techniques for temperature measurement of GaAs integrated circuit devices. The three measurement techniques considered were:

1. an electrical method using the temperature dependence of the forward voltage (V_f) of a GaAs Schottky diode,
2. liquid crystal thermography and
3. infrared microscopy.

Struble and Witte (1991) described the use of power transistors as both the heat source and temperature measurement device in liquid immersion cooling experiments. Steady-state heating was produced in the emitter-base region, and then current was removed for a period of 1 ms during which the voltage across this region was measured as a temperature sensitive parameter. A paper by Geeraerts et al (1993) described the use of a diode matrix to produce a known power dissipation on a chip at a specific location and sample the temperature distribution on the chip. Another paper by Wirth (1992) discussed the use of an IGBT's peak rate of decrease of collector current at turn-off to provide a convenient means of measuring its junction temperature. Two papers by Lee et al (1992c, 1992d) discussed two different techniques for transient thermal measurment of semiconductor device temperature. One technique employs the change of optical index of refraction with temperature which, in optics, is usually referred to as the thermo-optic effect. The other technique utilizes the heater resistor change with temperature. The intrinsic response time of both methods is a few nanoseconds making it possible to measure the rapid initial temperature rise caused by a step power input.

Measurements of temperature and power dissipation are often used to characterize component thermal performance in terms of thermal resistance. This thermal resistance may be expressed as a total resistance from junction-to-air (R_{TOT} or Θ_{JA}) or it is separated into an internal resistance from junction-to-case (R_{INT} or Θ_{JC}) plus an external resistance from case-to-air (R_{TOT} or Θ_{JA}). Two papers by Siegal (1991, 1993) discussed the need for a better methodology for thermal characterization of surface mount devices. Both papers covered several measurement setup examples including those for determining θ_{JA} and θ_{JC} as well as application oriented examples. In many instances, internal thermal resistance is treated as constant and independent of external cooling conditions such as air flow or the packaging density on the PCB. Unfortunately, this is only true for an isothermal package surface and, in many cases, internal thermal resistance measured under one cooling or packagaing condition will exhibit a different value under other conditions. An earlier paper by Bar-Cohen et al (1989) discussed this problem and proposed a new approach providing "a universal, though modified, θ_{JC} for use with a particular package in every possible packaging configuration." In a more recent paper, Krueger and Bar-Cohen (1992) applied the expanded methodology to a plastic leaded chip carrier (PLCC) package.

Lemczyk et al (1992a) discussed an approach to the thermal characterization problem based upon the optimum thermal resistance (that for the isothermal case package) and an optimization factor "which can be used by a thermal designer to measure the variation of the performance under given system constraints." Experimental data and the results of a three-dimensional thermal simulation were used to determine eleven sensitivity coefficients which may be used to determine a weighted average case temperature.

A study conducted by Kozarek (1991) investigated the effect of case temperature measurement errors with thermocouples on the measured junction-to-case thermal resistance (Θ_{JC}) of a 132 pin ceramic pin grid array (PGA) package. This study utilized the water-cooled heat sink approach specified in SEMI and MIL standards to measure Θ_{JC}. The results of this study indicated the potential for large errors in determining Θ_{JC} with this approach unless an accurate method of determining case temperature is used. A paper by

Sweet and Cooley (1990) described measurements and calculations for ceramic leadless chip carriers illustrating the ranges of accuracy which can be obtained in determining Θ_{JC} or Θ_{JA} for different test conditions. The study examined both the heat sink and fluid bath measurement test approaches. Two papers by O'Mathuna (1992) and O'Mathuna and Daguang (1992) reported on the development of an automated test system for the thermal characterization of IC packages. The system is designed to provide oven calibration of on-chip temperature sensors as well as thermal resistance measurements in an oven or on a temperature controlled heat sink.

It is sometimes necessary to determine the heat transfer coefficient acting upon electronic module or card surfaces. To do so, it is necessary to determine the temperature of the surface, the heat flow from the surface to the air, the area of the surface and the temperature of the cooling air. In many cases, the heat flow from the surface may not be accurately known, introducing a large uncertainty in the "measured" value of the heat transfer coefficient. Such uncertainties may be avoided by using the napthalene sublimation technique to experimentally determine heat transfer coefficients. As explained by Souza Mendes (1991), in a review paper on the technique, only mass transfer experiments are performed in the laboratory, and then, heat transfer results are obtained be employing the concept of the analogy between heat and mass transfer. The theoretical background and use of the technique are reported in the paper, along with many examples of successful applications and a comprehensive list of references.

In addition to measurements of temperature, an understanding of flow patterns over and around modules on a PCB is important for a complete understanding of the associated heat transfer performance. Wang and Ghajar (1991) conducted a study to investigate the effect of component dimension, geometry, orientation and placement (inline vs. staggered) on flow distribution. Experiments were conducted in a wind tunnel with smoke injected for vizualization of the flow over and around the simulated components. Another study examining the impact of component layout and geometry on flow distribution over simulated components was reported by Azar and Russell (1991). The experiments were conducted in a water tunnel with diluted ink injected to vizualize the flow structure around the components.

Thermal sensing can also play an important role in protecting electronic components and packages from excessive temperatures during operation in product environments. As a result of the high power densities used in many components and packages today, a cooling failure (such as loss of air flow) could result in physical damage to the affected component, making thermal protection an essential element of thermal management. An article by Berardinis (1990) discussed the types and operation of electromechanical temperature switches as well as some of the improvements which have been made in their design. A paper by O'Hanlon (1991) described the use of a thermistor within the IBM TCM to directly measure the purity of the helium gas which was used as a thermally conductive medium within the module. It was reported that it was possible to detect as little as 15 percent air contamination of the helium charge.

Another important aspect of thermal measurments is the determination of thermal transport properties of semiconductore and electronic packaging materials. An accurate determination of property values is important both for selection of materials and for use in package thermal analysis. A paper by Tye and Maesano (1992) described a technique based on ac calorimetry, developed to measure the thermal diffusivity and specific heat of specimens less than 0.3 mm thick over a broad temperature range.

4.8 THERMAL ANALYSIS AND MODELING

As the interval between successive generations of electronic equipment decreases, the time available for performing prototype tests on next generation equipment has decreased. This has led to an increased emphasis on the use of analytic and numerical models for the analysis and prediction of package temperatures and thermal performance. Sophisticated thermal analysis of proposed electronic packages is rapidly becoming the rule, rather than the exception. In many cases automated thermal modeling tools are being integrated into the Computer Aided Engineering (CAE) and Computer Aided Design (CAD) design systems used for electrical design and physical layout of electronic packages. A paper by Babus'Haq et al (1990) provided an overview of the role of thermal analysis in thermal management and some of the modeling techniques employed. Another paper by Babus'Haq et al (1992) reviewed the use of finite difference, finite element and network analysis for the thermal analysis of electronic packages and systems. The paper also describes the use of a thermal analysis template to facilitate computer-aided analyses of thermal systems. A paper by Nigen and Amon (1991) proposed a thermal design methodology suitable for concurrent design of cost-driven electronic systems. The design methodology is based upon an evolutionary concept, with the analysis tools capable of adjusting their level of complexity as the design evolves. The level of approximation at each stage is selected based upon a consideration of both time and accuracy requirements. A paper by Edwards (1992) described the integral role of thermal analysis in the design of the IBM thermal conduction module (TCM).

Articles by Garfinkel (1990), Maudgal (1991), Goodman et al (1990) and Tokola and Wolff (1990) discussed the continuing trend towards greater automation of thermal analysis and design for printed circuit boards. A paper by DeMey (1990) discussed thermal simulation in relation to CAD for hybrid circuits. Avaritsiotis and Eleftheriades (1991) described an approach, implemented on a PC/XT combining device layout and thermal analysis. In another paper, Almgren and Almgren (1990) provided a general discussion of the accuracy of PCB thermal analyzers with suggestions for assessing their validity and accuracy.

Closely associated with the thermal analysis of PCB's is the optimum placement of components on a PCB to enhance reliability. Papers by Osterman and Pecht (1990) and Osterman (1992) discussed the development of a placement methodology to maximize reliability by minimizing the total hazard rate of components on a PCB with forced convection cooling. A paper by Eliasi et al (1990) also presented the methodology and results of a Monte Carlo optimization of thermal component placement on a PCB to minimize the total failure rate. Two papers by Cahlon et al (1991, 1992) described a mathematical model to calculate the temperature of electronic components on a board. The model is used in conjunction with an annealing algorithm to determine the optimal placement of components in order to minimize the maximum temperature of the system. A paper by Queipo et al (1994) presented and discussed genetic algorithms within the framework of an adaptive solution methodology for investigating otherwise intractable optimization problems in the thermosciences. The discussion focused on an example application to find optimal and nearly optimal arrangements of convectively cooled components placed in-line on the bottom wall two dimensional channel.

In another paper, Miner (1990) discussed the use of Lagrange multipliers to determine the optimum flow distribution in a parallel flow cooling system to minimize the total component failure rate. An article by Yao (1990) emphasized the need to integrate thermal and reliability analyses to obtain meaningful results for the reliability of an electronic system.

Relatively simple thermal resistance networks are often used to model the heat flow paths in an electronic package. Malhammer (1991) discussed component thermal resistance network models which included:

1. a junction-to-ambient model,
2. a junction-to-pin model,
3. a sensitivity model and
4. a three parameter model.

Guenin and Mahulkar (1993) discussed a thermal resistance model developed to characterize the thermal performance of an MQUAD package on a test board cooled by natural convection. As discussed in the paper, and experimentally determined heat transfer coefficient and board thermal conductivity were utilized in the model. Hadim et al (1990) used a thermal network analyzer called SINDA (System Improved Numerical Differencing Analyzer) to create a thermal model of a PCB cooled by natural convection. The authors described an iterative technique which compared experimentally measured temperatures with computed temperatures in order to refine the correlation used to compute natural convection heat transfer to air. Csanky (1990) described the application of a series resistance thermal model to calculate the temperature of IMPATT diodes under life test. The model incorporated temperature-dependent thermal conductivities of silicon and diamond by means of simple empirical relationships.

A paper by Gautier (1991) discussed the reduction of larger thermal network models to produce reduced models focusing on particular nodes. In another paper, Kelly (1990) presented thermal equivalent circuits derived from finite element analyses of three different surface mount packages.

With the appropriate addition of capacitive elements, thermal network models may also be used to analyze transient thermal behavior. Guenin (1994) used such a model to analyze the transient behavior of a 160 lead, cavity down, MQUAD package for situations in which power to the chip changes. Fukuoka and Ishizuka (1991) discussed the use of a thermal resistance and capacitance network to analyze the transient response of a package containing a low melting point alloy. The alloy was used as a phase change material (PCM) for heat absorption. Estes (1991) demonstrated the use of a simple thermal analog network to model heat flow in a sealed outdoor electronics enclosure containing a thin PCM layer on the inside walls. Exact solutions were obtained for internal temperature response to daily sinusoidal variations in outside temperature and solar load. A paper by Sofia (1994) discussed a technique for synthesizing dynamic models comprised of discrete thermal resistances and capacitances directly from packaged semiconductor device thermal step-response data.

Papers by Baker and Shortt (1990) and Awkward and Lewis (1991) described applications of SPICE (Simulation Program with Integrated Circuit Emphasis) to model device or component thermal characteristics with electrical analog circuits. A paper by Van Petegem et al (1990) described an electrothermal simulator (ETS) for combined electrical and thermal modeling of integrated circuits. Electrical analysis is performed with a "SPICE-like" subprogram and thermal analysis is performed with a subprogram employing finite element methods. Each subprogram is executed consecutively with the output of one program providing input to the other until a converged solution is achieved. Another paper by Hebrard et al (1992) discussed modeling of electrothermal interactions using a program called SETIPIC (Simulation of Electrothermal Interatcions in Power Integrated Circuits). As described by the authors, SETIPIC is organized around the SPICE-3 electrical simulator and a 3-D thermal simulator called PICMOST. Gough et al (1991) also described a computer program to simulate the coupled electrothermal behavior of semiconductor power devices in 2 and 3 dimensions. In this case, an essentially hybrid Newton method is used to obtain a coupled solutiion of the semiconductor device equations, followed by a solution to the heat equation. The scheme is then iterated until a self consistent solution is obtained. A discussion of the application of the technique to investigate the effect of temperature on the apportioning of current between the emitter fingers is provided. A paper by Hefner and Blackburn (1994) discussed a procedure for developing thermal component models for electrothermal network simulation. The methods used to implement the models into a circuit simulator are also discussed.

Relatively simple flow resistance networks are also sometimes used to model airflow through electronic cabinets or enclosures. A paper by Juds and Chan (1992) described the CABINET computer program based on a lumped parameter resistance technique. The CABINET program was designed for thermal analysis of a complex electronic cabinet with many internal compartments, wall heat, sinks, vents and fans. The solution capabilities include air, wall and heat sink temperatures, fan forced and buoyancy driven air flow and multisurface radiation echange with participating gases. Another paper by Ishizuka (1992) described a model comprised of a fluid nodal point network and a temperature nodal point network used to analyze the thermal performance of a laptop computer.

Finite element modeling (FEM) techniques are often used for thermal analysis of packages with greater geometric complexity, or to obtain a more detailed definition of temperatures. Several papers covered in this section address the application of FEM techniques and programs, A paper by Tatara (1991) described the use of FEM programs in general, and a program called TLAYERS in particular, for thermal analysis of semiconductor packages. A paper by Buschbom and Blanton (1990) discussed the importance of thermal management for all aspects of next generation systems and the use of FEM as a thermal design management tool. Godfrey et al (1993) described the development of an efficient procedure for the determination of PCB temperatures using an available finite element program. Ibrahim (1991) described the development of an integrated thermal/structural model of an MMIC (Monolithic Microwave Integrated Circuit) transmitter module using the ANSYS code. In another paper, Cooley and Razani (1991) discuss the application of finite element codes to perform a thermal analysis of a 40 pin leadless chip carrier (LCC) mounted on an alumina substrate. Included in the discussion is a technique (which the authors term homogenization) for simplification of a complex region by introducing an effective homogeneous material. A paper by Simon et al (1990) discussed a FEM analysis of a plastic dual-in-line package (DIP) to assess the influence of thermal contact resistance with possible material interfacial delamination.

A paper by Demmin (1991) discussed the use of finite element models to study the thermal performance of MCM's and a paper by Handa et al (1993) discussed the use of finite element models to simulate the effects of thermal vias and a heat sink on chip-to-air thermal resistance. Sankaran et al (1990) reported using the FIDAP code to model the spatial and temporal distribution of temperatures in high power silicon controlled rectifiers (SCR's). A paper by Nye and Brook (1991) describes the use of statistical experimental design techniques in concert with finite element analysis (FEA) results to create polynomial expressions to predict package response (Θ_{JA} for example) to design factors (such as mold compound, die size, leadframe, pin count, package size, and air flow). The effectiveness of the technique is illustrated by comparions of Θ_{JA} versus airflow based on the fitted polynomial expression and the FEA model results.

Another technique which is sometimes used in thermal modeling is the boundary element method (BEM). Four papers referred to in this section reported on BEM applications to thermal analysis. A paper by Vick and Nelson (1990) reported on the development of a new BEM thermal model for sliding contact problems of the type found in head/disk interfaces. The authors noted that a unique aspect of their work was the use of a moving Green's function to incorporate the convective effect of the moving region, as opposed to a stationary, diffusion Green's function used in standard boundary element methods. A paper by Driscart and DeMay (1990) provided a discussion and example of the application of the boundary element method to the thermal analysis of simple hybrid circuit packages. The use of the BEM approach for analysis of a VLSI PGA package was illustrated in a paper by Liu et al (1990). It was reported that a main advantage of the BEM approach is that the time and effort needed to prepare a model is dramatically reduced. The paper also included an analysis of the same package using the FEM approach. The authors noted that "no accuracy differences are apparent between BEM and FEM models." A paper by Wright et al (1991) also provided an example of the use of the BEM approach for thermal analysis and a comparison with other methods. Four different techniques were used to predict GaAs MMIC junction temperatures:

1. a series solution and

2. finite difference,

3. finite element and

4. boundary element methods.

The authors concluded that the comparative study indicated that the boundary element method is best suited for accurately predicting MMIC channel temperatures.

Although one of the oldest techniques for the numerical solution of differential equations, finite difference techniques are still widely used in thermal analysis of electronic packages. Five papers reported in this section discussed thermal analyses based upon the finite difference approach. Webb and Russell (1991) used a three dimensional finite difference simulator to model the transient thermal response of microwave devices for particular duty cycles. The method of analysis described used a combination of steady state and single transient thermal simulation to simplify the procedure and reduce the required computer CPU time required to obtain a solution. Landram (1991) described the use of the finite difference method in concert with an integral method to create a hybrid computational model of longitudinal fin heat transfer to a laminar internal flow. The model was used to identify the optimum coolant flow channel geometry between each fin in relation to coolant to wall thermal conductivity ratio and the fin to channel width ratio.

Nelson and Sayers (1992) used a control volume based finite difference method to calculate thermal spreading resistance based upon planar and axisymmetric models. The results were compared with 3-D spreading resistances calculated for devices with various aspect ratios using the TAMS code. Based upon the comparisons, conclusions were made as to when the two-dimensional approximations may be accurately applied to three-dimensional problems. Weisberg and Bau (1992) discussed an analysis of micro-channel heat exchangers as might be used to cool microcircuit chips. The analysis was performed using a central differencing scheme for simultaneous determination of the temperature fields in both the solid walls surrounding a fluid channel and in the fluid passing through the channel. Wilson and Decker (1994) used a finite difference scheme to perform detailed modeling of a Gallium Arsenide (GaS) power amplifier monolitic microwave integrated circuit (MMIC) to obtain temperatures used to correlate reliability life test results.

The use of analytic techniques, as discussed by papers in this section, continues to be an important tool for thermal analysis. Krane (1991) obtained Fourier series solutions to the steady state thermal conduction equations for two and three-dimensional rectangular slabs with a uniform base temperature and heat input over a portion of the opposite face. The results were presented graphically in the form of a thermal spreading resistance ($\bar{R} = \Delta T_{\max}/q$) from the peak temperature in the heated region to the isothermal temperature boundary. Nakwaski (1991) also discussed the solution of the two-dimensional thermal conduction equation to obtain the thermal spreading resistance of a stripe-geometry diode laser. A paper by Lavine and Bai (1991) discussed the analysis of heat transfer in Josephson junction devices. The problem was effectively reduced to that of obtaining a Fourier series solution for temperature distribution in the bottom electrode of a device. Funk et al (1991) described a semi-analytical model developed to predict steady state temperatures on PCB's subject to heating by single or multiple heat sources. The model used an iterative scheme to couple separate analytic solutions for temperatures of the PCB and chips mounted on it.

Ortega and Kabir (1991) described an analytical model developed for conduction from a heated surface element to a conductive substrate or board. As explained by the authors, "the model utilizes a rigorous linear superposition technique to decompose the element temperature into the isolated element base contribution and into the contribution of adjacent elements." A paper by Lall et al (1994) discussed the derivation of a new first order thermal design for determining the temperatures on conducting substrates and for determining the thermal "zone of influence" or "footprint" associated with a component on a board. Exact solutions were presented for sources at the center, edge and corner of a rectangular board and compared with 2-D and 3-D finite element calculation results for rectangular sources on a board. Another paper by Ortega et al (1993b) presented analytical solutions for the temperature field resulting from the application of a heat source on an adiabatic plate or board when the fluid is represented as a uniform flow with an effective turbulent diffusivity.

A paper by Min et al (1990a) discussed the derivation of a three-dimensional transient temperature solution for a two-layer semi-infinite plate structure with imbedded heat sources. A computer program was written based upon the solution and the method of images, and a number of application examples are given in the paper.

Another paper by Min et al (1990b) presented a systematic approach to obtain the thermal response of a five-layer plate structure with heat sources of arbitrary shape. Chien et al (1992) also discussed temperature solutions for a five-layer plate structure. However, a circular source instead of a square source was used as a unit source to produce a unit thermal profile. It was stated that the CPU time for a temperature calculation was reduced by a factor of 10 compared to that for rectangular source solutions. Albers (1994) describes a recursive technique developed to provide the surface temperature solution for a multi-layer steady state heat flow problem. An example of the use of the technique to predict surface temperature of a buried oxide structure containing several thicknesses of surface silicon layers was shown.

A paper by Lee and Jay-Min (1991) provides further discussion of thermal modeling of multi-layer structures using analytical methods. The use of the method of images to obtain a solution for a finite plate geometry from an infinite plate solution is described and applications to printed circuit board thermal modeling are given. Another paper by Lin and Fang (1991) also discussed the use of the image method to solve for the temperature distribution in a semiconductor laser array mounted on a copper heat sink with a diamond film for thermal spreading. A paper by Sherif et al (1992) presented a thermal analysis to assess the thermal gradients across the solder inteconnections of a chip mounted on a substrate during the power-on transient. Closed form solutions were developed based upon a 1-D model. Finite elements solutions and experimental data were also obtained to verify the validity of the 1-D model.

Brewster and Sherif (1992) presented an analysis of the temperature field resulting from Joule heating in vias surrounded by electrical insulating materials. Electrical power dissipation in the vias, as well as thermal conduction in the vias and surrounding substrate material were included in the model. It was shown that for an example application, neglect of the Joule heating in the vias can result in significant underprediction of the temperature. Lemczyk et al (1992b) also presented a thermal analysis to determine the temperature rise of a surface or buried current carrying lines (or traces) on a fiberglass PCB. A two-dimensional thermal conduction analysis performed on a multilayered cell model with a finite heat source to estimate the effective thermal conductivity for a typical PCB was also described in the same paper.

Six papers discussed the use of META (MicroElectronic Thermal Analyzer) to solve conjugate heat transfer problems. As described by Culham et al (1991), META is an analytical/numerical conjugate fluid-solid model for thermal analysis of air-cooled PCB's with arbitrarily located component heat sources. In another paper, Culham et al (1994) decribed the use of META to model isothermal cuboids and rectangular heat sinks cooled by natural convection. Lee et al (1991a, 1991b) used META to investigate the effect of common design parameters on PCB component temperature under natural and forced convection cooling conditions respectively. Some of the parameters studied included board thermal conductivity, board emissivity, upstream heating, and air velocity under forced convection conditions. Other papers by Lee et al (1991c, 1992a) extended the results of the previous papers to include mixed convection. A paper by Culham et al (1993) discussed the use of META to discuss the conjugate effects associated with a heated cubical object attached to a board and cooled by natural convection. The results of the study demonstrated the importance of using high conductivity lands as a means for lowering heat source temperatures.

Three papers discussed the use of TAMS (Thermal Analyzer for Multilayer Structures). TAMS is a computer code based on the

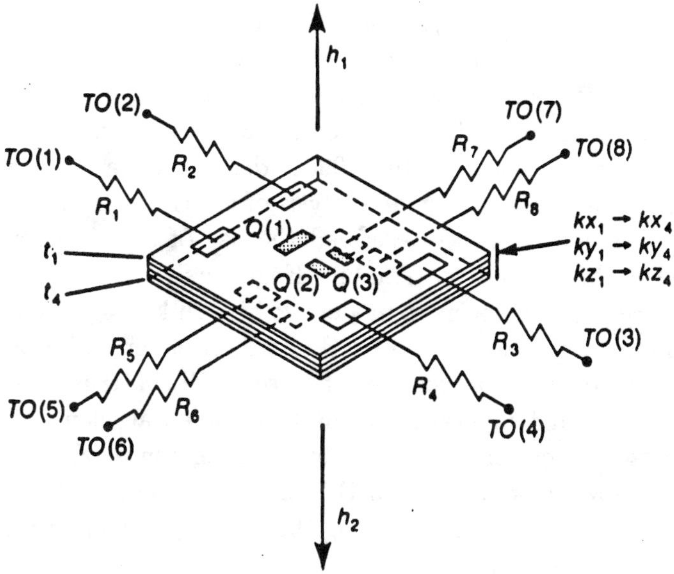

Figure 4.10: Example of geometry and relevant elements representing heat sources and lumped parameter thermal resistances on a multi-layer substrate [Ellison (1984) - Reprinted with permission of the International Society for Hybrid Microelectronics, Reston, VA].

Fourier-series solution in an anisotropic plate with a thermal conductivity k_{xi}, k_{yi} and k_{zi} for each layer, subsurface heat sources, and lumped thermal resistances intermixed with both the top and bottom surfaces as depicted in Fig 4.10. Ellison (1990) provided an overview of the theoretical foundation and methodology used in TAMS to permit "the analysis of hybrid circuit geometries that can have up to four layers of differing thermal conductivity (including orthotropy) and thickness." This paper also included an example illustrating the application of TAMS to the analysis of a hypothetical multi-chip hybrid package. Another paper by Ellison (1991) provided a further investigation and discussion of the use of lumped parameter thermal resistances in TAMS models to simulate heat flow paths from source-to-substrate and source-to ambient. A paper by Hussein et al (1991b) discussed TAMS models used to study the thermal interaction effects of two heat generating devices mounted on the same substrate.

Computational fluid dynamic (CFD) modeling of flow over heated components, across printed circuit boards, and in boxes containing electronic assemblies is becoming an increasingly important component of thermal analysis of electronic equipment. Three articles by Reynell (1990a, 1990b, 1990c) discussed the application of CFD modeling to the thermal analysis of electronic equipment. An article by Leibson (1994) described the use of a commercial code (FLOTHERM) by Silicon Graphics Engineers to model a workstation and Sequent engineers to model two heat sink designs to cool two Pentium microprocessors on a CPU board. A paper by Lee and Mahalingham (1994) also discussed the use of FLOTHERM to model the velocity and temperature fields to a forced air flow in a computer system enclosure. A paper by Temmerman et al (1992) also described how a commercial CFD tool is used to evaluate the cooling of printed circuit boards during the design phase of an advanced communication system.

Linton (1991) described an approach to modeling flow and temperatures in electronics enclosures using a commercial finite control volume code (PHOENICS). Techniques for modeling grilles, fans, heat sinks, circuit boards and other factors affecting thermal performance were discussed. Lee and Kim (1991) discussed a PHOENICS model developed to account for the effects on heat transfer of the swirling motion of air across a fan. Agonafer and Moffatt (1990) described the use of PHOENICS to model flow over blocks simulating electronic modules. Comparison of computed heat transfer coefficients with experimental data in the fully developed flow region showed agreement within 8 percent. In another paper, Linton and Agonafer (1994) described the use of PHOENICS for detailed modeling of a parallel plate heat sink. Results from the detailed model in terms of pressure drop and thermal resistance versus average flow velocity were used to create a coarse numerical model of the heat sink for use in higher level models with more components. Kamath (1994) discussed the use of PHOENICS to model the cooling performance of a pin fin heat sink for two different flow schemes. The first scheme consisted of a conventional forced convection air flow perpendicular to the longitudinal axis of the pin fins. The second scheme consisted of an impinging forced convection flow parallel to the longitudinal axis of the pin fins.

A paper by Lasance (1991) discussed numerical modeling of natural convection air flow in a television set using both THEBES and FLOTHERM. The paper also discussed experimental validation of calculated air temperature increase at various locations within the television set. In several other papers researchers discussed numerical flow and heat transfer models which utilized the SIMPLER[1] algorithm detailed by Patankar (1980). Smith et al (1991) modeled combined heat transfer by conduction, natural convection and radiation in a sealed electronic package consisting of 3 PCB's carrying a variety of heat generating components. Shaw et al (1991) modeled mixed convection heat transfer phenomena in a three-dimensional channel with a single heating element representing an LSI package. Asako and Faghri (1991) modeled periodic fully developed fluid flow and heat transfer characteristics for turbulent flow over three-dimensional arrays of heated blocks simulating electronic modules on a PCB in a cooling channel. In another paper, Faghri et al (1994) described a numerical study conducted to investigate both the fully developed periodic and developing regions for turbulent flow over an array of heated rectangular blocks.

Du and Bilgen (1990) created a two-dimensional laminar flow model to study natural convection heat transfer within a sealed enclosure containing three components on a PCB. Ahmed et al (1990) created a two-dimensional model of a single electronic component on a vertical, flat, rectangular plate heat sink located directly above a horizontal component board. The model accounted for heat transfer via fully-coupled natural convection, conduction and radiation processes. Papanicolaou and Jaluria (1991) conducted a numerical study of two-dimensional forced and mixed convection heat transfer within an enclosure with heat sources simulating electronic components on the two side walls and the bottom wall. The effects of forced ventilation were included by means of flow through openings between each of the side walls and the top wall. Liakopoulos et al (1991) used a spectral element method to obtain numerical solutions for convective flow in tall cavities under a variety of conditions. These conditions were intended to simulate natural convection flows due to heat dissipating electronic components mounted on vertical PCB's. Nigen

[1]SIMPLER is an acronym for Semi-Implicit Method for Pressure-Linked Equations Revised.

and Amon (1991) developed numerical simulations of the flow field and heat transfer characteristics of oscillatory and non-oscillatory flows in geometries similar to those found in printed circuit board configurations with surface mounted components. The extent of enhancement achieved due to the oscillations was gauged through direct comparison to results corresponding to the steady-flow regime.

A paper by Lee and Mahalingham (1992) discussed the development of software to predict the optimum operating conditions for a two-phase liquid cooling environment. The software consists of a data bank including the physical and thermodynamics properties of fluorocarbons and heat transfer correlations from single phase boiling incipience through nucleate boiling to the critical heat flux. The authors discuss the application of software to pool boiling, flow boiling and jet impingement boiling.

4.9 PACKAGE COOLING TECHNIQUES

Specific cooling techniques in electronic equipment applications can yield equipment sizes that vary from the very small to the very large. But, whether the application involves a single natural convection-cooled heat producing component, or a high power, liquid cooled, multi-chip module, the objective remains the same: to allow thermal energy to flow from source to sink within the constraints of the specified temperature, space and cost. This section lists papers which describe a variety of package cooling applications.

At the component level, a number of papers discussed thermal performance and evaluation of air-cooled components and packages. An article by Shedden et al (1990) addressed basic concepts of utilizing air-cooling for thermal management of PCB's in telecommunication systems. The major topics discussed included thermal paths, conductive interfaces, thermal planes, thermal vias, component placement, heat sink application, reliability and series/parallel air flow networks. A paper by Dean (1991) provided a general discussion of thermal design for components on PCB's. Manno et al (1993) described a series of experiments conducted to assess the impact of circuit board thermal conductivity on the performance of air cooled electronic components. The results were reported to show that the use of moderate thermal conductivity ($k = 1.1\,\mathrm{W/m\text{-}K}$) boards increases the effective heat transfer area of a component by a factor of 3 or more. Shaukatullah and Gaynes (1994) also conducted an experimental study to determine the effect of PCB thermal conductivity on the thermal performance of surface mount packages. It was reported that for packages with good internal heat spreading, thermal performance is relatively more sensitive to the amount of copper in the cards. Wang and Saulnier (1993) performed a conjugate heat transfer analysis of four uniformly heated integrated circuit modules on a PCB in a two-dimensional channel. The effects of the materials constituting the PCB's and the IC's were studied. The results indicated that the heat exchange with air was less sensitive to component materials than to the PCB material, assuming perfect contact between the component and the PCB.

A paper by Manno et al (1990) describes an experimental study of the dependence of component thermal resistance on typical design variables such as component arrangement, power configuration, board placement and cooling mode (natural or forced convection). It was found that neighboring components can have a significant effect on the temperature of an individual component and can increase the effective component thermal resistance by more than 25 percent. Another paper by Manno and Azar (1990) also discussed the effects of neighboring components on thermal performance. A paper by Azar et al (1994) discussed experimental and numerical simulations conducted to study the effects of circuit board and component parameters on thermal performance under natural convection cooling conditions. The parameters considered included board thermal conductivity, component spacing and component height.

A paper by Ortega et al (1993) discussed a study of heat transfer in low aspect ratio horizontal enclosures typical of laptop computers. Experiments were preformed to characterize the combined conduction, convection and radiation heat transfer from a horizontal PCB in a shallow enclosure and compared with simulation results. It was noted that "the magnitude of the heat transfer coefficient used in the predictive model does not have to be known extremely accurately to predict maximum board temperatures with good accuracy." A paper by Edwards et al (1994) discussed a study to evaluate thermal enhancement techniques for plastic packages. The study was carried out using test structure measurements and FEM thermal modeling.

Enhancement techniques studied included the use of additional metal tracks on the PCB, the use of heat slugs and heat spreaders inside the package and the use of high conductivity mold compounds.

Three papers discussed the thermal performance of pin grid array (PGA) packages. Tanaka et al (1990) discussed a 35 mm ×35 mm PGA package which utilized an aluminum nitride substrate and a staggered fin heat sink to obtain a low thermal resistance. At a cooling air velocity of 1 m/s the package exhibited a thermal resistance of 3.0°C/W, or about one-half that obtained using an alumina substrate. Hamzehdoost (1990b) conducted an experimental study of design factors affecting the thermal performance of high performance PGA's. Among the design factors studied were die size, substrate material (ceramic vs. plastic), surface area, heat spreader, heat sink, heat sinks and air velocity. Sulhan et al (1991a) also conducted a study of the thermal performance of pin grid arrays and multichip modules. The effects on thermal resistance of die attach thickness and thermal conductivity, substrate thickness and thermal conductivity, cavity orientation (up or down) and air velocity were reported.

A paper by Lee et al (1992c) presented an analytical approach for thermal modeling of via networks used to facilitate thermal design from chips to the substrate in multichip module designs. The complete thermal resistance between the die and substrate was determined utilizing unit cells in a combination of series and parallel paths. The thermal resistance of the components making up a typical via network was also determined by closed form expressions presented in the paper.

During the 1980's, surface mount technology (SMT) was the fastest growing package technology in small systems and consumer electronics and its growth continues today. An example of the continued growth and interest in SMT applications is provided by the papers in this section related to SMT thermal performance and characteristics. In an article on SMT cooling, Markstein (1991) noted that "surface mount technology has compounded the heat removal problem because devices (chips) are packaged more closely spaced on a printed circuit board - hence, further increasing the heat density." It was also noted that, "due to significant differences in physical configuration and size, SMT devices such as SOICs and quad flatpacks, are more difficult to cool than through-hole devices such as DIPs." One option that SMTs offer is the capability to efficiently package components on both sides of a PCB. A paper by VanLaanen (1990) provided an example of the thermal and reliability effect to be expected when two single-sided PCB packages are combined into one dual-sided PCB package. Papers by Aghazadeh and Mallik (1990), Hamzehdoost (1990a) and Moore (1991) reported the results of numerical and experimental studies to characterize the thermal performance of plastic quad flat pack (PQFP) packages. A paper by Mulganoker et al (1994) reported the results of a study to project the thermal resistance performance limits of a the QFP family of packages. The study was based upon simulation using a finite difference based thermal code for IC packages. The power dissipation limits of nominal and enhanced PQFP packages was projected in the range of 1 W to 1.9 W for natural convection and 1.2 W to 2.7 W for forced convection.The limits for CQFP packages was projected in the range of 1.1 W to 2.7 W for natural convection and 1.4 W to 4 W for forced convection.

A paper by Zimmerman et al (1990) discussed a FEM analysis and experiments to characterize the thermal performance of a molded multi-chip module (MMCM). As described in the paper, the MMCM package incorporates the low cost features of a plastic leaded chip carrier (PLCC) package with the improved thermal performance of a ceramic package. The silicon is mounted directly on a ceramic substrate, which is solder-attached to a leadframe and then molded. Another paper by Rodgers et al (1990) addressed the design, thermal modeling, and characterization of an enhanced small outline-j-lead (SOJ) package. This package was described as incorporating an aluminum or copper slug to which both the chip and leadframe are attached with epoxy. The slug provides a direct thermal path from the chip to cooling air, and results in as much as a 27 percent reduction in R_{JA}. A paper by Chambers and Mahalingham (1992) reported experimental data on the thermal characteristics of 23 lead single-in-line (SIP-23) power packages clamped to a cold plate with both dry and greased interface conditions. It was noted that internal thermal resistance (junction-to-heat-tab) constituted the majority of the total thermal resistance (junction-to-coldplate) over the range of

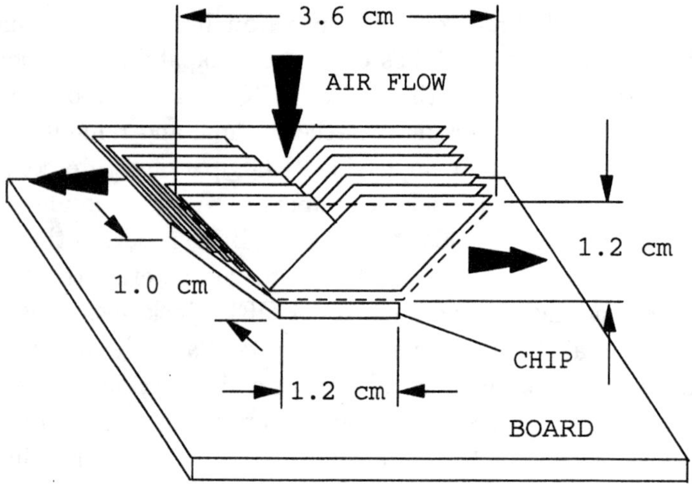

Figure 4.11: Perspective view of air-cooled microlaminar heat sink [Hilbert et al (1990a)].

conditions measured, with the die area representing the single most important parameter affecting the internal thermal resistance.

McNelis and Buller (1990a) reported the results of an investigation of the natural convection cooling performance of SMT components for several different heat sinks in horizontal and vertical orientations. A paper by Shaukatullah et al (1993) addressed the thermal performance of various surface mount packages with and without heat sinks for a range of air velocities from 0 to 2 m/s. It was shown that with an appropriate heat sink, the thermal performance of a plastic package may be comparable to that of a metal or plastic package with an exposed heat spreader. It was also concluded that "In terms of cost, heat sink attachment offers the most cost effective way of improving the thermal performance of the plastic package provided that there is some space available for the heat sink."

Another paper by McNelis and Buller (1990b) reported the results of a study conducted to evaluate the thermal bounds of tape automated bonding (TAB) components in relation to similar package structures in order to obtain insight into the application of TAB components. Among the conclusions drawn, it was noted that the conductive heat flow path for TAB components is not as beneficial due to the low thermal conductivity and small cross-sectional area of the supporting tape structure. It was further noted that there appears to be a greater sensitivity to component location and card orientation, as well as reduced benefits from increased airflow. Davis (1990) also reported the results of a numerical and experimental study to thermally characterize a 160 I/O TAB component. McMaye (1994) reported on a study to evaluate alternative designs for TAB packages. The three designs considered were:

1. thermal vias connected to a coooling ground plane,
2. thermal vias connected to a coooling ground plane and a heat sink attached to the backside and
3. a copper slug witha heat sink attached to the backside.

The methods used in the evaluation were

1. a one-dimensional thermal resistance analysis,
2. a numerical analysis using FLOWTHERM and
3. experiments.

Of the three design alternatives, the third option was found to offer the best performance. Reynolds et al (1992) described an effort to develop a thermal enhancement technique for single chip thin film carriers. The objectives were to develop the appropriate heat sink, thermal adhesive and assembly process. Results were presented based upon a test package consisting of a 7.5 mm $\times$ 7.5 mm chip bonded to a high lead count organic tape substrate using tape automated bonding.

Two papers by Hilbert et al (1990a, 1990b) described a high performance air cooling concept which combines a micro-laminar flow with an efficient heat sink design. As shown in Fig 4.11, the heat sink is designed to receive impinging air at the top and to exhaust heated air out of the sides. It was reported that "the thermal resistance of the micro-laminar heat sinks was measured to be less than 2 °C/W for a 1 cm^2 chip at an airflow rate of 2 CFM." It was also noted that the capability to remove in excess of 600 W from a multi-chip module (25 chips at 25 W) while limiting chip temperature rise to 55 ° C above ambient was demonstrated using this concept. A

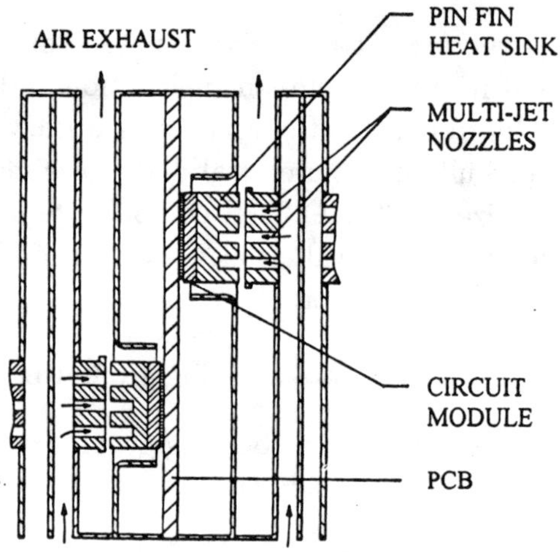

Figure 4.12: Cross-section view of apparatus for controlled air-impingement module cooling [Adapted from Hamadh et al (1991)].

paper by Nelson et al (1993) reported on a system application of this heat sink design, scaled to cool large PGA or MCM packages on PCB's in a card cage. A patent by Hamadh et al (1991) disclosed another impinging air cooling scheme. As shown in Fig 4.12, a supply plenum is provided with a multi-jet nozzle array facing electronic circuit modules with pin fin heat sinks. Cooling air impinges on the heat sinks and then flows out into an exhaust plenum created between the supply plenum and a baffle around the modules.

The packaging schemes used in avionics and for military electronics applications often differ from those used for commercial applications. A typical approach is to conduct heat from modules into a good thermal conductor that is attached to, or is a part of, the PCB. The heat is then conducted along the conductor, across a mechanical interface, and into the sidewalls of a chassis which contains finned air-cooling passages. Leatherman (1990) conducted a trade-off study to compare the typical conduction approach versus an approach using a hollow core board. Cross-sectional views of a solid core board and a hollow core board are shown in Fig 4.13. As shown in Fig 4.14, air is forced directly through the passages formed by the internal fins within the hollow core board. Two examples are provided in the

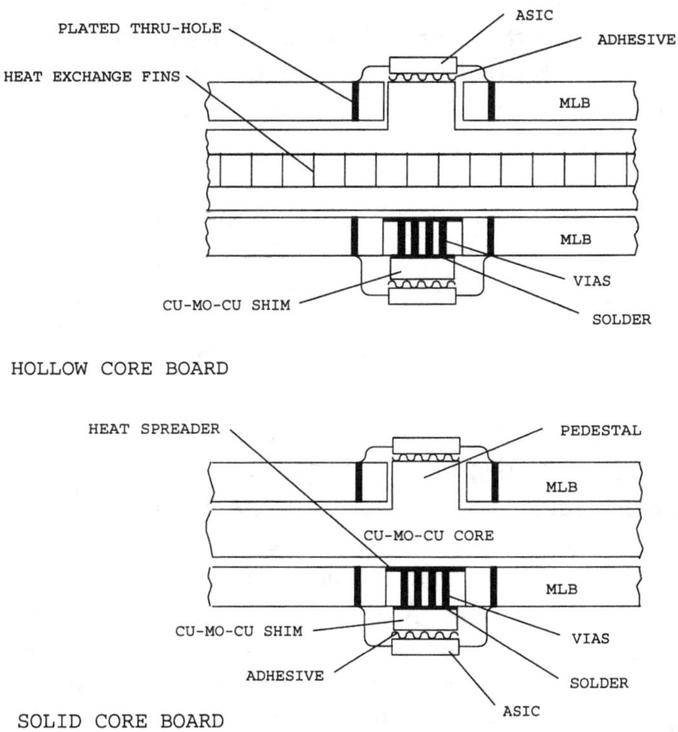

Figure 4.13. Cross-section views of solid core board and hollow core board [Leatherman (1990)].

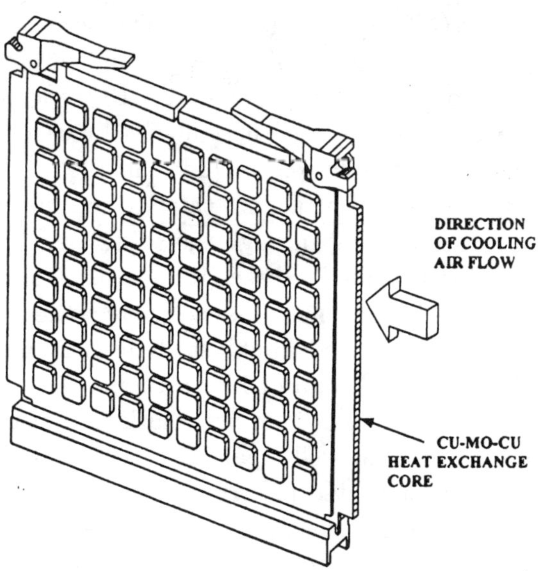

Figure 4.14. Typical hollow core air-cooled board [Leatherman (1990)].

paper illustrating the improved cooling performance obtained using the hollow core board. A paper by Herold et al (1992) also considered flow-through air cooling of military standard electronic module (SEM) boards in a format-E configuration. Gates and Hontz (1991) conducted a study of edge-cooled packages, as well as flow-through cooling with air or water. The study focused on determining the benefits of eliminating thermal interfaces and layers of materials by mounting chips directly on circuit boards rather than within hermetic modules. A paper by Seals (1991) discussed a packaging and cooling investigation conducted by AT&T Bell Laboratories under the Air Force Advanced Aircraft Avionics Packaging Technologies (AAAPT) program.

Ibrahim et al (1992) discussed the thermal design upgrade of a military electronics rack to accommodate 50 percent more power. The rack was described as consisting of a stack of horizontal trays providing air-cooled finned cold plates on which to mount the electronic module assemblies. A paper by Snyder (1991) considered the use of a phase change material to absorb heat generated by a small airborne electronics package over a short duration of time. The analysis presented in the paper illustrated the effectiveness of this approach over a short time duration (20 seconds) compared to simply using the thermal capacitance of a metal sink.

A growing number of applications are appearing which utilize multichip module (MCM) package designs. These designs offer higher chip density, reduced interconnect delays, and improved electrical performance. A paper by Sulhan et al (1991b) reported the results of a study to characterize the thermal resistance (R_{JA}) of MCM designs with 4, 5 and 9 chips. Included in the study was an assessment of the benefit of attaching chips to internal thermal slugs brazed into cavities passing through the substrate. The slugs made of copper-tungsten provided a direct thermal path from the chip to heat sink, and provided as much as a 28 percent reduction in R_{JA}. Keeley and Ryan (1991) conducted a study to evaluate the effectiveness of proposed thermal solutions for a copper-polyimide interconnect technology for MCM's. Thermal solutions considered included thermal cutouts, thermal wells, partial thermal wells and thermal vias. Results were presented in terms of thermal resistance versus electrical routing density available beneath the die (i.e. the chip).

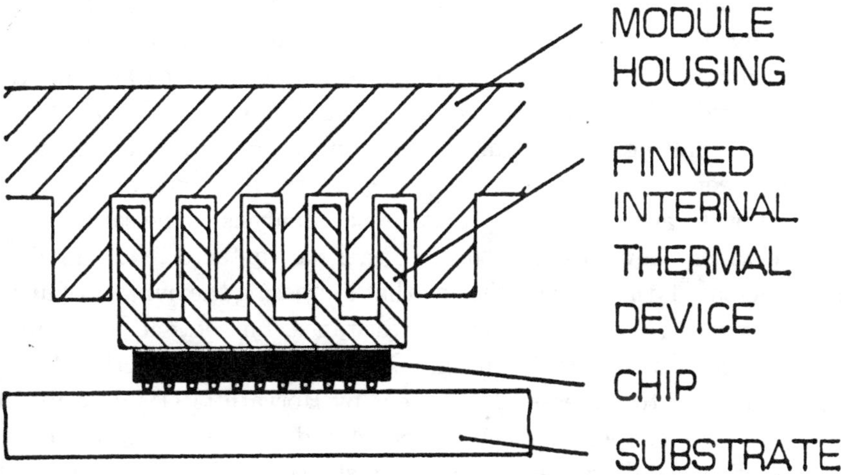

Figure 4.15: Internal fin-to-fin chip-to-housing cooling device [Adapted from Horvath et al (1991)].

A paper by Dolbear (1990) provided a discussion and comparison of "above-substrate" cooling versus "through-substrate" cooling. The "above-substrate" cooling approach was described as one where each chip is cooled by conduction to a cold plate in contact with it. The "through substrate" cooling approach was described as one where the substrate to which the chip is attached is used to conduct heat to a cold plate or heat sink at its back surface or edges. A paper by Mok (1994) describes a "through-substrate" approach which utilizes a spring-loaded folded fin heat sink to cool a silicon substrate with silicon chips. A grease loaded with ZnO is used to control the thermal interface resistance between the folded fin heat sink and the back of the silicon substrate. A paper by Snyder (1992) discussed a thermal performance analysis and modeling analysis of a copper-polymide thin-film-on-silicon multichip module. The emphasis was placed on examining heat removal from the active side of the chip through solder pads and vias to the silicon substrate.

A patent by Horvath et al (1991) disclosed a finned internal thermal device to conduct heat away from the back of the chip and into the module housing. As shown in Fig 4.15, the fins on the chip cooling device interleave between corresponding fins formed in the module housing. Thermal conduction between the chip and cooling device, and between fins on the cooling device and housing is facilitated by filling the module with oil. Hamburgen (1991) discussed the design and analysis of the same type of device which he termed "interleaved fin thermal connectors." Hamburgen noted that "the principal challenge in designing such connectors is minimizing the gap between the fins while maintaining the required compliance."

Ashiwake et al (1991) also provided an analysis of the same type device which they termed a "flexible thermal contactor."' As might be anticipated, the analysis showed that increasing the number of fins per unit area reduces the thermal resistance, but beyond a certain point, further increases in the number of fins will cause the thermal resistance to increase. The patent by Horvath (1991) described multiple disk shaped elements to be placed between a chip and a cold plate. The disks are progressively stacked and consist of bulge shapes with radial finger contacts to provide multiple compliant heat conduction paths between each chip and the cold plate. A patent by Chu et al (1992) decribed another variation of the IBM TCM piston cooling approach termed "super-position cooling." In this case. multiple pistons are used to contact the chip. Instead of conducting heat from the pistons to the surrounding wall via helium, jets of a dielectric coolant impinge upon and flow around the pistons to produce direct convective cooling.

Song and Bang (1990) investigated a circular piston geometry similar to that used in the IBM thermal conduction modules. In this case, however, an O-ring seal was placed near the bottom of each piston and its hole in the housing and water was circulated around the remainder of the piston. A patent by Flint et al (1991) described a means of cooling chips on a multichip substrate by using a thin copper sheet to contact all the chips and separate the chips from cooling water. Miniature fins are provided on the water-side opposite each chip and thermal grease is provided between each chip and the sheet. A paper by Watari (1990) described the computer packaging technology (including cooling) used in the NEC SX3 supercomputer. As shown in Fig 4.16, the liquid cooled module design (LCM) which houses 100 chips, incorporates an integral water-cooling structure that utilizes liquid jet impingement on the heat transfer block to

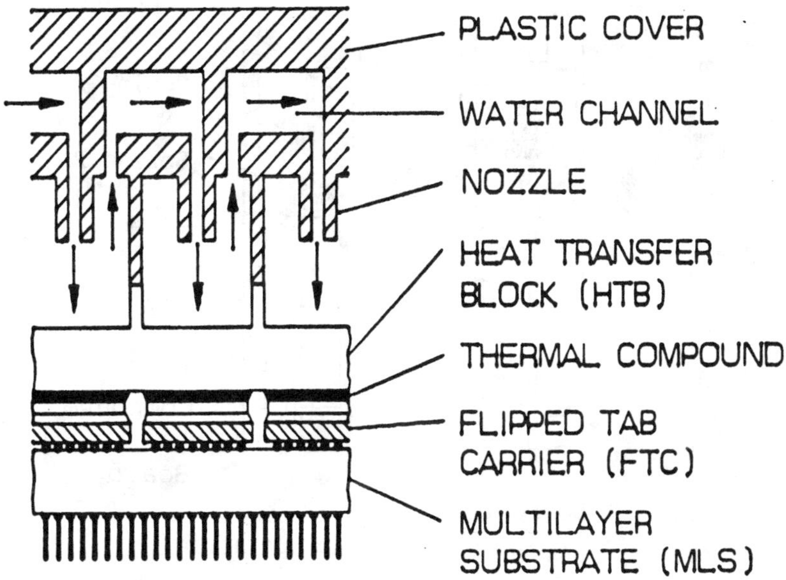

Figure 4.16: Cross-section of NEC liquid cooled module used for SX-3 [Adapted from Watari (1990)].

achieve high heat transfer rates. According to the author this design provides a thermal resistance to water (R_{JW}) of 0.6° C/W to support individual chip powers as high as 38 W and a total module power of 3.8 kW.

A paper by Nelson et al (1994) described an MCM package design to use direct liquid immersion cooling to transfer heat from chips on a substrate to a final heat transfer medium outside the package. A pin fin heat exhanger at the top of the module can be operated either as a submerged condenser or a vapor condenser. Fins on either side of the wall are used to transfer heat to coolig water. The fluid used within the module is FC-72. A paper by Bar-Cohen et al (1994) discussed a new thermal management technique disclosed in a patent by Cray and Sherwood (1992). The technique uses a high velocity flow of a liquid-gas mixture in the narrow channel between populated substrates to support both high heat fluxes and high heat densities. A prototype high packaging density module was successfully operated by Cray Computer Corporation using this approach and demonstrated the ability to support a chip heat flux of 20 W/cm^2 and a volumetric heat density of 18 W/cm^3. Results reported by Bar-

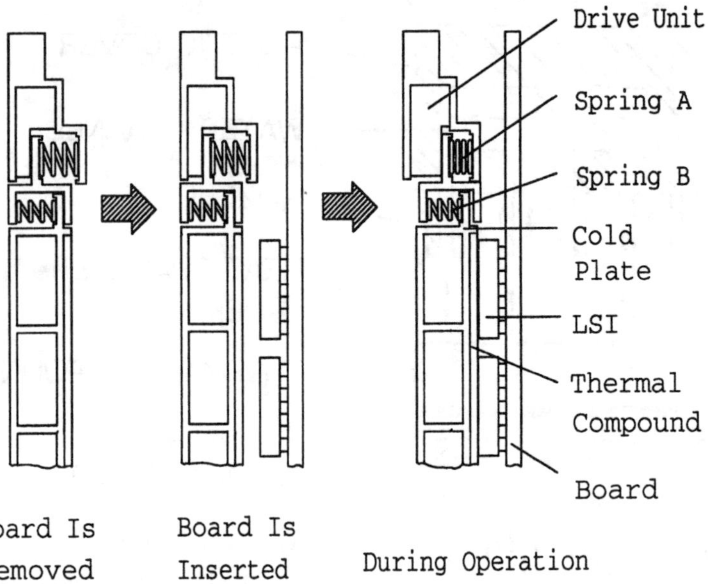

Figure 4.17: NEC removable conductive cooling module (RCCM) concept [Arai et al (1992)].

Cohen et al suggest that "in excess of 50 W/cm^3 could be removed in a 3-dimensional package by using helium or nitrogen assisted flow of Fluorinert^{TM} through a sub-millimeter channel.

A paper by Arai et al (1992) described a conductive cooling approach used to cool LSI chip modules in broadband integrated service digital network (BISDN) switching systems. As shown in Fig 4.17, a water cooled removable conductive cooling module (RCCM) is designed to contact LSI modules on a board when in position in the electronics frame. Heat is conducted from the LSI modules into the water cooled cold plate via a thermal compound at the interface.

A paper by Mathisen (1990) discussed theory, applications, and performance of a liquid heat sink (LHS). The LHS consists of a flexible plastic pouch containing a fluorocarbon liquid, that is placed between the heated surface of electronic modules and a cooler surface (such as a cold plate or other heat exchanger). Heat is transferred from one surface to the other by liquid natural convection within the LHS. Data is presented in the paper comparing actual vs. theoretical performance of the LHS in vertical and horizontal orientations.

Yeh (1990) performed a study of three different cold plate flow schemes for cooling an array of modules with 18 modules in a row. The various schemes studied included:

1. a single, unidirectional flow,
2. a side-by-side, bidirectional flow and
3. a top-to-bottom bidirectional flow.

Module temperature distributions were reported for each scheme studied. It was found that the top-to-bottom, bidirectional scheme was the most effective for reducing temperature and temperature variation. A paper by Hwang et al (1992) described the analysis and experiments supporting the design of the water cooled cold plates for TCM's used in the IBM ES/9000 processor. The topics discussed included cold plate design, thermal experiments, interface resistance (dry and with oil), cold plate thermal resistance, cold plate fouling and cooling limits. A paper by Soule (1992) discussed thermal management using water cooling with cold plates and liquid cooled bus bars.

A study conducted by Nakayama (1990) illustrated the effect of the requirement to provide coolant flow channels on the average wiring length for two- and three-dimensional module packaging configurations. The results of the study showed that the need for cooling channels will become a serious impediment to the reduction of wiring distances which could otherwise be realized in three-dimensional configurations. It was also concluded that for the cases studied, "immersion cooling by FC-77 invites a far larger penalty in the increase of wiring distance than indirect water cooling, if no measures are taken to enhance heat transfer to FC-77."

A paper by Hamburgen and Fitch (1993) described a thermosyphon or "wickless heta pipe" designed to cool a 12.6 mm × 15.44 mm microprocessor chip. As shown in Fig 4.18, the chip is joined with epoxy to a small finned heat sink projecting into the liquid space of the thermosyphon structure. Boiling of the sealed water-alcohol mixture accommodates high heat fluxes at the chip while heat is transferred by vapor condensation on the inner wall to air cooled fins on the outside of the pipe. It was reported that the junction-to-air ther-

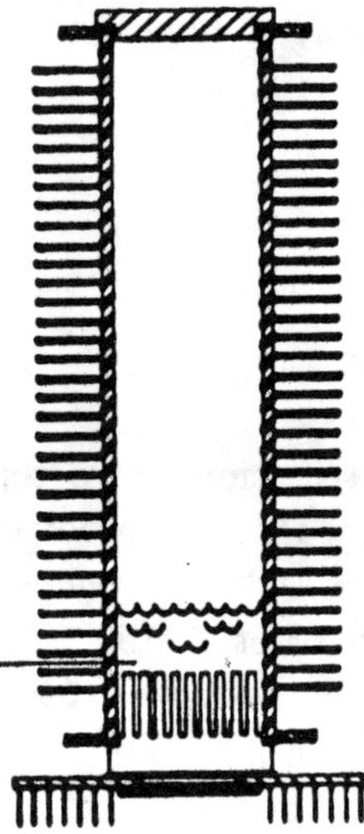

Figure 4.18: DEC air cooled thermosiphon ECL microprocessor package [Adapted from Hamburgen and Fitch (1993)].

mal resistance was about 0.4°C/W, making it possible to maintain junction temperatures below 100°C while dissipating 150 W.

Package cooling techniques encompass not only chip, module and PCB applications, but also applications providing temperature control of other electrical devices. A paper by Sato et al (1990) reported a study of the heat transfer characteristics in 5.25 inch disk enclosures at high speeds (1800 to 7200 rpm). Yuki et al (1990) discussed an experimental study to determine the effects of vent size and location on component cooling in a table model television receiver. Existing vents in the television cabinet were systematically blocked until the set was operated in a completely sealed condition. Component temperatures on the main circuit board were reported for different vent conditions. Typical flow patterns were reported based upon flow visualization tests with all vents open.

4.10 SYSTEM COOLING TECHNIQUES

As challenging as the package cooling problem may be, an equally challenging problem is that of the cooling system design. Usually, an electronic system will consist of many components or packages. The electronic system could be a small personal computer with a few natural convection cooled printed circuit boards, or a large mainframe computer containing both air-cooled and water-cooled package assemblies. In either case, the system cooling design is required to ensure that the cooling fluid can be supplied to each component or package at the proper flow rate and temperature needed to provide adequate package cooling. The papers covered in this section discuss some of the techniques used in air and liquid cooling systems.

Some air cooling system designs use only natural convection. The use of natural convection eliminates the need for fans or blowers, as well as related controls, flow sensing, and alarm circuitry. In addition, it obviates the requirements for air filters and cooling system maintenance. A paper by Gabrielli and Meli (1990) described a methodology to calculate the air temperature distribution within a cabinet containing electronic components cooled by natural convection. A paper by Ramos and Milanez (1993) discussed a study to assess the effects of various methods of front door attachment on natural cooled modular packages in a digital transmission equipment structure. The study considered hinges, hinges and tape and screws. The temperature rise on the critical component was determined for each scheme, along with that for the open door case.

As either component power, or the total power dissipation within a cabinet or system enclosure is increased, it often becomes necessary to provide a forced flow of air. Adding a forced flow of air provides the dual benefits of increasing the heat transfer coefficient at the component level, and reducing the average air cooling temperature. Several publications addressed aspects of forced flow cooling in electronic cabinets or enclosures. A trade press article by Paggen (1990) discussed the management of heat flow from units mounted in standard 19-inch racks. The discussion included cooling in sealed and vented racks. Another trade press article by Hill and Lind (1990) provided "rules-of-thumb" and guidelines for the design of ventilation grilles for electronic equipment. The discussion also addressed the need to consider possible electromagnetic interference (EMI) radiated through the grille openings and safety requirements pertaining to ventilation openings in electrical cabinets. A paper by Hill and Hill (1990) presented the results of experiments to investigate the effects of electronic enclosure cooling fan plenum size on the performance of typical small, axial flow fans. Data in the form of static pressure versus flow was provided for free inlet/outlet conditions, and for plenum depth to fan diameter ratios of 0.1, 0.2 and 1.0.

Harada et al (1990) described the design of a forced-air cooling system for a cabinet containing a telecommunications system. Because the system is installed in non-staffed offices, emphasis was placed upon design of an air delivery system to provide high availability. An optimal fan size was selected to provide sufficient cooling airflow even when two out of a total of eight fans failed. A paper by McPhee et al (1990) discussed the air cooling system for the DEC VAX 9000 computer. The cooling system provides impinging air flow to a 4 × 4 planar array of Multi Chip Units (MCU's). The discussion included a description of the design, early model predictions, and representative test results of isolated components. It was reported that the system is capable of delivering an average flow of 50 CFM to each MCU, more than satisfying the design target. The development of an impingement air jet cooling system for telecommunication multichip modules was discussed in a paper by Kishimoto et al (1991). A board level switching system was described consisting of eight 100 mm × 100 mm MCM's, each containing a 4 × 4 array of 12.4 mm square chips. Each module was cooled by a pinfin heat sink supporting a heat flux of 6.5 W/cm^2 at the chip level and 1.6 W/cm^2 at the module level.

In some instances the electronics cabinet or enclosure may be exposed to outdoor environments. Two papers included in this section address this situation. Volk (1990) discussed considerations in the design of sealed, outdoor electronic cabinets. The discussion focused on determining the impact of solar radiation and outside air temperature on the internal cabinet environment. McKay (1990) discussed a series of tests conducted to assess the effectiveness of using an air-to-air heat exchanger to reduce the peak temperature rise in an outdoor electronics cabinet. It was reported that with the heat exchanger used in the tests, the peak temperature was reduced about 10°F on average. It was also noted that use of the heat exchanger should allow a 40 to 50 percent increase in internal heat load within the cabinet for roughly the same peak temperature rise.

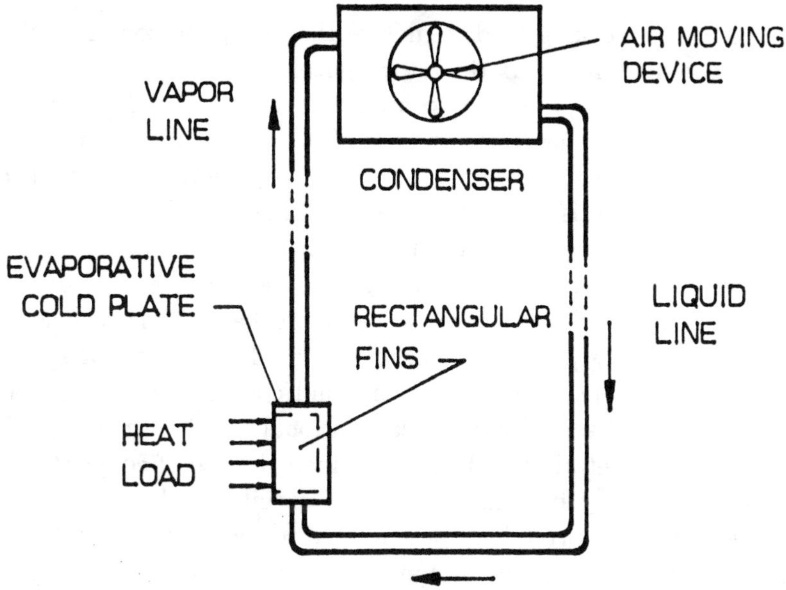

Figure 4.19: Schematic diagram of thermosyphon cooling system [Adapted from Heng and Pei (1990)].

A paper by Pei et al (1991) provided a description of a variety of system cooling concepts for computers. Both air cooled and liquid cooled system concepts were discussed with an emphasis on the advantages offered by closed-loop cooling systems. A paper by Lee et al (1992b) described the evaluation of a compact liquid cooling system as a benchmark for self-contained heat exchanger units that can be used in small moveable electronic equipment. Another paper by Heng and Pei (1990) presented thermal performance test results for single-tube and twin-tube remote condenser phase-change cooling systems. As shown in Fig 4.19, the twin-tube cooling system consisted of an evaporative cold plate connected by tubing to a separate air-cooled condenser, essentially constituting an extended, wickless heat pipe (or thermosyphon). As noted by the authors, "remote placement of the condenser enables the cooling air for the condenser to be provided by the traditional air flow management within the computer cabinet." A paper by Kishimoto and Harada (1992) discussed the development of a thermosyphon cooling system for telecommunication multichip modules to be used in a BISDN cooling system. Experiments were described evaluating water, FC-87 and Freon-142b as the thermosyphon working fluids. Another thermosiphon cooling system used to cool high heat flux thyristor devices was described in a paper by Kuwahara et al (1994). Two papers by Kojima et al (1992) and Sotani et al (1992) described the use of heat pipes to cool arrays of modules on PCB's in telecommunication equipment. Heat from each module on a board is conducted across a thin layer of silicone rubber to a heat-absorber plate containing 3 mm diameter heat pipes. The heat pipes are used to transport the heat beyond the edge of each board to be rejected to air flowing over microchannel fins attached to each heat absorber plate. The authors claimed that a cooling system using the micro heat pipe approach could cool heat loads that are several times higher compared to conventional direct air cooling. The paper by Sotani et al (1992) also described the use of copper/water micro heat pipes embedded in a printed circuit board.

Two papers reported here were related to the application of cryogenic cooling to electronic equipment. Novotny (1990) described a test program to evaluate a Gifford-McMahon cycle refrigerator. The test program was part of an overall feasibility study to develop a computer based on cryogenically cooled CMOS devices. Anderson et al (1992) describe an analysis conducted to assess the performance and cost of three types of liquid cooling systems for a 1 kW system heat load:

1. a "once-through system in which liquid nitrogen is allowed to boil at the CPU boards and is then vented to the surroundings,

2. a saturated pool boiling refrigeration system in which gaseous nitrogen generated by boiling on the CPU boards is condensed in a vapor space above the liquid nitrogen pool and

3. a subcooled pool boiling system in which the condenser is completely submerged in a subcooled liquid nitrogen pool surrounding the CPU boards.

It was recommended that when the board heat fluxes are less than 15 W/cm^2, a "once-through" should be selected, because such a system is more economical than a refrigerated system. It was also recommended that, if a refrigeration system is adopted, a subcooled system should be selected to provide 10 to 30 percent greater critical heat flux than a saturated system at a comparable cost.

The remaining five papers in this section each addressed installation level aspects of electronic equipment cooling. A paper by Nakao et al (1991) evaluated and compared the performance of four air flow distribution schemes for a high heat density ($600\,W/m^2$) room containing electronic equipment. The schemes considered included:

1. underfloor air supply/overhead return,
2. underfloor air supply/horizontal return,
3. overhead air supply/underfloor return and
4. overhead air supply/horizontal return.

Experimental results indicated that system-1, the underfloor air supply/overhead return, provided superior performance in terms of both lower and more uniform equipment inlet air temperatures. Hayama et al (1990) discussed air flow distribution design for telecommunications equipment rooms characterized by high heat densities. The design specifically addressed the cooling characteristics of natural cooling equipment and the uneven distribution of heat in equipment rooms. A method of calculating the air supply/return distribution to optimize air flow distribution was presented. A paper by Forsman and Zirath (1990) described a natural convection room cooling approach termed a "climate ceiling." In this approach, air-to-water finned tube heat exchangers were integrated into a false ceiling containing large square chimneys and forming a grill covering the whole area of the room. Prudhoe et al (1991) described the use of a "vacuum panel heat exchanger" (VPHE) mounted in thc ceiling of a telecommunication equipment shelter to provide passive cooling. As described by the authors, "the VPHE is a variation on the customary sealed pipe thermosyphon in that it has been configured as a panel rather than a pipe." Included in the paper was a proposed design for a high reliability energy efficient equipment shelter combining the use of VPHE units and a phase change material (PCM) thermal storage chamber. The shelter would also contain an air conditioning unit. The authors note, however, that "for most of the yearly operating cycle the VPHE in combination with PCM storage will provide all of the controlled cooling by passive means." A further discussion of the possible use of both VPHE's and PCM's for thermal control of

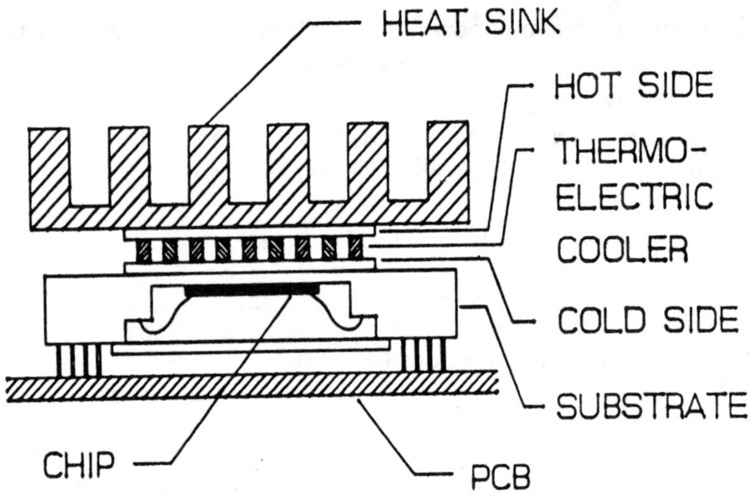

Figure 4.20: Cross-sectional view of a single chip module package with a thermoelectric and heat sink augmentation [Adapted from Mansuria (1991)].

electronic equipment shelters was provided in a paper by Prudhoe and Doukas (1990).

4.11 COOLING DEVICES AND HARDWARE

A variety of devices and hardware may be used to control temperatures and improve cooling capability in electronic equipment. Fans and blowers are used to provide cooling air flow over power dissipating components and packages. Heat sinks and heat pipes are used to transport heat more effectively from source to sink. Mechanical or thermoelectric refrigeration devices are used to pump heat or provide sink temperatures below ambient. The papers in this section discuss the design, evaluation, and application of some of these devices and hardware.

An article by Burke (1990) discussed the selection and application of fans and blowers to provide cooling airflow for electronic packages and systems. Topics covered in the article included fan curves and airflow parameters, choice of forced or induced flow and the use of brushless motors for variable flow cooling. Another article by Swager

(1990) addressed the use of thermoelectric devices and heat sinks to augment the cooling of the growing number of fast and dense integrated chips and circuits. An example of the effectiveness of using a thermoelectric device in concert with a heat sink was provided in a paper by Mansuria (1991). Mansuria conducted experiments to determine how much chip temperature could be reduced, using a commercially available thermoelectric module in the package configuration shown in Fig 4.20. With an air flow velocity of 300 fpm and an inlet temperature of 25°C a chip temperature of 22°C was obtained with the combined thermoelectric-heat sink package, compared to a chip temperature of 45°C for a package with only the heat sink. A paper by Van Tongeren and Thijs (1992) discussed the application of a thermoelectric cooler to control the temperature of a pump-laser diode.

An article by Kiley and Soule (1990) addressed the variety of heat sinks and methods of heat sink fabrication that are available today for electronic cooling applications. The topics discussed included standard heat sinks and extrusion technology, stamped sheet metal parts, folded fins, brazed/soldered fins, epoxy bonded fins, casting technologies, liquid-cooled heat sinks and phase change heat transfer. A paper by Pei et al (1990) discussed the fans and heat sinks used to cool the MCU's in the DEC VAX 9000 computer. A paper and article by Heng and Pei (1991a, 1991b) provided a further discussion of the heat sinks. The aluminum heat sink was described as consisting of more than 600, 0.75 inch tall pin fins on a 4 × 4.1 inch base and exhibiting a thermal resistance of 0.14°C/W for an airflow rate of 31 CFM. Both publications discussed four different manufacturing processes investigated for fabrication of the heat sink. A paper by Bartilson (1991) discussed the development of a miniature heat sink (9.45 mm × 9.45 mm × 2.54 mm) with 256 pin fins. The air jet impingement cooled heat sink was designed to be attached to the base of a 28.95 mm × 28.95 mm ceramic chip carrier and project through a square hole in the PCB to which the chip carrier is attached. The heat sink was reported to provide a total thermal resistance (R_{JA}) of 5.0°C/W for an air flow rate of 0.028 m^3/min.

The remainder of the papers discussed in this section are related to the design and application of heat pipes for electronic cooling. Pruzan et al (1991) reported on an analytical and experimental study of dryout in heat pipe wicks and a parametric design study to achieve high-performance wicks. The authors noted that the study was motivated by the needs of the electronics industry. It was shown in the paper, that high-performance flat plate and cylindrical wicks, 10 centimeters in length, are theoretically capable of dissipating 50 and 100 W/cm^3. Babin et al (1990b) reported the results of a combined analytical and experimental investigation to identify and understand phenomena that govern operating characteristic and performance limits of micro heat pipes. It was explained that "in practical terms, a micro heat pipe is a wickless, noncircular channel with an approximate diameter of 100 to 1000 μm and a length of about 10 to 60 mm." The paper provided a detailed description of the methodology used to develop the steady-state model and a comparison of predicted and experimental results.

Two papers by Wu and Peterson (1991a, 1991b) reported a continuation of the study to investigate the transient behavior of micro heat pipes during startup, or variation in the evaporator heat load. The first paper (1991a) reported on the development of an analytical model, and the second paper (1991b) reported on the experimental investigation. A paper by Fox et al (1991) described an experimental multi-chip test vehicle built to evaluate direct cooling of chips by means of a heat pipe integral to the package. In this scheme the chips are cooled by evaporation of the working fluid and the resulting vapor is condensed on the inner surface of the cap. A porous wick is provided between the cap and each chip to facilitate return of the coolant to the chips by capillary action. A paper by Mallik et al (1992) discussed a three-dimensional numerical model to determine the potential advantages of constructing an array of micro (100 μm) heat pipes as an integral part of the semiconductor chips. Miyagi et al (1990) described the use of heat pipes in an MCM containing a 3 $\times$ 4 array of gallium arsenide chips. As shown in Fig 4.21, a spring-loaded copper piston is used to conduct heat from the back of each chip carrier and into the housing. The heat is rejected from the housing via four heat pipes with air-cooled fins at the far end.

North and Avedesian (1993) reported the results of an experimental study which demonstrated the ability of heat pipes to dissipate high heat fluxes and high total power at low surface temperatures for application to semiconductor chip modules. Both air-cooled and

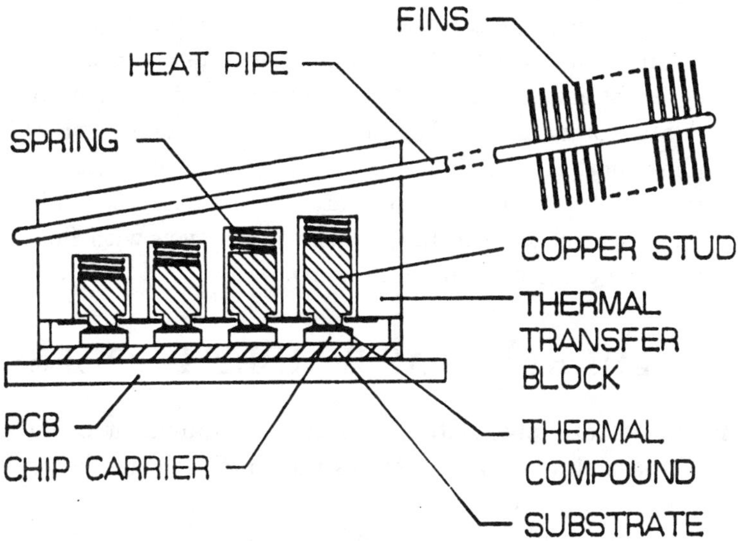

Figure 4.21: Cross-sectional view of a heat pipe cooling module package [Adapted from Miyagi et al (1990)].

water-cooled designs were studied. The air-cooled design consisted of a base plate with an array of holes drilled in it and lined with sintered copper powder to serve as a wick. An array of wick lined tubes with a series of stacked fin plates was attached normal to the base plate to serve as a condenser. The water-cooled heat pipe was disk shaped with a sintered wick structure and water cooling channels over the periphery of the housing. Water was employed as the working fluid for the air-cooled version. and ethanol for the water-cooled version. A heat flux of 47 W/cm^2 and a maximum surface temperature of 93 °C with 27°C cooling air was reported for the air-cooled heat pipe. Similarly, a heat flux of 31 W/cm^2 and a maximum surface temperature of 75 °C with 25°C cooling water was reported for the water-cooled heat pipe.

A paper by Toth and Garner (1990) described the operation of variable conductance (gas controlled) heat pipe heat exchangers (VCHX) and discussed recent advances in VCHX technology. As noted in the paper, variable conductance heat pipe heat exchangers are passive heat transfer devices that maintain electronic temperatures in an optimum range while power levels and coolant temperatures vary. A paper by Reese (1990) described tests of a variable conductance heat pipe for a shipboard electronics cooling application. While in operation, the heat pipe assembly was subjected to ± 30 degrees of motion relative to the horizontal at a rate up to 5 cpm to simulate worst case ship's motion. The tests demonstrated that the rolling motion of a ship does not have a detrimental effect on heat pipe startup and that evaporator temperature control can be maintained within the specified requirements even with fluctuating heat load and coolant supply temperature.

4.12 THERMAL ASSEMBLY PROCESSES

Although not directly related to electronic equipment cooling, the topic of thermal assembly processes warrants inclusion here. Many of the steps and processes in the manufacture and assembly of semiconductor chips and packages require controlled application of heat and temperature. Failure to achieve the requisite degree of control may be reflected in poor quality and degraded reliability.

A common technique used in the electronics industry for joining components to PCB's is infrared reflow soldering. A typical configuration in which PCB assemblies are supported on their edges and carried by a belt through an oven equipped with IR heating elements is shown in Fig 4.22. Four papers covered in this section addressed computer modeling of infrared reflow soldering processes. Tang et al (1990) discussed the development of a numerical model to investigate the effects of various components on the thermal profiles of a PCB during the soldering process. In the model, the PCB was fixed at the center of a furnace of twice the length of the actual furnace and the temperatures of the individual heating panels were varied to simulate motion through the furnace. The modeling results demonstrated that the thermal profiles of PCBs during soldering can be tailored by the incorporation of thermal components to act as heat sinks or thermal barriers. Whalley et al (1991) described recent results of a project to develop an evolving model of the infrared reflow soldering process within commercial reflow furnaces. Modeling improvements discussed were the addition of convective cooling of the PCB as it exits the furnace muffle and the addition of realistic component structures to the PCB assembly. Initial results from a second model

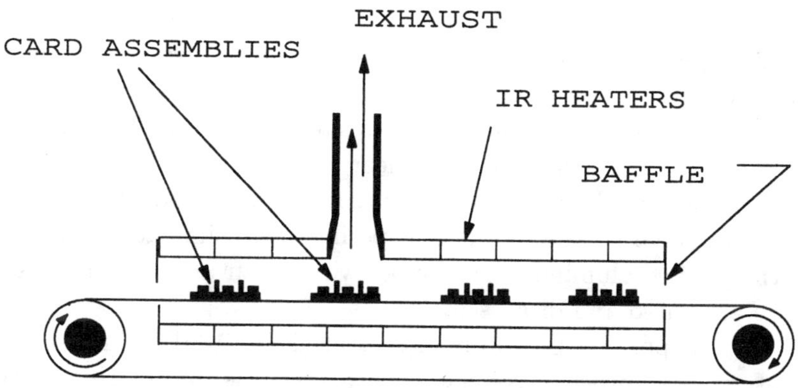

Figure 4.22: Schematic of a typical IR solder reflow process for joining components to a PCB [Eftychiou et al (1993)].

of a high production volume, high quality, mass manufacture oven were also presented. Fernandes et al (1992) discussed an experimental study conducted to reveal the relevant heat transfer mechanisms within an infrared reflow oven. The transient temperature response of simulated card assemblies to combined radiative and convective heating was measured and a simple numerical model was developed. Eftychiou et al (1993) discussed development of a thermal model of the infrared reflow soldering process and reported sample predictions. The authors noted that "the model, which accounts for coupled radiation, mixed convection and conduction heat transfer" is capable of predicting relevant thermal effects ranging from the convection characteristics within the IR oven to the detailed thermal response, including solid-liquid phase change, of individual solder connections.

A paper by Wang et al (1991) discussed a numerical thermal model of the laser heating process to create SMT microsoldered joints. The model was reported to include the influence on the thermal process of solder thermal conductivity variation with temperature, the light reflection coefficient of the lead wire surface and heat transfer on the surface of SMT materials. Another paper by Mok and Kang (1991) discussed a FEM model developed to predict chip temperature response during inner-lead bonding of TAB packages. The results presented in the paper included the effects of applied bonding force, thermode oxide layer thickness, plating materials on the leads, length and thickness of the leads, bump pitch and chip size.

A paper by Huang et al (1991) reported on an investigation of the effect of polarity on heat transfer during the ball formation process in wire bonding of semiconductor chips to lead frames. The electrical discharge process, the associated heat transport, and the wire material combine to determine the nature of the ball formed during the bonding process. The results of experiments to simulate the process with upscaled aluminum and copper wires in air at reduced pressure were presented and discussed.

Two papers by Grigoropoulos et al (1990, 1991) addressed the heat transfer process associated with laser melting of thin silicon films. As explained in the papers, the use of lasers to melt and subsequently recrystallize thin semiconductor layer on insulators (such as SOI structures) has shown good potential for application to commercial VLSI technology. Discussions of a computational conductive heat transfer model and a comparison with experimental results were provided.

Two papers by Hirasawa et al (1992, 1994) discussed modeling efforts to predict the transient temperature distribution of semiconductor wafers in diffusion furnaces used for semiconductor device processing. The first paper considered a vertical diffusion furnace with a stack of horizontal wafers surrounded by a quartz tube, a SiC tube and a heating coil. The second paper considered a hot-wall type furnace with a vertically mounted wafer facing a heated wall.

4.13 NOMENCLATURE

Roman Letter Symbols

a	constant, m-K/W
B	heat sink height, m
b	constant, m/W
D	board spacing, m
	jet surface spacing, m
	lateral heat sink spacing, m
H	height of module, m
h	coefficient of heat transfer, W/m^2-K
k	thermal conductivity, W/m-K
Nu	Nusselt number, dimensionless
Pr	Prandtl number, dimensionless
q	heat flow, W
R	thermal resistance, °C/W or K/W
$\bar{R}$	thermal spreading resistance, °C/W or K/W
U	velocity m/s
V	voltage, V
Z	jet exit spacing, m

Greek Letter Symbols

α	thermal diffusivity, m^2/s
Δ	change in, dimensionless
θ	thermal resistance, °C/W or K/W
μ	dynamic viscosity, kg/m-s
ν	kinematic viscosity, m^2/m-s

Roman Letter Subscripts

ad	adiabatic condition
EB	emitter-to-base
f	forward
in	inlet condition
int	internal condition
JA	junction-to-ambient condition
JC	junction-to-case condition
JW	junction-to-water
m	mean condition

Roman Letter Subscripts (Cont'd)

max	maximum condition
opt	optimum condition
TOT	total condition
∞	free stream condition

4.14 REFERENCES

Aghazadeh, M and Mallik, D (1990). "Thermal characteristics of single and multi-layer high performance PQFP packages," IEEE Trans Comp, Hybrids, Manuf Tech, CHMT-13 (4), Dec, 975-979.

Agonafer, D, and Moffatt, D F (1990). "Numerical modeling of forced convection heat transfer for modules mounted on circuit boards," ASME J Elec Packaging, 112 (4), Dec, 333-337.

Ahmed, I, Krane, R J, and Parsons, J R (1990). "A preliminary investigation of the cooling of electronic components with flat plate heat sinks," presented at ASME Winter Annual Meeting, Dallas, TX, Nov, ASME 90-WA/EEP-25.

Albers, J (1994). "An exact solution of the steady-state surface temperature for a general multilayer structure," Proc 10th Semiconductor Thermal Measurement and Management Symp, San Jose, CA, Feb, 129-137.

Almgren, D W, and Almgren, D W (1990). "The accuracy of PCB thermal analyzers," Proc Natl Elec Packaging and Prod Conf - NEPCON East, Boston, MA, Jun, 131-140.

Anderson, A M, and Moffat, R J (1990). "A new type of heat transfer correlation for air cooling of regular arrays of electronic components," ASME, HTD-153, Nov, 27-40.

Anderson, A M, and Moffat, R J (1991). "Direct air cooling of electronic components: reducing component temperatures by controlled thermal mixing," ASME J Heat Transfer, 113, Feb, 56-62.

Anderson, A M, and Moffat, R J (1992a). "The adiabatic heat transfer coefficient and ther superposition kernel function: part 1 - data for arrays of flatpacks with deifferent flow consitions," ASME J Elec Packaging, 114, Mar, 14-21.

Anderson, A M, and Moffat, R J (1992b). "The adiabatic heat transfer coefficient and ther superposition kernel function: part 1 - modeling flatpack data as a function of channel turbulence," ASME J Elec Packaging, 114, Mar, 22-29.

Anderson, J C, Krane, R J, Parsons, J R, Bar-Cohen, A (1992). "Optimization of the thermal design of a cryogenically cooled computer," IEEE Trans Components, Hybrids, Manuf Tech, CPMT-15 (2), Apr, 786-793.

Antonetti, V W, and Simons, R E (1985). "Bibliography of heat transfer in electronic equipment," IEEE Trans Components, Hybrids, Manuf Tech, CHMT-8 (2), 289-295.

Arai, T, Kurisaka, M, and Ishi H (1992) "Removeable conductive cooling technology of a BISDN switching system," Proc. of the Joint ASME - JSME Conference on Electronic Packaging, Milpitas, CA, Apr, 67-74.

Asako, Y and Faghri, M (1991). "Parametric study of turbulent three-dimensional heat transfer of arrays of heated blocks encountered in electronic equipment," ASME, HTD-171, 135-141.

Ashiwake, N, Daikoku, T, Kawamura, K, and Zushi, S (1991). "A flexible thermal contactor for the cooling of electronic components," Proc 3rd ASME - ASME Thermal Engineering Conf, 2, 357-364.

Avaritsiotis, J N, and Eleftheriades, G (1990). "Layout and thermal analysis of power devices using a PC/XT," Active and Passive Elec Comp, 14 (2), 95-109.

Awkward, K W and Lewis, R D (1991). "Model for determining thermal profiles of bond wires using 'PSPICE, analysis," Proc.

7th Semiconductor Thermal Measurement and Management Symp, Phoenix, AZ, Feb, 86-90.

Azar, K, Benson, J R, and Manno, V P (1991a). "An experimental investigation of microjet impingement cooling," ASME, HTD-171, 1-10.

Azar, K, Benson, J R, and Manno, V P (1991b). "Liquid crystal imaging for temperature measurement of electronic devices," Proc. 7th Semiconductor Thermal Measurement and Management Symp, Phoenix, AZ, Feb, 23-33.

Azar, K, Mandrone, C D and Segelken, J M (1993). "Liquid crystal imaging for temperature measurement of electronic devices," Proc. 9th Semiconductor Thermal Measurement and Management Symp, Austin, TX, AZ, Feb, 19-27.

Azar, K, McLeod, R S and Caron, R E (1992). "Narrow channel heat sink for cooling high powered electronic components," Proc. 9th Semiconductor Thermal Measurement and Management Symp, Austin, TX, AZ, Feb, 19-27.

Azar, K, and Moffat, R J (1991). "Heat transfer coefficient and its estimation in electronic enclosures," Proc Natl Elec Packaging and Prod Conf - NEPCON East, Boston, MA, Jun, 361-371.

Azar, K, Pan, S S, Parry, J and Rosten, H (1994). "Effect of circuit board parameters on thermal performance of electronic components," Proc. 10th Semiconductor Thermal Measurement and Management Symp, San Jose, CA, Feb, 163-169.

Azar, K, and Russell, E T (1991). "Effect of component layout and geometry on the flow distribution in electronics circuit packs," ASME J Elec Packaging, 113, Mar, 50-57.

Azimi-Rashti, M, Jaeger, R C, Baginski, M E, and Goodling, J S (1990). "Liquid freon immersion cooling for microelectronic applications, Proc Natl Elec Packaging and Prod Conf - NEPCON West, 2, Anaheim, CA, Mar, 1643-1649.

Babin, B R, Leiner, W, and Peterson, G P (1990a). "Cooling electronic components with thermosyphons," ASME, HTD-153, 89-94.

Babin, B R, Peterson, G P, and Wu, D (1990b). "Steady-state modeling and testing of a micro heat pipe," ASME J Heat Transfer, 112, Aug, 595-601.

Babus'Haq, R F, George, H E, and O'Callaghan, P W (1992). "A thermal analysis template," Intl J of Computer Applications in Technology (UK), 5 (1), 67-71.

Babus'Haq, R F, George, H E, O'Callaghan, P W, and Constant, B A (1990). "Thermal management of electronics: problems and analytical techniques," Computer-Aided Engineering Journal (UK), Feb, 23-26.

Babus'Haq, R F, O'Callaghan, P W, and Probert, S D (1991). "Reducing the thermal resistance of a pressed contact by employing an interfacial filler," Heat Transfer Engineering, 12 (1), 28-36.

Baker, W E, and Shortt, D J (1990). "Integrated electrical/thermal modeling," Proc of the Intersociety Energy Conversion Engineering Conf, 78-83.

Bar-Cohen, A, Elperin, T, and Eliasi, R (1989). "Rjc characteristics of chip packages - justification, limits and future," IEEE Trans Components, Hybrids, Manuf Tech, CHMT-12 (4), Dec, 724-731.

Bar-Cohen, A, Sherwood, G, and Hodes, M (1994). "Gas-assisted cooling of high density electronic modules," Proc. of Intersociety Conf on Thermal Phenomena on Electronic Systems (I-Therm IV), Washington, DC, 32-40.

Bartilson, B W (1991). "Air jet impingement on a miniature pin-fin heat sink," ASME 91-WA-EEP-41, ASME Winter Annual Meeting, Atlanta, GA, Dec.

Bazydola, S, and Taslim, M E (1993). "An experimental investigation of a staggered array of heat sinks in the hydrodynamic and thermal entrance regions of a duct," ASME J Elec Packaging, 115, Mar, 106-111.

Bejan A, and Lee, S W (1993). "Optimal geometry of convection cooled electronic packages," Proc of 6th Intl Symp on Transport Phenomena, 1, Seoul, Korea, May, 106-111.

Bejan A, and Morega, A M (1993). "The optimal stacking of a stack of plates cooled by turbulent convection," Intl J of Heat and Mass Transfer, 37 (6), Apr, 1045-1050.

Berardinis, L A (1990). "Electromechanical temperature switches survive in a high-tech world," Machine Design, Feb, 112-119.

Bergles, A E (editor) (1990). Heat Transfer in Electronic and Microelectronic Equipment, Hemisphere Pub Corp, New York, NY.

Besserman, D L, Ramadhyani, S, and Incropera, F P (1991). "Numerical simulation of laminar flow and heat transfer for liquid jet impingement cooling of a circular heat source with annular collection of the spent fluid," Numerical Heat Transfer, 20, 263-278.

Besserman, D L, Ramadhyani, S, and Incropera, F P (1992). "Heat transfer from a square source to an impinging liquid jet confined by an annular wall," ASME J Heat Transfer, 114, 284-287.

Bhavnani, S H, Tsai, C-P, and Jaeger, R C (1991). "Pool boiling characteristics of enhanced hybrid silicon surfaces, ASME, HTD-171, 19-27.

Bjorneklett, A, Tuhus, T, Halbo, L, and Kristiansen, H (1993). "Thermal resistance, thermomechanical stress and thermal cycling endurance of silicon chips bonded with adhesives," Proc. 9th Semiconductor Thermal Measurement and Management Symp, Austin, TX, Feb, 136-143.

Bowers, M B and Mudawar, I (1994). "High flux boiling in low flow rate, low pressure drop mini-channel and micro-channel heat sinks," Intl J Heat and Mass Transfer, 37 (2), Jan, 321-332.

Brewster, R A and Sherif, R A (1992). "Thermal analysis of a substrate with power dissipation in the vias," IEEE Trans Components, Hybrids, Manuf Tech, CHMT-15 (5), Oct, 667-674.

Burke, T S (1990). "Forced air keeps electronics cool," Electronic Packaging and Production, Feb, 181-185.

Buschbom, M, and Blanton, P (1990). "Thermal management issues for high speed/high power components," Proc Natl Elec Packaging and Prod Conf - NEPCON East, Boston, MA, Jun, 1001-1013.

Cahlon, B, Gertsbakh, I, Schochetman, I E, and Shillor, M (1991). "A model for the convective cooling of electronic components with application to optimal placement," Mathematical Computer Modelling (UK), 15 (2), 59-75.

Cahlon, B, Schochetman, I E and Shillor, M (1992). "Optimal placement of heat sources on a rectangular grid," Mathematics and Computers in Simulation (Netherlands), 34(3-4), Sep, 351-364.

Carvalho, R D, and Bergles, A E (1990). "The influence of subcooling on the pool nucleate boiling and critical heat flux of simulated electronic chips," Proc of the 9th Intl Heat Transfer Conf, 2, Jerusalem, Israel, 289-293.

Cengel, Y A, and Ngai, T H (1991). "Cooling of vertical shrouded-fin arrays of rectangular profile by natural convection: an experimental study," Heat Transfer Engineering, 12 (4), 27-39.

Chadwick, M L, Webb, B W, and Heaton, H S (1991). "Natural convection from two-dimensional discrete heat sources in a rectangular enclosure," Intl J Heat and Mass Trans, 34 (7), Jul, 1679-1693.

Chambers, B. and Mahalingham, M (1992). "Thermal performance of power packages - single-inline package as a case study," Proc. 8th Semiconductor Thermal Measurement and Management Symp, Austin, TX, Feb, 83-88.

Chan, K F, Leung, C W, and Probert, S D (1991). "Natural convection cooling of a housed, simulated printed-circuit board," Applied Energy, 38, 245-252.

Chapman, C L, Lee, S, and Schmidt, W L (1994). "Thermal performance of an elliptical pin fin heat sink," Proc. 10th Semiconductor Thermal Measurement and Management Symp, San Jose, CA, Feb, 24-31.

Chen, L, Keyhani, M, and Pitts, D R (1991). "Convection heat transfer due to protruded heat sources in an enclosure," J of Thermophysics, 5 (2), Apr-Jun, 217-223.

Chien, D H, Wang, C Y and Lee, CC (1992). Temperature solution of five layer structure with a circular embedded source and its applications," IEEE Trans Components, Hybrids, Manuf Tech, CHMT-15 (5), Oct, 707-714.

Cho, C H, Lee, S R, and Seong, H J (1990). "Boiling heat transfer from inclined channels in a natural circulation loop," Proc of Intersociety Conf on Thermal Phenomena in Electronic Systems (I-Therm II), Las Vegas, NV, 86-89.

Chrysler, G M, Chu, R C, and Simons, R E (1994). "Jet impingement boiling of a dielectric coolant in narrow gaps," Proc. of Intersociety Conf on Thermal Phenomena on Electronic Systems (I-Therm IV), Washiongton, DC, 1-8.

Chrysler, G M and Simons, R E (1990). "An experimental investigation of the forced convection heat transfer characteristics of fluorocarbon liquid flowing through a packed-bed for immersion cooling of microelectronic heat sources", ASME. HTD-131, Aug, 21-27.

Chu, R C and Simons, R E (1990). "Review of thermal design for multi-chip modules," Proc Natl Elec Packaging and Prod Conf - NEPCON West, 2, Anaheim, CA, Mar, 1633-1642.

Chu, R C and Simons, R E (1993). "Recent development of computer cooling technology," Proc. 6th Intl Symp. on Transport Phenomena, 1, Seoul, Korea, May, 17-25.

Chu, R C, Simons, R E, and Vader, D T (1992). "Super-position cooling," US Patent 5,097,385.

Chung, K, Boyle, A, and Sager, J (1991). "MCM die attachment using low-stress, thermally conductive epoxies," Proc Intl Elec Packaging Soc (IEPS) Conf, 1, San Diego, CA, Sep, 167-175.

Churchill, S W (1977). "A comprehanesive correlating equation for laminar, assisting, forced and free convection," AIChE J, 23 (1), Jan, 10-16. technique

Cooley, W T, and Razani, A (1991). "Thermal analysis of surface mounted leadless chip carriers," ASME J Elec Packaging, 1132, Jun, 156-163.

Copeland, D W (1994). "Single-phase and boiling cooling of small pin fin arrays by multiple slot nozzle suction and impingement," Proc. of Intersociety Conf on Thermal Phenomena on Electronic Systems (I-Therm IV), Washington, DC, 16-22.

Copeland, D W and Powell, D O (1990). "Aluminum nitride metallized ceramic substrate performance," Proc. 6th Semiconductor Thermal Measurement and Management Symp, Phoenix, AZ, Feb, 104-107.

Crall, R F (1990). "The use of thermal imaging to verify thermal management in new product designs," Proc. of the Surface Mount Technology Association (SMTA) Technical Program: Surface Mount 90 Exposition and Conference, Boston, MA, 42-49.

Cray, S and Sherwood, G (1992). "Gas-liquid forced turbulence cooling," US Patent 5,131,233.

Cripps, S C (1990). "A new technique for screening and measuring channel temperature in RF and microwave hybrid circuits," Proc. 7th Semiconductor Thermal Measurement and Management Symp, Phoenix, AZ, Feb, 40-42.

Csanky, G (1990). "Reliability critical thermal model for double-drift IMPATT diodes on diamond heat sinks," Quality and Reliability Engineering International, 6, 73-84.

Culham, J R, Lee, S, and Yovanovich, M M (1993). "Conjugate heat transfer from a raised isothermal heat source attached to a vertical board," Proc. 9th Semiconductor Thermal Measurement and Management Symp, Austin, TX, Feb, 151-159.

Culham, J R, Lemczyk, T F, Lee, S, and Yovanovich, M M (1991). "META- a conjugate heat transfer model for air cooling of circuit boards with arbitrarily located heat sources," ASME, HTD-171, 117-126.

Culham, J R, Yovanovich, M M, and Lee S (1994). "Thermal modeling of isothermal cuboids and rectangular heat sinks cooled by natural convection," Proc. of Intersociety Conf on Thermal Phenomena on Electronic Systems (I-Therm IV), Washington, DC, 73-82.

Davis, T L (1990). "Carrier impact on the thermal performance of tape-automated-bonding components," IEEE Trans Comp, Hybrids, Manuf Tech, CHMT-13 (4), Dec, 998-1005.

Dean, D J (1990). "Circuit board thermal design or how to succeed at the impossible," Cicuit World (UK), 22, May, 16-17.

De May, G (1990). "Thermal analysis in CAD of hybrid circuits," Hybrid Circuits, 22, May, 16-17.

Demmin, J C (1991). "Thermal Modeling of multi-chip modules," Proc Natl Elec Packaging and Prod Conf - NEPCON East, Boston, MA, Jun, 413-424.

Dolbear, T (1990). "Thermal management of multi-chip modules," Proc Natl Elec Packaging and Prod Conf - NEPCON West, 2, Anaheim, CA, Mar, 1186-1208.

Driscart, M, and De May, G (1990). "The boundary element method for thermal analysis of electronic equipment in a multi dimensional environment," First Intl Conf on Advanced Computational Methods in Heat Transfer, Southampton, UK, Jul, 343-354.

Du, Z-G, and Bilgen, E (1990). "Effects of heat intensity, size, and position of the components on temperature distribution within an electronic PCB enclosure," ASME J Elec Packaging, 112 (3), Sep, 249-254.

Edwards, D L (1992). "Integral thermal analysis in TCM design," Proc. 8th Semiconductor Thermal Measurement and Management Symp, Austin, TX, Feb, 1-4.

Edwards, D L, Hwang, M, and Stearns, W (1994). "Thermal enhancement of IC packages," Proc. 10th Semiconductor Thermal Measurement and Management Symp, Austin, TX, Feb, 1-4.

Eftychiou, M A, Bergman, L T, and Masada, G Y (1993). "A detailed thermal model of the infrared soldering process," ASME J Elec Packaging, 115 (1), Dec, 55-62.

Eliasi, R, Elperin, T, and Bar-Cohen, A (1990). "Monte Carlo thermal optimization of populated printed circuit board," IEEE Trans Comp, Hybrids, Manuf Tech, CHMT-13 (4), Dec, 953-960.

Ellison, G N (1984). "A review of a thermal analysis computer program for microelectronic devices," in Thermal Management Concepts in Microelectronic Packaging, S S Furkay, R F Kilburn, and G Monti (eds), International Society for Hybrid Microelectronics, Reston, VA.

Ellison, G N (1990). "TAMS: a thermal analyzer for multilayer structures," Electrosoft, 1 (2), 85-97.

Ellison, G N (1991). "Extensions of a closed form method for substrate thermal analyzers to include thermal resistances from source-to-substrate and source-to-ambient," Proc 7th Semiconductor Thermal Measurement and Management Symp, Phoenix, AZ, Feb, 140-148.

Enck, R C, Harris, R D, and Fields, J L (1991). "High thermal conductivity substrate materials," Proc Natl Elec Packaging and Prod Conf - NEPCON West, 1, Anaheim, CA, Feb, 523-531.

Eseteki, M H, Reizes, J A, and Behnia, M (1993). "Natural convection heat transfer from electronic components in an enclosure." Proc of 6th Intl Symp on Transport Phenomena, 1, Seoul, Korea, May, 249-254.

Estes, R C (1991). "The effect of thermal capacitance and phase change on outside plant enclosures," Proc 7th Semiconductor Thermal Measurement and Management Symp, Phoenix, AZ, Feb, 91-96.

Faghri, M, Ray, A, Sridhar, S, and Schmidt, R (1991). "Entrance heat transfer correlation for air cooling of rectangular arrays of rectangular blocks," ASME, HTD-183, 19-23.

Faghri, M, Molki, M, and Chrupcala, J (1994). "Entrance analysis of turbulent flow in an array of heated rectangular blocks," Proc. of Intersociety Conf on Thermal Phenomena on Electronic Systems (I-Therm IV), Washington, DC, 198-206.

Fernandes, N J, Bergman, T L, and Masada, G Y (1992). "Thermal effects during infrared solder reflow - Part I: Heat transfer mechanisms," ASME J Elec Packaging, 114, Mar, 41-47.

Fletcher, L S (1990). "A review of thermal enhancement techniques for electronic systems," IEEE Trans Comp, Hybrids, Manuf Tech, CHMT-13 (4), Dec, 1012-1021.

Flint, E B, Grebe, K R, Gruber, P A, and Zingher, A R (1991). "Flexible finned heat exchanger," U.S. Patent 5,022,462.

Forsman, B, and Zirath, H (1990). "Heat absorption in telephone exchanges through an optimized climate ceiling," Proc 12^{rmth} Intl Telecom Energy Conf, Orlando, FL, Oct, 202-205.

Fox, L R, Kromann, G, and Hannemann, R J (1991). "Multichip package cooling with integral heat pipe," Proc Natl Elec Packaging and Prod Conf - NEPCON West, Anaheim, CA, Feb, 673-689.

Fukuoka, Y, and Ishizuka, M (1991). "Thermal analysis of a new high density package cooling technology using low melting point alloys," Japanese Journal of Applied Physics, 30 (6), Jun, 1313-1319.

Funk, J N, Menguc, M P, Tagavi, K A and Cremers, C J (1991). "A semi-analytical method to predict printed circuit board package temperatures," Proc 7th Semiconductor Thermal Measurement and Management Symp, Phoenix, AZ, Feb, 7-14.

Gabrielli, F, and Meli, G (1990). "Thermal simulation for power equipment using natural cooling," Proc 12th Intl Telecom Energy Conf, Orlando, FL, Oct, 556-562.

Garfinkel, G A (1990). "Simplifying thermal and structural analysis," Circuit Design, May, 50-57.

Garimella, S V, and Eibeck, P A (1990). "Heat transfer characteristics of an array of protuding elements in single phase forced convection," Intl J Heat and Mass Trans, 33 (12), Dec, 2659-2669.

Garimella, S V, and Eibeck, P A (1991a). "Effect of spanwise spacing on the heat transfer from an array of protuding elements in forced convection," Intl J Heat and Mass Trans, 34 (9), Sep, 2427-2430.

Garimella, S V, and Eibeck, P A (1991b). "Enhancement of single phase convective heat transfer from protuding elements using vortex generators," Intl J Heat and Mass Trans, 34 (9), Sep, 2431-2433.

Gates, L E, and Hontz, M P (1991). "Design analysis of a high power sealed chip-on-board circuit," Proc Intl Elec Packaging Soc (IEPS) Conf, 1, San Diego, CA, Sep, 109-116.

Gautier, T (1991). "Construction and validation of thermal models of packages," Proc 7th Semiconductor Thermal Measurement and Management Symp, Phoenix, AZ, Feb, 74-79.

Gaynes, M A and Shaukatullah, H (1994). "Adhesives for thermal enhancement of surface mount plastic electronic packages," Proc Natl Elec Packaging and Prod Conf - NEPCON West, 2, Anaheim, CA, Mar, 1560-1569.

Geeraerts, B, Van Petegem, W and Sansen, W (1993). "A BI-MOS diode matrix for the characterization of static and transient thermal performance on silicon," Proc 9th Semiconductor Thermal Measurement and Management Symp, Austin, TX, Feb, 108-111.

Gersey, C O and Mudawar, I (1992). "Effects of orientation on critical heat flux from chip arrays during flow boiling," Proc. of the Joint ASME - ASME Conference on Electronic Packaging, Milpitas, CA, Apr, 123-134.

Gersey, C O and Mudawar, I (1993). "Nucleate boiling and critical heat flux flux from protruded chip arrays during flow boiling," ASME J Elec Packaging, 115, Mar, 78-88.

Gersey, C O, Willingham, T C, and Mudawar, I (1992). "Design parameters and practical considerations in the two phase forced convection of multi-chip modules," ASME J Elec Packaging, 114, Sep, 280-289.

Ghodbane, M, and Holman, J P (1991). "Experimental study of spray cooling with Freon-113," Intl J Heat and Mass Trans, 34 (4/5), 1163-1174.

Giammaruti, R J, Farouk, B, and Carroll, J P (1991). "Impingement cooling of simulated electronic components in the presence of a crossflow," ASME, HTD-163, 19-25.

Godfrey, W M, Tagavi, K A, Cremers, C J, and Menguc, M P (1993). "Interactive thermal modeling of electronic circuit boards," IEEE Trans Components, Hybrids, Manuf Tech, CHMT-16 (8), Dec, 978-985.

Goodman, T, Snyder, D, and Weiss, J (1990). "Predicting circuit board hot spots," Computer-Aided Engineering (UK), Sep, 71-72.

Goosey, M T, Lodge, K J, and Logan, E A (1991). "Aluminum nitride for packaging high performance electronic circuits," GEC Journal of Research, 8 (3), 137-144.

Gough, P A, Walker, P and Whight, K R (1991). "Electrothermal simulation of power semiconductor devices," Proc. 3rd Intl Symp on Power Semiconductor Devices and IC's, Baltimore, Md, Apr, 89-94.

Goyal, A, Jaeger, R C, Bhavnani, S H, Ellis, C D, Phadke, N K, Azimi-Rashte, M, and Goodling, J S (1993). "Formation of silicon reentrant cavity heat sinks using anisotropic etching and direct water bonding." IEEE Electron Device Letters, 14 (1). 29-32.

Grigoropoulos, C P, Dutcher, W E, and Emery, A F (1991). "Experimental and computational analysis of laser melting of thin silicon films," ASME J Heat Transfer, 113, 21-29.

Grigoropoulos, C P, Emery, A F, and Wipf, E O (1990). "Heat transfer in thin silicon film melting by laser line sources," Intl J Heat and Mass Trans, 33 (5), May, 797-803.

Guenin, B M (1994). " Transient thermal model for the MQUAD microelectronic package," Proc 10th Semiconductor Thermal Measurement and Management Symp, San Jose, CA, Feb, 86-95.

Guenin, B M and Mahulkar, D (1993). "Methodology for the thermal characteristics of the MQUAD electronic package," Proc 9th Semiconductor Thermal Measurement and Management Symp, Austin, TX, Feb, 176-185.

Hadim, A, Chang, A T, and DiEduardo, T (1990). "Numerical and experimental heat transfer analysis of a printed circuit board," presented at ASME Winter Annual Meeting, Dallas, TX, Nov, ASME 90-WA/EEP-3.

Hamadh, T T, Ryder, D N, and Friedman, H S (1991). "Apparatus for controlled air-impingement module cooling," U.S. Patent 5,063,476, Nov.

Hamburgen, W R (1991). "Interleaved fin thermal connectors for multichip modules," Proc Intl Elec Packaging Soc (IEPS) Conf, 1, San Diego, CA, Sep, 419-427.

Hamburgen, W R and Fitch, J S (1993). "Packaging a 150 W bipolar ECL microprocessor, IEEE Trans Components, Hybrids, Manuf Tech, CHMT-16 (1), Feb, 28-38.

Hamzehdoost, A (1990a). "Thermal performance of high lead count plastic quad flat packs," Proc Intl Elec Packaging Soc (IEPS) Conf, Marlbourough, MA, Sep, 68-76.

Hamzehdoost, A (1990b). "Thermal performance of high performance pin grid arrays," Proc Intl Elec Packaging Soc (IEPS) Conf, Marlbourough, MA, Sep, 634-641.

Handa, T, Iida, S, and Utsunomiya, J (1993). "Thermal design for high-speed, high-density multi-chip module," Proc of 11th IEEE/CHMT Intl Electr Manuf Tech Symp, 266-269.

Harada, A, Kaneko, Y, and Ibaraki, O (1990). "Forced air convection cooling for telecommunication systems installed in nonstaffed offices," Proc Intl Elec Packaging Soc (IEPS) Conf, Marlbourough, MA, Sep, 1164-1178.

Harpole, G M and Eninger, J E (1991). "Micro-channel heat exchanger optimization," Proc 7th Semiconductor Thermal Measurement and Management Symp, Phoenix, AZ, Feb, 59-63.

Hayama, H, Nakao, M, and Sanbe, M (1990). "Airflow distribution in telecommunications equipment rooms," Proc 12th Intl Telecom Energy Conf, Orlando, FL, Oct, 206-212.

Hebrard, L, Jacquemod, G, Boutherin, B and LeHelly, M (1992). "Simulation of electrothermal interactions in power integrated circuits." Proc 8th Semiconductor Thermal Measurement and Management Symp, Austin, TX, Feb, 76-82.

Hefner, A R and Blackburn D L (1994). "Thermal component models for an electrothermal network simulation," IEEE Trans Components, Hybrids, Manuf Tech, CHMT-17 (3), Sep, 413-424.

Heng, S and Pei, J (1990). "Remote condenser phase-change cooling system," Proc Intl Elec Packaging Soc (IEPS) Conf, Marlbourough, MA, Sep, 1147-1163.

Heng, S and Pei, J (1991a). "Air impingement cooled pin-fin heat sink for multi-chip unit," Proc Natl Elec Packaging and Prod Conf - NEPCON West, 2, Anaheim, CA, 1156-1165.

Heng, S and Pei, J (1991b). "Cooling MCMs with pin in heat sinks," Electronic Packaging and Production, Oct, 54-57.

Herold, K E, Sridhar, S, and Hu, S (1992). " Cooling of electronic boards using internal fluid flows," Proc. of the Joint ASME - ASME Conference on Electronic Packaging, Milpitas, CA, Apr, 285-290.

Hilbert, C, Sommerfeldt, S, Gupta, O, and Herrell, D J (1990a). "High performance air cooled heat sinks for integrated circuits," IEEE Trans Comp, Hybrids, Manuf Tech, CHMT-13 (4), Dec, 1022-1031.

Hilbert, C, Sommerfeldt, S, Gupta, O, and Herrell, D J (1990b). "A low pressure air cooling system for high performance modules," Proc Natl Elec Packaging and Prod Conf - NEPCON West, 2, Anaheim, CA, Mar, 1667-1675.

Hill, T B and Hill, C C (1990). "Effects of electronic enclosure layout on fan performance," presented at ASME Winter Annual Meeting, Dallas, TX, Nov, ASME 90-WA/EEP-6.

Hill, T B and Lind, J (1990). "Designing ventilation grilles for electronic equipment," Machine Design. Aug, 73-78.

Hirasawa, S, Kieda, S, Watanabe, T, Torii, T, Takagaki, T, and Uchino, T (1992). " Temperature distribution in semiconductor wafers heated in a vertical diffusion furnace," Proc. of Intersociety Conf on Thermal Phenomena in Electronic Systems (I-Therm III), Austin, TX, 276-280.

Hirasawa, S, Watanabe, Takagaki, T, and Uchino, T (1994). " Temperature distribution in semiconductor wafers heated in a hot-wall type diffusion furnace," Proc. of Intersociety Conf on Thermal Phenomena in Electronic Systems (I-Therm IV), Washington, DC, 253-260.

Hollworth, B R, Maxwell, K A, and Ukeiley, L S (1990). "Impingement cooling of electronics: effects of venting through circuit board," ASME, HTD-131, 29-36.

Horvath, J L, Biskeborn, R G, and Harvilchuck, J M (1991). "High conduction cooling module having internal fins and compliant interfaces for VLSI chip technology," U.S. Patent 5,052,481.

Huang, L J, Jog, M A, Cohen, I M, and Ayyaswamy, P S (1991). "Effect of polarity on heat transfer in the ball formation process," ASME J Elec Packaging, 113 , Mar, 33-39.

Hussein, M M, Elshabini-Riad, A, and Nelson, D J (1991a). "Thermal management of hybrid circuits: experiment versus prediction," Proc of 4th Electronics Materials and Processing Congress, Montreal, Canada, Aug, 389-397.

Hussein, M M, Nelson, D J, and Elshabini-Riad, A (1991b). "Thermal interaction of semiconductor devices on copper-clad ceramic substrates," Proc 7th Semiconductor Thermal Measurement and Management Symp, Phoenix, AZ, Feb, 117-122.

Hwang, U P, Moran, K P, and Kemink, R G (1992). "Cold plate design for IBM ES/9000 TCM Electronic Modules," Proc. of the Joint ASME - ASME Conference on Electronic Packaging, Milpitas, CA, Apr, 75-81.

Ibrahim, M S (1991). "Thermal design of electronic modules in high ambient temperature environments," Proc Natl Elec Packaging and Prod Conf - NEPCON East, Boston, MA, Jun, 341-347.

Ibrahim, M S, Paradis, L R, and Paterson, D (1992). "Finite element modeling of a MMIC transmitter module for thermal/structural design optimization," IEEE Trans Comp, Hybrids and Manuf Tech, CHMT-15 (5), Oct, 723-729.

Ishizuka, M (1992). "Thermal design of a laptop computer by a personal computer," Proc. of Intersociety Conf on Thermal Phenomena in Electronic Systems (I-Therm III), Austin, TX, 184-189.

Iwasaki, H, Sasaki, T, and Ishizuka, M (1994). Cooling performance of plate fins for multichip modules," Proc. of Intersociety Conf on Thermal Phenomena in Electronic Systems (I-Therm IV), Washington, DC, 144-147.

Johnson, W E, Spexarth, F A, and Forster, J A (1992). "Copper clad invar as a thermal management material, "Proc. Intl Elec Packaging Soc (IEPS) Conf, 1, Austin, TX, Sep, 403-419.

Jomard, T, Eckes, U, and Lallemand, M (1991). "Two-phase cooling evolution of high power semiconductors," Proc of 4th European Conf on Power Electronics and Applications, Firenze, Italy, Sep, 2, 491-496.

Joshi, Y K (1992). "Liquid immersion cooling of electronic equipment," Naval Research Review, 44 (1), 35-42.

Joshi, Y, Kelleher, M D, Powell, M, and Torres, E I (1991). "Natural convection heat transfer from an array of rectangular protusions in an enclosure filled with dielectric liquid," ASME, HTD-183, 9-18.

Joshi, Y and Knight, D L (1990). "Natural convection from a column of flush heat sources in a vertical channel in water," ASME J Elec Packaging, 112, Dec, 367-374.

Joshi, Y and Paje, R A (1991). "Natural convection cooling of a ceramic substrate mounted leadless chip carrier in dielectric liquids," Intl Comm Heat Mass Transfer, 18, 39-47.

Juds, M A and Chan, S H (1992). "Cabinet: a lumped parameter computer program for multi-compartment vented enclosures," Proc. of the Joint ASME - ASME Conference on Electronic Packaging, Milpitas, CA, Apr, 89-101.

Kakac, S, Yuncu, H, and Hijikata, K (eds) (1994). Cooling of Electronic Systems, Kluwer Academic Publishers, Dordrecht, The Netherlands.

Kamath, V (1994). " Air injection and convection cooling of multichip modules: a computational study," Proc. of Intersociety Conf on Thermal Phenomena in Electronic Systems (I-Therm IV), Washington, DC, 83-90

Kang, B H and Jaluria, Y (1990). "Mixed convection transport from a protuding heat source module on a vertical surface," J of Thermophysics, 4 (3), Jul, 384-390.

Keeley, A and Ryan, C (1991). "Achieving a balance of thermal performance and routing density in a multichip module substrate," Proc Natl Elec Packaging and Prod Conf - NEPCON East, Boston, MA, Jun, 551-561.

Kelly, E G (1990). "Thermal characteristics of surface mount packages," Proc Surface Mount Technology Conf (SMTCON), Atlantic City, NJ, Apr, 127-137.

Keyhani, M, Chen, L, and Pitts, R (1990). "The aspect ratio effect on natural convection in an enclosure with protruding heat sources," ASME, HTD-131, 1-11.

Kiley, R F and Soule, C A, (1990). "Engineered heat sinks," Powertechnics Magazine, Jul, 16-20.

Kishimoto, T, Harada, A, and Kaneko, Y (1991). "Air cooling system for telecom multichip modules in broadband ISDN," Proc Intl Elec Packaging Soc (IEPS) Conf, 1, San Diego, CA, Sep, 405-418.

Kishimoto, T and Harada, A (1992). "Two-phase thermosyphon cooling for telecom multichip modules," Proc. of the Joint ASME - ASME Conference on Electronic Packaging, Milpitas, CA, Apr, 136-141.

Kitajo, S, Takeda, Y, Kurokawa, Y, Ohta, T, and Mizunashi, H (1992). "Development of a high performance air cooled heat sink for multi-chip modules," Proc 12th Semiconductor Thermal Measurement and Management Symp, Austin, TX, Feb, 119-124.

Knight, R W, Goodling, J S, and Hall R W (1991), "Optimal thermal design of forced convection heat sinks," ASME J Elec Packaging, 113 (3), Sep, 313-321.

Knight, R W, Goodling, J S, and Gross, B E (1992a), "Optimum thermal design of air cooled forced convection finned heat sinks - experimental verification," IEEE Trans Components, Hybrids, Manuf Tech, CHMT-15 (5), Oct, 754-760.

Knight, R W, Hall, D J, Goodling, J S, and Jaeger, R C (1992b), "Heat sink optimization with application to microchannels," IEEE Trans Components, Hybrids, Manuf Tech, CHMT-15 (8), Apr, 832-842.

Kozarek, R L (1991). "Effect of case temperature measurement errors on the junction-to-case thermal resistance of a ceramic

PGA," Proc 7th Semiconductor Thermal Measurement and Management Symp, Phoenix, AZ, Feb, 44-51.

Krane, M J (1991). "Constriction resistance in rectangular bodies," ASME J Elec Packaging, 113, Dec, 392-396.

Kristiansen, H, Bjorneklett, A, Ohlckers, P, and Gleditsch, T (1990). "Thermal management evaluation of evaporation cooling with fluorocarbon liquids using naked test chips," Proc Intl Elec Packaging Soc (IEPS) Conf, Marlbourough, MA, Sep, 1133-1146.

Krueger, W, and Bar-Cohen A (1992) "Thermal characteristics of a PLCC - expanded R_{JC} methodology," IEEE Trans Components, Hybrids, Manuf Tech, CHMT-15 (5), Oct, 691-698.

Kuhlman, M A and Sehitoglu, H (1992). "A thermal evaluation of multichip module (MCM) materials and designs," Proc 8th Semiconductor Thermal Measurement and Management Symp, Austin, TX, Feb, 110-118.

Kunkel, S H and Peck W M (1994), "Predicting thermal contact resistance in circuit card assemblies," Proc. of Intersociety Conf on Thermal Phenomena on Electronic Systems (I-Therm IV), Washington, DC, 91-100.

Kuwahara, H, Takahashi, K, Nakajima, T, Suzuki, O, and Takasaki, T (1994). "Enhancement of two-phase thermosyphon for cooling high flux power devices," Proc. of Intersociety Conf on Thermal Phenomena on Electronic Systems (I-Therm IV), Washington, DC, 184-189.

Lall, B S, Ortega, A and Kabir, H (1994). "Thermal design rules for electronic components on conducting boards in passively cooled enclosures," Proc. of Intersociety Conf on Thermal Phenomena on Electronic Systems (I-Therm IV), Washington, DC, 50-61.

Landram, C S (1991). "Computational model for optimizing longitudinal fin heat transfer in laminar internal flows," ASME, HTD-171, 127-134.

Lasance, C J M (1991). "Numerical simulation of natural convection heat transfer in a television set" ASME, HTD-171, 101-108.

Lavine, A S and Bai, C (1991). "An analysis of heat transfer in Josephson junction devices," ASME J Heat Transfer, 113, 535-543.

Leatherman, R A (1990). "Innovative thermal design concepts for surface mount multilayer boards," Proc Intl Elec Packaging Soc (IEPS) Conf, Marlbourough, MA, Sep, 613-627.

Lee, S W and Kim, S J (1991). "Numerical and experimental analysis of fan-induced flow and heat transfer in electronic systems," ASME, HTD-171, 109-115.

Lee, T Y and Mahalingham, M (1992). "Thermal design space prediction in two-phase liquid cooling," IEEE Trans Components, Hybrids, Manuf Tech, CHMT-15 (2), Oct, 778-785.

Lee, T Y and Mahalingham, M (1994). "Applications of a CFD tool for system-level thermal simulation," IEEE Trans Components, Hybrids, Manuf Tech, CHMT-17 (4), Dec, 564-572.

Lee, C C and Jay-Min, Y (1991). "Thermal modeling of semiconductor devices and packages using analytical methods," Proc Natl Elec Packaging and Prod Conf - NEPCON West, 1, Anaheim, CA, Feb, 147-160.

Lee, R S, Huang, H C, and Chen, W Y (1990). "A thermal characteristic study of extruded-type heat sinks in considering air flow bypass phenomenom," Proc 6th Semiconductor Thermal and Temperature Measurement Symp, Phoenix, AZ, Feb, 95-102.

Lee, S, Culham, J R, and Yovanovich, M M (1991a). "The effect of common design parameters on the thermal performance of microelectronic equipment: part I - natural convection," ASME, HTD-171, 47-54.

Lee, S, Culham, J R, and Yovanovich, M M (1991b). "The effect of common design parameters on the thermal performance of microelectronic equipment: part II - forced convection," ASME, HTD-171, 55-62.

Lee, S, Culham, J R, and Yovanovich, M M (1991c). "Parametric investigation of conjugate heat transfer from microelectronic circuit boards under mixed convection cooling," Proc Intl Elec Packaging Soc (IEPS) Conf, 1, San Diego, CA, Sep, 428-446.

Lee, S, Lemczyk, T F, and Yovanovich, M M (1992a). "Analysis of thermal vias in high density interconnect technology," Proc 8th Semiconductor Thermal Measurement and Management Symp, Austin, TX, Feb, 55-61.

Lee, T Y, Andrews, J A, Chow, P, and Saums, D (1992b). "Compact cooling system for small moveable electronic equipment," IEEE Trans Components, Hybrids, Manuf Tech, CHMT-15 (5), Apr, 786-793.

Lee, C C, Su, T, and Chao, M (1992c). "Transient thermal measurments using the index of refraction as a temperature sensitive parameter," IEEE Trans Components, Hybrids, Manuf Tech, CHMT-15 (5), Oct, 625-631.

Lee, C C, Su, T, and Chao, M (1992d). "Transient thermal measurments using thermooptic and thermoelectric effects." Proc 8th Semiconductor Thermal Measurement and Management Symp, Austin. TX, Feb, 41-46.

Lee, T Y, Mahalingham, M, and Normington, P J (1993). "Subcooled pool boiling critical heat flux in dielectric liquid mixtures," ASME J Elec Packaging, 115, Mar, 134-137.

Lehmann, G L, and Huang, Y (1991). "Enhanced direct air cooling of electronic components using secondary flow mixing," ASME, HTD-171, 11-17.

Lehmann, G L, and Kosteva, S J (1990). "A study of forced convection direct air cooling in the downstream vicinity of heat sinks," ASME J Elec Packaging, 112, Sep, 234-240.

Lehmann, G L, and Pembroke, J (1991). "Forced convection air cooling of simulated low profile electronic components: part 1- base case," ASME J Elec Packaging, 113 , Mar, 21-26.

Leibson, S H (1994). "Beating processor heat through thermal simulation," EDN, Oct, 53-58.

Lemczyk, T F, Mack, B, Culham, J R, and Yovanovich, M M (1992a). "Fopt - a thermal optimization factor for a microelectronic package." Proc 8th Semiconductor Thermal Measurement and Management Symp, Austin, TX, Feb, 89-94.

Lemczyk, T F, Mack, B, Culham, J R, and Yovanovich, M M (1992b). "PCB trace thermal analysis and effective conductivity," ASME J Elec Packaging, 114, Dec, 413-438.

Liakopoulos, A, Huang, X, Blythe, P A, and Simpkins, P G (1991). "Buoyancy driven motions due to a vertical array of heat sources," ASME, HTD-171, 63-69.

Lin, T-Y, and Hsieh, S-S (1990). "Natural convection of opposing/assisting flows in vertical channels with asymmetrically discrete heated ribs," Intl J Heat and Mass Trans, 33 (10), 2295-2309.

Lin, Y, and Fang, Z (1991). "Thermal characteristics of semiconductor laser phase-locked arrays with diamond film heatsink," Electronics Letters, 27 (1), Jan, 18-19.

Linnander, B (1993). "When it's too hot to touch, use infrared thermography," IEEE-circuits and devices, 9 (4), Jul, 35-37.

Linton, R L (1991). "CFD modelling of electronic enclosures," ASME, HTD-171, 95-100.

Linton, R L and Agonafer, D (1994). "Coarse and detailed CFD modeling of a finned heat sink," Proc. of Intersociety Conf on Thermal Phenomena on Electronic Systems (I-Therm IV), Washington, DC, 156-161.

Liu, J C, Hunter, D E, and Kozarek, R L (1990). "Electronic package thermal design using boundary elements," Proc 6th Semiconductor Thermal and Temperature Measurement Symp, Phoenix, AZ, Feb, 65-69.

Lu, C Z, Wang, M Z, Gui, X, and Gao, G B (1991). "Fast measurement of the peak junction temperature of power transistors using electrical method, Proc 7th Semiconductor Thermal Measurement and Management Symp, Phoenix, AZ, Feb, 39-43.

Ma, C-F and Bergles, A E (1983). "Boiling jet impingement cooling of simulated microelectronic chips," *Heat Transfer in Electronic Equipment*, 1983, ASME-HTD, 28, 3-10.

Maddox, D E and Bar-Cohen, A (1991). "Thermofluid design of submerged-jet impingement cooling for electronic components," HTD-171, 71-80.

Mahaney, H V, Incropera, F P, and Ramadhyani, S (1990a). "Comparison of predicted and measured mixed convection heat transfer from an array of discrete sources in a horizontal rectangular channel," Intl J Heat and Mass Trans, 33 (6), Jun, 1233-1245.

Mahaney, H V, Incropera, F P, and Ramadhyani, S (1990b). "Measurement of mixed-convection heat transfer from an array of discrete sources in a horizontal rectangular channel with and without surface augmentation," Experimental Heat Transfer, 3, 215-237.

Mahaney, H V, Ramadhyani, S, and Incropera, F P (1990c). "Mixed convection heat transfer from an array of discrete sources in a horizontal rectangular channel," ASME, HTD- 131, 13-19.

Malhammar, A (1991). "Thermal models for circuit boards and microcircuits, today and tommorrow," Proc of 2nd Components Engineering, Reliability and Test Conference, 22-59.

Mallik, A K, Peterson, G P, and Weichold, M H (1992). "On the use of heat pipes as an integral part of semicvonductor devices," ASME J Elec Packaging, 114, Dec, 436-442.

Manno, V P and Azar, K (1990). "The effect of neighboring components on thermal performance of air-cooled circuit packs," Proc Natl Elec Packaging and Prod Conf - NEPCON East, Boston, MA, Jun, 121-130.

Manno, V P and Azar, K (1991). "Impact of heat transfer coefficient models on the simulation of electronic enclosure thermal response," Proc Natl Elec Packaging and Prod Conf - NEPCON East, Boston, MA, Jun, 349-360.

Manno, V P and Azar, K (1993). "Thermal-fluid interactions of neighboring components on air-cooled circuit boards," ASME J Elec Packaging, 113, 374-381.

Manno, V P, Kurita, N R, and Azar, K (1993). "Experimental characterization of board conduction effects," Proc 9th Semiconductor Thermal and Temperature Measurement Symp, Austin, TX, Feb, 127-136.

Manno, V P, Leisk, G, and Azar, K (1990). "Effect of system orientation and cooling mechanism on component thermal impedances in an electronic enclosure," Proc 6th Semiconductor Thermal and Temperature Measurement Symp, Phoenix, AZ, Feb, 17-24.

Mansuria, M (1991). "Application of the thermoelectric cooler for electronic packaging cooling," Proc Intl Elec Packaging Soc (IEPS) Conf, 1, San Diego, CA, Sep, 251-265.

Markstein, H W (1991). "Design tips for SMT cooling," Electronic Packaging and Production, Jul, 49-55.

Mathisen, T T (1990). "Performance of liquid heat sink for cooling electronics," Proc Intl Elec Packaging Soc (IEPS) Conf, Marlbourough, MA, Sep, 602-612.

Maudgal, V K (1991). "Computer-aided thermal analysis," Hybrid Circuit Technology, May, 19-21.

Maudgal, V K, Kliman, J, and Sunderland, J E (1990a). "A thermal measurement method to test circuit boards in production," Intl Jour for Hybrid Microelectronics, 13 (3), Sep, 53-60.

Maudgal, V K, Kliman, J, Miles, J, and Sunderland, J E (1990b). "Experimental determination of thermal contact in a diode-heat sink assembly," ASME J Elec Packaging, 112, Dec, 350-356.

Maudgal, V K, Kliman, J, Miles, J, and Sunderland, J E (1990c). "Transient thermal contact resistance in a diode heat- sink assembly," Applied Energy (UK), 39, 31-43.

McGillis, W R, and Carey, V P (1990). "Immersion cooling of an array of heat dissipating elements - an assessment of different flow boiling methodologies," ASME, HTD-131, 37-44.

McGillis, W R, Carey, V P, Fitch, J S, and Hamburgen, W R (1991a). "Pool boiling enhancement techniques for water at low pressures, Proc 7th Semiconductor Thermal Measurement and Management Symp, Phoenix, AZ, Feb, 64-72.

McGillis, W R, Carey, V P, and Strom, B D (1991b). "Geometry effects on critical heat flux for subcooled convective boiling from an array of heated elements," ASME J Heat Transfer, 113, May, 463-471.

McKay, J R (1990). "Use of an air-to-air heat exchanger to reduce peak temperatures in outdoor cabinets," Proc 12th Intl Telecom Energy Conf, Orlando, FL, Oct, 584-588.

McKeown, S A and LeVasseur, R D (1991). "High performance heat sink for surface mount applications," Proc of 12th IEEE/AIAA

Digital Avionics Systems Conf, Los Angeles, CA, Oct, 153-157.

McMaye, F (1994). "Evaluation of alternative cooling techniques for TAB packages," Proc 10th Semiconductor Thermal Measurement and Management Symp, San Jose, CA, Feb, 95-102.

McNelis, B and Buller, L (1990a). "Enhancing thermal performance of SMT components," Proc Surface Mount Technology Conf (SMTCON), Atlantic City, NJ, Apr, 29-37.

McNelis, B and Buller, L (1990b). "Thermal characteristics of TAB for small systems," IEEE Trans Comp, Hybrids, Manuf Tech, CHMT-13 (4), Dec, 989-997.

McPhee, J M, O'Toole, T S, and Yedvabny, M (1990). "Cooling the VAX 9000," Electro/90 Conference Record, Boston, MA, May, 28-292.

Mereu, S, Sciubbe, E, and Bejan A (1993. "The optimal cooling of a stack of heat generating boards with fixed pressure drop, flow rate or pumping power," Intl J Heat and Mass Trans, 36 (12), Oct, 3677-3686.

Michael, M and Nguyen, L (1992). "Effect of mold compound thermal conductivity on IC package thermal performance," Proc. of Intersociety Conf on Thermal Phenomena on Electronic Systems (I-Therm IV), Washington, DC, 246-252.

Min, Y J, Palisoc, A L, and Lee, C C (1990a). "Transient thermal study of semiconductor devices," IEEE Trans Comp, Hybrids, Manuf Tech, CHMT-13 (4), Dec, 980-988.

Min, Y J, Palisoc, A L, and Lee, C C (1990b). "Point source response of five-layer anisotropic plate structures," Proc of Intersociety Conf on Thermal Phenomena in Electronic Systems (I-Therm II), Las Vegas, NV, 149-153.

Miner, S M (1990). "Maximizing electronic system reliability through optimized distribution of system coolant," ASME J Elec Packaging, 112, Dec, 345-349.

Miyagi, T, Itoh, K, and Kimijima, S (1990). "GaAs multichip module for a parallel processing system," Proc 40th Elec Comp and Tech Conf, Las Vegas, NV, May, 580-585.

Moffat, R J, and Anderson, A, (1990). "Applying heat transfer coefficient data to electronic cooling," ASME J Elec Packaging, 112, Nov, 882-890.

Mok, L, S (1994). "Thermal management of silicon-based multichip modules," Proc 10th Semiconductor Thermal Measurement and Management Symp, San Jose, CA, Feb, 59-63.

Mok, L S, and Kang, S K (1991). "Chip temperature responses during inner-lead bonding of TAB packages," Proc of 4th Electronics Materials and Processing Congress, Montreal, Canada, Aug, 193-197.

Moore, T (1991). "The design and characterization of a high thermal efficiency leadframe with an integral heatspreader," Proc 7th Semiconductor Thermal Measurement and Management Symp, Phoenix, AZ, Feb, 98-104.

Morega, A M and Bejan, A (1994). "Optimum spacing of parallel boards with discrete heat sources cooled by laminar forced convection," Numerical Heat Transfer - Part A: Applications, 25 (2), Apr, 373-392.

Morihara, A, Naganuma, Y, and Yokoyama, H (1990). "High performance line contact cooling module for multichip packaging," Proc of Intersociety Conf on Thermal Phenomena in Electronic Systems (I-Therm II), Las Vegas, NV, 161-164.

Moy, P, Hannigan, M, Allen, E, Bradley, S, and Choi, J (1992). "Design and process considerations for heat sink attachment," Proc Intl Elec Packaging Soc (IEPS) Conf, Marlbourough, MA, Sep, 178-190.

Mudawar, I (1992). "Direct immersion cooling for high power electronic chips," Proc of Intersociety Conf on Thermal Phenomena in Electronic Systems (I-Therm III), Austin, TX, 74-84.

Mudawar, I, and Anderson, T M (1990). "Parametric investigation into the effects of pressure, subcooling, surface augmentation and choice of coolant on pool boiling in the design of cooling systems for high-power-density electronic chips," ASME J Elec Packaging, 112, Dec, 375-382.

Mudawar, I, and Anderson, T M (1993). "Optimization of enhanced surfaces for high flux chip cooling by pool boiling, " ASME J Elec Packaging, 115, Mar, 89-101.

Mudawar, I, and Maddox, D E (1989). "Critical heat flux in subcooled flow boiling of fluorocarbon liquid on a simulated electronic chip in a vertical rectangular channel," Intl J Heat and Mass Transfer, 32 (2), Feb, 379-393.

Mudawar, I, and Maddox, D E (1990). "Enhancement of critical heat flux from a high power microelectronic heat sources in a flow channel," ASME J Elec Packaging, 112, Sep, 241-248.

Mudawar, I, and Wadsworth, D C (1991). "Critical heat flux from a simulated chip to a confined rectangular impinging jet of dielectric liquid," Intl J Heat and Mass Trans, 34 (6), Jun, 1465-1479.

Mulganoker, S, Chambers, B, Mahalingham, M, Ganesan, G, Hause, V, and Berg, H (1994). "Thermal performance limits of the QFP family," IEEE Trans Comp, Hybrids, Manuf Tech, CHMT-17 (4), Dec, 573-582.

Naganuma, Y, Norihara, A, and Yokoyama, H (1990). "Effect of radius of cylinder on thermal resistance of line contact cooling block for semiconductor," Proc 6th Semiconductor Thermal and Temperature Measurement Symp, Phoenix, AZ, Feb, 114-117.

Nakao, M, Hayama, H, and Nishioka, M (1991). "Which cooling air supply system is better for a high heat density room: underfloor or overhead?" 13th Intl Telecom Energy Conf, Kyoto, Japan, Nov, 393-400.

Nakayama, W (1990). "On the accommodation of coolant flow paths in high-density packaging," IEEE Trans Comp, Hybrids, Manuf Tech, CHMT-13 (4), Dec, 1040-1049.

Nakayama, W and Bergles, A E (1990). "Cooling electronic equipment - past, present, and future," Bulletin Intl Centre for Heat and Mass Transfer, 4, 1-67.

Nakwaski, W (1991). "Spreading thermal resistance of a diode- laser heat sink," Optical and Quantum Electronics, 23, 427-432.

Nelson, D J and Sayers, W A (1992). "A comparison of two dimensional planar, axisymmetric and three-dimensional spreading resistance," Proc 8th Semiconductor Thermal and Temperature Measurement Symp, Austin, TX, Feb, 62-68.

Nelson, R D, Dolbear, T, Chen, K, and Chao, C (1993), "Application of air-cooled microchannel heat exchangers to card cage systems," Proc 9th Semiconductor Thermal and Temperature Measurement Symp, Austin, TX, Feb, 50-62.

Nelson, R D, Sommerfeldt, S, and Bar-Cohen, A (1994), "Thermal performance of an integral immerson cooled multichip module package," IEEE Trans Comp, Hybrids, Manuf Tech, CHMT-17 (3), Sep, 405-412.

Nigen, J S and Amon, C H (1991). "Forced convective cooling enhancement of surface-mounted electronic package configurations through self-sustained oscillatory flows," ASME, HTD-171, 39-46.

Nishiguchi, M, Fujihara, M, Miki, A, and Nishizawa, H (1991). "Precisional Comparison of Surface Temperature Measurement Techniques for GaAs ICs," Proc 7th Semiconductor Thermal Measurement and Management Symp, Phoenix, AZ, Feb, 34-38.

Normington, P J, Mahalingham, M, and Lee, T Y (1992). "Thermal management control without overshoot using combinations of boiling liquids," IEEE Trans Comp, Hybrids, Manuf Tech, CHMT-15 (2), Apr, 806-814

North, M T and Avedesian, C T (1993). "Heat pipes for cooling high flux/high power semiconductor chips," ASME J Elec Packaging, 115, Mar, 112-117.

Novotny, S (1990). "Performance evaluation of a Gifford-McMahon refrigerator for cryogenically cooled computer applications," Proc of Intersociety Conf on Thermal Phenomena in Electronic Systems (I-Therm II), Las Vegas, NV, 97-100.

Nye, L W, and Brook, R (1991). "Thermal performance analysis using orthogonal arrays for designing integrated circuit packages," Proc Intl Elec Packaging Soc (IEPS) Conf, 1, San Diego, CA, Sep, 457-474.

O'Hanlon, J F (1991). "Detection of air in helium filled thermal conduction modules," IEEE Trans Comp, Hybrids, Manuf Tech, CHMT-14 (1), Mar, 211-213.

Ohara, T, Aihara, T, and Hayashi, C (1992). "Cooling performance of compact heat sinks with semi-microchannels," Proc. of the Joint ASME - ASME Conference on Electronic Packaging, Milpitas, CA, Apr, 249-252.

O'Mathuna, S C (1992). "Development and analysis of an automated test system for the thermal characterization of IC packaging technologies," IEEE Trans Components, Hybrids, Manuf Tech, CHMT-15 (5), Oct, 615-624.

O'Mathuna, S C and Daguang, L (1992). "Development and application of test chips and test system for the thermal characterization of IC packaging technologies," Proc 7th Semiconductor Thermal Measurement and Management Symp, Phoenix, AZ, Feb, 47-54.

Ortega, A and Kabir, H (1991). "Substrate conduction mechanisms in convectively cooled simulated electronic packages," Proc

7th Semiconductor Thermal Measurement and Management Symp, Phoenix, AZ, Feb, 1-14.

Ortega, A, Lall, B S, Chicci, J, Aghazedeh, M, and Kiang, B (1993a). "Heat transfer in a low aspect ratio horizontal enclosure for forced air cooling of electronic components," Proc 9th Semiconductor Thermal Measurement and Management Symp, Austin, TX, Feb, 42-49.

Ortega, A, Ramathana, S, Chicci, J D, and Prince, J L (1993b). "Thermal wake models for forced air cooling of electronics components," Proc 9th Semiconductor Thermal Measurement and Management Symp, Austin, TX, Feb, 63-74.

Osterman, M D, (1992). "A physics of failure approach to component placement," ASME J Elec Packaging, 114, Sep, 305-309.

Osterman, M D, and Pecht, M G (1990). "Placement for reliability on physics of failure concepts," Proc of 1990 IEEE Natl Aerospace and Electronics Conf (NAECON), Dayton, OH, 1021-1027.

Paggen, W (1990). "Managing heat flow in 19 in. rack," Electronic Product Design (UK), May, 39-41.

Papanicolaou, E and Jaluria, Y (1991). "Convective cooling of multiple components in an enclosure," ASME, HTD-171, 29-37.

Parsley, M (1991). "The use of thermochromic liquid crystals in research applications, thermal mapping and non-destructive testing," Proc 7th Semiconductor Thermal Measurement and Management Symp, Phoenix, AZ, Feb, 53-58.

Patankar, S V (1980). Numerical Heat Transfer and Fluid Flow, McGraw-Hill, New York, NY.

Pei, J, Heng, S, Charlantini, R, and Gildea, P (1990). "Cooling components used in VAX 9000 family of computers," Proc Intl Elec Packaging Soc (IEPS) Conf, Marlbourough, MA, Sep, 587-601.

Pei, H-S, Heng, S, and Watson, J (1991). "Closed-loop cooling for computers - opportunities for the 90's," Proc Natl Elec Packaging and Prod Conf - NEPCON East, Boston, MA, Jun, 373-386.

Phadke, N K, Bhavnani, S H, Goyal, A, Jaeger, R C, and Goodling, J S (1992a). "Re-entrant cavity surface enhancements for immersion cooling of silicon multichip packages," IEEE Trans Components, Hybrids, Manuf Tech, CHMT-15 (2), Apr, 815-822.

Phadke, N K, Bhavnani, S H, Goyal, A, Jaeger, R C, and Goodling, J S (1992b). "Enhanced surfaces for cooling high flux microelectronic components," Proc. of the Joint ASME - ASME Conference on Electronic Packaging, Milpitas, CA, Apr, 253-261.

Proctor, O and Solc, J (1991). "Improved thermal conductivity in microelectronic encapsulants," IEEE Trans Comp, Hybrids, Manuf Tech, CHMT-14 (4), Dec, 708-713.

Prasad, V, Keyhani, M , and Shen, R (1990). "Free convection in a discretely heated vertical enclosure: effects of Prandtl number and cavity size," ASME J Elec Packaging, 112, 63-68.

Prudhoe, R K, and Doukas, L (1990). "The use of phase change materials (PCMs) and vacuum panel heat exchangers for energy conservation and thermal stability of electronic equipment enclosures," Proc 12th Intl Telecom Energy Conf, Orlando, FL, Oct, 186-193.

Prudhoe, R K, Henshaw, R, Doukas, L, and Snow, R (1991). "Passive cooling of electronic equipment rooms - thermal diode room construction," 13th Intl Telecom Energy Conf, Kyoto, Japan, Nov, 401-409.

Pruzan, D A, Klingensmith, L K, Torrance, K E, and Avedisian, C T (1991). "Design of high-performance sintered-wick heat pipes," Intl J Heat and Mass Trans, 34 (6), Jun, 1417-1427.

Quiepo, N, Dakarakonda, R, and Humphrey J A C (1994). "Genetic algorithms for thermosciences research: application to the optimized cooling of electronic components," Intl J Heat and Mass Trans, 37 (6), Apr, 893-908.

Ramos, R A and Milanez, L F (1993). "Effect of the module front door closing procedure of a digital transmission equipment on the temperature of the components," Proceedings of 6th Intl Symp on Transport Phenomena, Seoul, Korea, May, 255-260.

Reese, R M (1990). "Shipboard application of variable conductance heat pipes," ASME, HTD-153, 95-101.

Remsburg, R (1991). "New routes to cooler electronics," Machine Design, Jul, 73-76.

Reynell, M (1990a). "Advanced thermal analysis of packaged electronic systems using computational fluid dynamics techniques," Computer-Aided Engineering Journal (UK), Aug, 104-106.

Reynell, M (1990b). "Fluid dynamics algorithms for advanced thermal analysis," Electronic Product Design (UK), Mar, 53-56.

Reynell, M (1990c). "Thermal analysis using computational fluid dynamics," Electronic Packaging and Production, Oct, 82-85.

Reynolds, S D, Sammakia, B G, and Carden T F (1992). "Thermal enhancements for a thin film chip carrier," IEEE Trans Components, Hybrids, Manuf Tech, CHMT-15 (5), Oct, 699-706.

Riddle, R A, Cantolini, R J, and Bernhardt, A F (1991). "Design calculations for the microchannel heatsink," Proc Natl Elec Packaging and Prod Conf - NEPCON West, 1, Anaheim, CA, 161-171.

Robinson, G (1990). "Smart ways to beat the heat," Design News, May, 113-116.

Rodgers, A J, Casto, J, Davis, D, and McShane, M (1990). "A thermally enhanced surface mount package for BICMOS FS-

RAM applications," Proc Natl Elec Packaging and Prod Conf - NEPCON East, Jun, 141-148.

Rodgers, P and Davis, M (1994). "Measurements of air velocity and temperature around three naturally convecting fin arrays," Proc. of Intersociety Conf on Thermal Phenomena on Electronic Systems (I-Therm IV), Washington, DC, 190-197.

Sankaran, V A, Hudgins, J L, Rhodes, C A, and Portnoy, W M (1990). "Transient thermal analysis of thryistors using finite element method," Proc 21st IEEE Power Electronics Specialists Conf, San Antonio, TX, Jun, 422-428.

Sato, I, Otani, K, Mizukami, M, Oguchi, S, and Hoshiya, K (1990). "Characteristics of heat transfer in small disk enclosures at high rotation speeds," IEEE Trans Comp, Hybrids, Manuf Tech, CHMT-13 (4), Dec, 1006-1011.

Schafer, D, Incropera, F P, and Ramadhyani, S (1991). "Planar liquid jet impingement cooling of multiple discrete heat sources," ASME J Elec Packaging, 113, Dec, 359-366.

Schenone, C and Tanda, G (1990). "Forced convection heat transfer from shrouded fin arrays," Intl Comm Heat Mass Transfer, 17, 747-758.

Seals, J D (1991). "Next-generation avionics packaging and cooling - test results from a prototype system," Proc of 10th IEEE/AIAA Digital Avionics Systems Conf, Los Angeles, CA, Oct, 164-169.

Sefcik, D M, Webb, B W, and Heaton, H S (1991). "Natural convection in vertically vented enclosures," ASME J Heat Transfer, 113, 912-918.

Sharan, A, Huang, C C, and Hamezehdoost, A (1991). "High strength copper alloys for EIAJ PQFPs - thermal and electrical performance without compromising lead integrity," Proc Natl Elec Packaging and Prod Conf - NEPCON West, Anaheim, CA, Jun, 173-190.

Shaukatullah, H and Gaynes, M A (1994). "Experimental determination of the effect of printed circuit card conductivity on the thermal performance of surface mount electronic packages," Proc 10th Semiconductor Thermal Measurement and Management Symp, San Jose, CA, Feb, 44-52.

Shaukatullah, H, Gaynes, M A, and White, L H (1993). "Thermal enhancement of surface mount electronic packages with heat sinks," Proc 43rd IEEE Electronic Technology and Components Conf, Orlando, FL, Jun, 256-263.

Shaukatullah, H, Gaynes, M A, and White, L H (1994). "A nondimensional correlation for the external thermal characteristics of surface mount quad flat packs," Proc. of Intersociety Conf on Thermal Phenomena on Electronic Systems (I-Therm IV), Washington, DC, 237-244.

Shaw, H-J, Chen, W-H, and Chen, C-K (1991). "Study on the laminar mixed convective heat transfer in three-dimensional channels with a thermal source," J Elec Packaging, 113, Mar, 40-49.

Sherif, R A, Schwartz, H B, and Brewster, R A (1992). "Transient thermal gradients across solder interconnections in electronic systems," IEEE Trans Comp, Hybrids, Manuf Tech, CHMT-15 (5), Oct, 865-890.

Shedden, R, Teymouri, J, and Morris, B (1990). "Natural thermal management of PCB's for telecommunications systems," Electronic Manufacturing, Feb, 12-17.

Siegal, B S (1991). "Methodology for thermal characterization of surface mount devices," Proc of Surface Mount Conf and Exposition, San Jose, CA, Aug, 715-724.

Siegal, B S (1993). "Thermal characterization of surface mount devices," Proc 9th Semiconductor Thermal Measurement and Management Symp, Austin, TX, Feb, 99-107.

Simon, B R, Yuan, Y, Umaretiya, J R, Bavirisetty, R, and Prince, J L (1990). "Thermal and mechanical finite element analysis of a VLSI package including spatially varying thermal contact resistance," Proc 6th Semiconductor Thermal and Temperature Measurement Symp, Phoenix, AZ, Feb, 74-81.

Simons, R E (1990). "Bibliography of heat transfer in electronic equipment - 1987 and 1988," in Advances in Thermal Modeling of Electronic Components and Systems - Volume 2, A Bar-Cohen and A D Kraus (eds), ASME Press, New York, NY.

Simons, R E (1993). "Bibliography of heat transfer in electronic equipment - 1989," in Advances in Thermal Modeling of Electronic Components and Systems - Volume 3, A Bar-Cohen and A D Kraus (eds), ASME/IEEE Press, New York, NY.

Simons, R E (1994). "Microelectronics cooling and semitherm: a look back," Proc 10th Semiconductor Thermal Measurement and Management Symp, San Jose, CA, Feb, 1-16.

Smith, T F, Beckermann, C, and Weber, S W (1991). "Combined conduction, natural convection, and radiation heat transfer in an electronic chassis," ASME J Elec Packaging, 113, Dec, 382-391.

Snyder, K W (1991) "An investigation of using a phase-change material to improve the heat transfer in a small electronic module for an airborne radar application," Proc Intl Elec Packaging Soc (IEPS) Conf, 1, San Diego, CA, Sep, 276-303.

Sofia, J W (1994). "Analysis of thermal transient data with synthesized dynamic models for semiconductor devices," Proc 10th Semiconductor Thermal Measurement and Management Symp, San Jose, CA, Feb, 78-85.

Song, T-H and Bang, S Y (1990). "Performance of a conduction cooling module," ASME J Elec Packaging, 112 (1), Mar, 52-56.

Song, S, Moran, K P, and Lee, S (1993). "Thermal and electrical resistances of bolted joints between plates of unequal thickness," Proc 9th Semiconductor Thermal Measurement and Management Symp, Austin, TX, Feb, 28-34.

Sotani, J, Yamamoto, M, Tanaka, S, Nanba, K, Yamazaki, N, and Kojima, Y (1992). " Feasibility study of printed circuit board using heat pipes," Proc. of the Joint ASME - ASME Conference on Electronic Packaging, Milpitas, CA, Apr, 83-88.

Soule, C A (1992). "Water-cooled heat sinks improve thermal efficiency," Power Conversion and Intelligent Motion, Nov, 18-22.

Souza Mendes, P R (1991). "The napthalene sublimation technique," Experimental Fluid Science, 4. 510-523.

Strom, B D, Carey, V P, and McGillis, W R (1989). "An experimental investigation of the critical heat flux conditions for subcooled convective boiling from an array of simulated microelectronic devices," ASME, HTD-111, Aug, 135-142.

Struble, C L, and Witte, L C (1991). "An electrical method for measurement of jet nucleate boiling heat transfer on silicon chip heaters," ASME, HTD-183, 1-8.

Sulhan, R, Fredholm, M, Monaghan, T, and Agarwal, A (1991a). "Thermal modeling and analysis of pin grid arrays and multichip modules," Proc 7th Semiconductor Thermal Measurement and Management Symp, Phoenix, AZ, Feb, 110-116.

Sulhan, R, Monaghan, T, Agarwal, A, Kozarek, R, and Kromann, G (1991b). "Thermal modeling and analysis of multichip modules," Proc Natl Elec Packaging and Prod Conf - NEPCON East, Boston, MA, Jun, 209-218.

Sullivan, P F, Ramadhyani, S, and Inrcropera, F P (1992). "Extended surfaces to enhance impingement cooling with single circular liquid jets," Proc. of the Joint ASME - ASME Conference on Electronic Packaging, Milpitas, CA, Apr, 207-215.

Sun, H, Ma, C P, and Nakayama, W (1993). "Local characteristics of convective heat transfer from simulated microelectronic chips to submerged round jets," ASME J elec Packaging, 115, Mar, 71-77.

Swager, A W (1990). "Methods converge to cool fast and dense circuits," EDN, Dec, 162-168.

Sweet, J N and Cooley, W T (1990). "Thermal resistance measurement and finite element calculations for ceramic hermetic packages," Proc 6th Semiconductor Thermal and Temperature Measurement Symp, Phoenix, AZ, Feb, 10-16.

Takada, J, Kimura, Y, Ogushi, T, Fujii, M, Kohara, M, and Nakao, S (1990). "Development of a new air-cooling heatsink for a high density computer system," Proc 40th Elec Comp and Tech Conf (ECTC), Las Vegas, NV, May, 534-539.

Tanaka, A, Okamoto, M, Oohashi, M, Arakawa, H, and Yamada, K (1990). "Low thermal resistance AlN PGA with low inductance of power lines," Proc 40th Elec Comp and Tech Conf (ECTC), Las Vegas, NV, May, 327-332.

Tang, A C, Whalley, D C, and Williams, D J (1990). "Tailoring thermal profiles for the reflow of SMD assemblies," Proc of 9th IEEE/CHMT Intl Electr Manuf Tech (IEMT) Symp, Washington, DC, Oct, 77-81.

Tatara, R A (1991). "Thermal modeling previews electronic device performance." Power Conversion and Intelligent Motion, Oct, 9-21.

Temmerman, W, Nelemans, W, Hendrickx, R, Ryckebush, M, and Rottiers, L (1992). "Application of a numerical tool for thermal evaluation of printed circuit boards in communications systems," Proc. of Intersociety Conf on Thermal Phenomena in Electronic Systems (I-Therm III), Austin, TX, 197-201.

Terada, M, Komine, T, Katahira, A, and Takahashi, T (1990). "New high heat-sink metal cored multilayer PCB for surface mount technology (SMT)," Proc of the 1990 Printed Circuit World Convention, Glasgow, UK, Jun.

Tewari, S S and Jaluria, Y (1991). "Convective heat transfer with buoyancy effects from thermal sources on a flat plate," J of Thermophysics, 5 (2), Apr-Jun, 199-207.

Tokola, M and Wolff, D (1990). "Concurrent design permits early thermal analysis," Electronic Packaging and Production, Nov, 66-70.

Tong, W, Bar-Cohen, A, Simon, T W, and You, S M (1990). "Contact angle effects on boiling incipience of highly-wetting liquids," Intl J Heat and Mass Trans, 33 (1), Jan, 91-103.

Toth, J, and Garner, S (1990). "Heat pipe heat exchangers for electronic cabinet cooling," Proc Intl Elec Packaging Soc (IEPS) Conf, Marlbourough, MA, Sep, 1179-1190.

Tye, R P and Maesano, A (1992). "Determination of appropriate transport properties for electronics applications," Proc 8th Semiconductor Thermal Measurement and Management Symp, Austin, TX, Feb, 37-40.

VanLaanen, J B (1990). "SMT single- versus dual-sided modules - a case study," Proc 6th Semiconductor Thermal and Temperature Measurement Symp, Phoenix, AZ, Feb, 43-48.

Van Petegem, W, De Wachter, D, and Sansen, W (1990). "Electrothermal simulation of integrated circuits," Proc 6th Semiconductor Thermal and Temperature Measurement Symp, Phoenix, AZ, Feb, 70-73.

Van San, R (1991). "Liquid Immersion Cooling of High Heat Dissipating Microelectronics," in Reliability '91, R H Matthews (ed), Elsevier Applied Science, New York, NY.

Van Tongeren, H and Thijs, P J (1992). "Thermal aspects of pump laser packaging." Proc 42rd IEEE Electronic Technology and Components Conf, San Diego, CA, May, 848-852.

Vick, B and Nelson, D J (1990). "A boundary element thermal model of sliding contact: application to head/disk interfaces," Proc of Intersociety Conf on Thermal Phenomena in Electronic Systems (I-Therm II), Las Vegas, NV, 121-129.

Vogel, M R (1994). "Thermal performance of air-cooled hybrid heat sinks for a low velocity environment," Proc 10th Semiconductor Thermal Measurement and Management Symp, San Jose, CA, Feb, 17-23.

Volk, T G (1990). "Thermal requirements for electronic equipment cabinets exposed to outdoor environments," Proc 12th Intl Telecom Energy Conf, Orlando, FL, Oct, 571-576.

Wadsworth, D C, and Mudawar, I (1990). "Cooling of a multichip electronic module by means of confined two-dimensional jets of dielectric liquid," ASME J Heat Transfer, 112, 891-898.

Wadsworth, D C, and Mudawar, I (1992). "Enhancement of single phase heat transfer and critical heat flux from an ultra-high-flux simulated microelectronic heat source to a rectangular impinging jet of dielectric fluid," Proc. of the Joint ASME - ASME Conference on Electronic Packaging, Milpitas, CA, Apr, 143-151.

Walcutt, J (1991a). "Determining the thermal characteristics of hybrid packages," Electronic Packaging and Production, Feb, 102- 104.

Walcutt, J (1991b). "Technology choices in infrared thermal imaging," Evaluation Engineering, Nov, 42-47.

Wang, C Q, Qian, Y H, Jiang, Y H (1991). "SMT laser microsoldering thermal process," Soldering and Surface Mount Technology (UK), 8, Jun, 29-31.

Wang, H Y and Saulnier, J B (1993). "A sensitivity study of material properties for coupled convective-conductive heat transfer generated in an electronic component," Intl J of Heat and Mass Trans, 36 (12), Oct, 3831-3839.

Wang, X S, Dagan, Z, and Jiji, L M (1990). "Prediction of surface temperature and heat flux of a microelectronic chip with jet impingement cooling," ASME J Elec Packaging, 112 (2), Mar, 57-62.

Wang, Y and Ghajar, A J (1991). "Effect of component geometry and layout on flow distribution for surface mounted electronic components: a smoke visualization study," ASME, HTD-183, 25-31.

Wang, Z G and Mayinger, F (1993). "Natural convective air cooling of the PCB's array in a closed casing," Proc of 6th Intl Symp on Transport Phenomena, 1, Seoul, Korea, May, 243-248.

Watari, T (1990). "Computer packaging technology for system performance," NEC Research and Development, Jul, 49-59.

Webb, P W and Russell, I A (1991). "Thermal simulation of transients in microwave devices," IEEE Proceedings, Jun, 329-334.

Weisberg, A and Bau, H H (1992). "Analysis of microchannels for integrated circuit cooling," Intl J of Heat and Mass Trans, 35 (10), 2465-2474.

Whalley, D C, Ogunjimi, A O, Conway, P P, and Williams, D J (1991). "A process model of the infrared reflow soldering of printed circuit board assemblies," Proc of 11th IEEE/CHMT Intl Electr Manuf Tech Symp, 122-125.

Wilson, J and Decker, K (1994). "GaAs MMIC thermal modeling for channel temperatures in accelerated life test fixtures and microwave modules," Proc 10th Semiconductor Thermal Measurement and Management Symp, San Jose, CVA, Feb, 121-128.

Wirth, F (1992). "IGBT junction temperature measurement aids for circuit design," Power Conversion and Intelligent Motion. Jan, 37-42.

Wirtz, R A and McAuliffe, W (1990). "Experimental modeling of convection upstream from a row of electronic packages," Proc of Intersociety Conf on Thermal Phenomena in Electronic Systems (I-Therm II), Las Vegas, NV, 165-170.

Wirtz, R A and Chen, W (1991). "Laminar-transitional convection from repeated ribs in a channel," ASME HTD-171, 89-94.

Womac, D J, Aharoni, G, Ramadhyani, S, and Incropera, F P (1990). "Single phase liquid jet impingement cooling of small heat sources," Proc of the 9th International Heat Transfer Conference, 2, Jerusalem, Israel, 149-154.

Wright, J L, Marks, B W, and Decker, K D (1991). "Modeling of MMIC devices for determining MMIC channel temperatures during life tests," Proc 7th Semiconductor Thermal Measurement and Management Symp, Phoenix, AZ, Feb, 131-139.

Wu, D and Peterson, G P (1991a). "Investigation of the transient characteristics of a micro heat pipe," J of Thermophysics, 5 (2), Apr-Jun, 129-134.

Wu, D and Peterson, G P (1991b). "Transient experimental investigation of micro heat pipes," J of Thermophysics, 5 (4), Oct-Dec, 539-544.

Yao, S C (1990). "Thermal analysis for reliability predictions," Evaluation Engineering, Oct, 114-118.

Yeh, L T (1990). "Comparisons of various cooling schemes within a confined space," Proc of Intersociety Conf on Thermal Phenomena in Electronic Systems (I-Therm II), Las Vegas, NV, 113-119.

You, S M, Simon, T W, Bar-Cohen, A, and Tong, W (1990a). "Experimental investigation of nucleate boiling incipience with a highly-wetting dielectric fluid (R-113)," Intl J Heat and Mass Trans, 33 (1), Jan, 105-117.

You, S M, Bar-Cohen, A, and Simon, T W (1990b). "Boiling incipience and nucleate boiling heat transfer of highly -wetting dielectric fluids from electronic materials," IEEE Trans Comp, Hybrids, Manuf Tech, CHMT-13 (4), Dec, 1032-1039.

You, S M, Simon, T W, and Bar-Cohen, A (1990c). "Experiments on nucleate boiling heat transfer with a highly-wetting dielectric fluid: effects of pressure, subcooling and dissolved gas content," ASME, HTD-131, Jun, 45-52.

You, S M, Simon, T W, Bar-Cohen, A, and Tong, W (1990d). "Experimental investigation of nucleate boiling incipience with a highly-wetting dielectric fluid (R-113)," Intl J Heat and Mass Trans, 33 (1), Jan, 105-117.

You, S M, Simon, T W, and Bar-Cohen, A (1992a). "Pool boiling heat transfer with an array of flush mounted square heaters," ASME HTD 206-2, 105-117.

You, S M, Simon, T W, and Bar-Cohen, A (1992b). "A technique for enhancing boiling heat transfer with application tomcooling of electronic equipment," IEEE Trans Components, Hybrids, Manuf Tech, CPMT-15 (2), Apr, 823-831.

Yuki, M S, Parsons, J R, and Krane, R J (1990). "An investigation of the passive cooling of table model television receivers," ASME J Elec Packaging, 112 (4), Dec, 279-287.

Zimmerman, M A, Azar, K, and Mandrone, C D (1990). "Thermal characteristics of molded multi-chip modules," presented at ASME Winter Annual Meeting, Dallas, TX, Nov, ASME 90-WA/EEP-21.

Zumbrunnen, M L (1991). "Materials and processing approaches for high performance electronic cooling," Proc of 4th Electronics Materials and Processing Congress, Montreal, Canada, Aug, 399-403.

ndex

B

C

E

F

G

H

J

K

L

I

N

O

S

V

W

X

Y

Z

Index

A

B

C

E

F

G

H

T

U

V

W

X

Y

Z